# STATICS FOR ARCHITECTS

# STATICS FOR ARCHITECTS

**Jafar Vossoughi**

School of Engineering and Architecture
The Catholic University of America

VNR VAN NOSTRAND REINHOLD COMPANY
New York

Library of Congress Catalog Card Number: 86-9130
ISBN: 0-442-29111-6



Manufactured in the United States of America

Published by Van Nostrand Reinhold Company Inc.
115 Fifth Avenue
New York, New York 10003

Van Nostrand Reinhold Company Limited
Molly Millars Lane
Wokingham, Berkshire RG11 2PY, England

Van Nostrand Reinhold
480 La Trobe Street
Melbourne, Victoria 3000, Australia

Macmillan of Canada
Division of Canada Publishing Corporation
164 Commander Boulevard
Agincourt, Ontario M1S 3C7, Canada

15 14 13 12 11 10 9 8 7 6 5 4 3 2 1

**Library of Congress Cataloging-in-Publication Data**

Vossoughi, Jafar.
Statics for architects.

Includes index.
1. Structures, Theory of. I. Title.
TA645.V67 1986 624.1'7 86-9130
ISBN 0-442-29111-6

*To*
*Forrest Wilson*
*Excellent engineer*
*Splendid architect*
*Perfect teacher*
*and Good friend*

# PREFACE

This book was designed to cover the materials needed for the first course in a series of required courses entitled ''Building Science'' and ''Structures for Architects'' that forms part of the architecture curriculum all over the country. It is the outcome of several years of experience in teaching such courses to architectural students at The Catholic University of America. Although the author has taught ''Statics for Architects'' and a similar course for engineering students, ''Statics for Engineers,'' it is his belief that there is no comparison between them even though both cover essentially the same materials. The basis for this belief is as follows:

1. In ''Statics for Engineers,'' the teacher enjoys the use of advanced mathematics, whereas in ''Statics for Architects,'' more or less the same material has to be covered in terms of very low-level math and definitely without direct use of the calculus.
2. Engineering students taking statics have undoubtedly already studied the subject in a freshman physics course called Mechanics, where they had to apply even more advanced mathematics.
3. Engineering students enrolled in this course have already taken about eight credits of advanced college mathematics and are concurrently enrolled in yet another math course. Architectural students in many universities, however, are not required to take any college math courses, and in the few architectural programs where math is required, they take math courses with social science and art students. Such courses involve quite low-level mathematics.
4. Students choosing engineering are usually mathematically and technically oriented, whereas those choosing architecture are artistically oriented. As a consequence, most architecture students not only do not like math, they *hate* math.
5. Statics is a fundamental course for engineers and what they learn in it is frequently germane to their other courses, whereas it does

not fit in with the other courses of architects, and what they learn here is infrequently applied elsewhere.
6. Usually, a noticeable gap exists between students of architecture and their statics or structures teachers. In most schools, these courses are taught to architects as well as engineers by engineering professors, and usually the engineering faculty does not understand architects, and, in fact, their attitude toward them is very sarcastic.
7. Finally, if we assume that students do indeed get more than half of their learning from textbooks, then, since most books in statics are written with engineers in mind, engineering students have the advantage over architecture students. Particularly because of their mathematical handicaps, architecture students cannot get much help from such books. Seeing this lack of balance was the main motivation for writing this long overdue book. Its main contribution, therefore, is to fill this gap. In fact, if the author were asked to write a statics book for engineers, he wouldn't do it, since enough excellently written ones for engineers already exist.

The author believes that the major contributions of this book are the following:

1. To the best of his knowledge, it is the first book with such a title and application.
2. It is written in a congenial manner specifically intended for architectural students as readers.
3. The coverage of a highly technical subject is achieved by using quite low-level math.
4. The subject is introduced in very simple, readily comprehensible language, with step-by-step explanations.
5. Elementary but important problems are used as teaching aids. The problems are solved in a thoroughgoing manner and, in most cases, are even "overexplained."
6. No prior knowledge of math is required. All the mathematics needed is not merely reviewed but systematically taught in the first chapter. In fact, Chap. 1 can be considered as a textbook in itself, rehearsing most of the mathematics covered in high school and first-year college that is related to the present subject. Teachers should consider this chapter as a quarter of the course and follow it step by step.

It should be mentioned that writing such a book, although a lot of fun, was extremely difficult. Precise treatment of mechanics as well as mathematics was intentionally avoided in the belief that doing so might confuse the issue. Instead, the applied aspects of the various subjects were emphasized. As far as abstract mathematics and mechanics are concerned, many concepts will be seen to have been defined and treated very loosely. This practice was followed throughout the book in view of its intended audience.

This book would not have come about without the encouragment of Dr. Forrest Wilson, who initially came up with the idea of the need for such a volume and encouraged and helped the author to get started. Many thanks to him. The book would not have been finished, moreover, without the enormous effort of Ms. Madeline Sapienza, who computerized it in its entirety. Having heard over and over that the latest version would be the last, she patiently undertook the handling of each revision. My gratitude goes to Messieurs O. E. Kia and S. Maiorisi for their invaluable services in preparing the drawings. The author would also like to thank the editors of Van Nostrand Reinhold and the School of Engineering and Architecture of The Catholic University of America for their cooperation and support.

Finally, the author would like to thank his wife, Laili, for her continuing love and support. She took pride and interest in reading every page, offering many invaluable comments as well as checking all calculations.

J. VOSSOUGHI

Washington, D.C.

# CONTENTS

# STATICS FOR ARCHITECTS

# 1

# BASIC MATHEMATICS, UNITS, AND ACCURACY

## MATHEMATICAL PRELIMINARIES

In this chapter all the mathematics needed to understand this book will be reviewed. This review is particularly important for those who have either forgotten or perhaps never learned high school math. Such students must go through this chapter very thoroughly.

There is no way to understand statics and subsequently the strength of materials and structures without being familiar and comfortable with the mathematics involved. Notice that you do not have to remember all the mathematics you studied at high school and in the first year of college; the principles reviewed here will suffice.

### 1.1 STRAIGHT LINES AND ANGLES

Basic concepts of plane geometry involving lines and angles will be reviewed in this section. Following are a number of facts about straight lines and angles that will be frequently used in solving various problems:

1. *Parallel* lines are lines that never intersect, regardless of how far they are extended at both ends.
2. Two lines are said to be *normal* (or *perpendicular*) to each other if the angle between the two is a *right angle*, or 90°.
3. Concurrent or intersecting lines are lines that meet or intersect at a point.
4. Two lines that are both parallel to a third line are themselves parallel.
5. Two lines that are perpendicular to a third line are themselves parallel.
6. If a third straight line, MN, crosses two parallel lines, AB and CD,

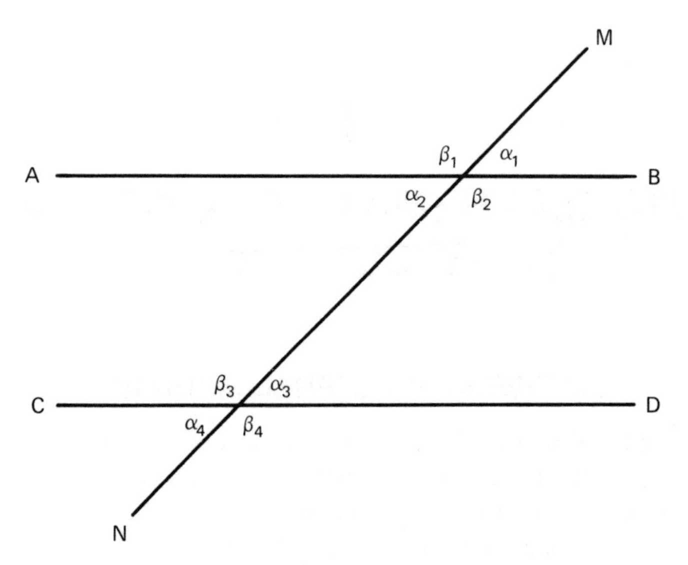


**Figure 1.1**

as shown in Fig. 1.1, the following equalities between the various angles hold true:

$$\alpha_1 = \alpha_2 = \alpha_3 = \alpha_4$$

Similarly,

$$\beta_1 = \beta_2 = \beta_3 = \beta_4$$

Angles between two lines are measured along a circular *arc* centered at the intersection of the two lines. The angle is considered *positive* when it is measured counterclockwise and *negative* when measured clockwise.

Angles are frequently measured by using a unit called a *degree*. A degree is subdivided into smaller units such as *minutes* and *seconds*. One degree (1°) is equal to 60 minutes (60′), and one minute is equal to 60 seconds (60″). As an example, an angle of 28 degrees, 15 minutes, and 47 seconds is written as

$$28°15'47''$$

For calculation purposes, however, this has to be converted to its decimal equivalent in the following manner:

$$47'' = \frac{47}{60} = 0.78 \text{ minutes}$$

$$15'47'' = 15 \text{ minutes} + 0.78 \text{ minutes} = 15.78 \text{ minutes}$$

$$28°15'47'' = 28°15.78'0''$$

$$15.78' = \frac{15.78}{60} = 0.26 \text{ degree}$$

$$28°15.78' = 28 \text{ degrees} + 0.26 \text{ degree} = 28.26 \text{ degrees}$$

Thus, 28°15′47″ = 28.26°.

Angles are also measured using *radians* and *grads*, but in this book angles will be measured in degrees. Going around a full circle, one complete turn will consist of 360 degrees (360°), which is also equal to $2\pi$ radians or 400 grads.* An angle of one-half a full circle, or 180°, is known as a *straight angle*, and an angle of one-fourth of a full circle (or 90°) is known as a right angle.

To convert angles from degrees to radians or vice versa, the equality, 180 degrees = $\pi$ radians, is used. To convert from degrees to radians, therefore, multiply the value of the angle by $\pi/180$. To convert from radians to degrees, multiply the value of the angle by $180/\pi$.

EXAMPLE 1: Convert 30°, 59°, 90°, 215°, and 356° to radians.

SOLUTION:

$$30° \times \frac{\pi}{180} = 0.524 \text{ radians, or } \frac{\pi}{6} \text{ radians}$$

$$59° \times \frac{\pi}{180} = 1.030 \text{ radians}$$

$$90° \times \frac{\pi}{180} = 1.571 \text{ radians, or } \frac{\pi}{2} \text{ radians}$$

$$215° \times \frac{\pi}{180} = 3.752 \text{ radians}$$

$$356° \times \frac{\pi}{180} = 6.213 \text{ radians}$$

*Pi ($\pi$) is a constant value equal to 3.1415, approximately.

EXAMPLE 2: Convert $\pi/6$ radians, 2 radians, and $2\pi/3$ radians to degrees.

SOLUTION:

$$\frac{\pi}{6} \text{ radians} \times \frac{180}{\pi} = 30°$$

$$2 \text{ radians} \times \frac{180}{\pi} = 114.6°$$

$$\frac{2\pi}{3} \text{ radians} \times \frac{180}{\pi} = 120°$$

Note that in Fig. 1.1 we can also write expressions such as the following:

$$\alpha_1 + \beta_1 + \alpha_2 + \beta_2 = 360° = 2\pi \text{ radians}$$

and since $\alpha_1 = \alpha_2$ and $\beta_1 = \beta_2$, then $\alpha_1 + \beta_1 = 180°$. In this case, we use the Greek letters $\alpha$ (alpha) and $\beta$ (beta) to indicate the angles.

Next, consider two angles, $x\hat{O}y$ and $x'\hat{O}'y'$, as shown in Fig. 1.2(a).* If the corresponding sides of these two angles are parallel, then the two angles are equal, that is: $\alpha = \beta$.

Now consider Fig. 1.2(b). As in the preceding case, these two angles

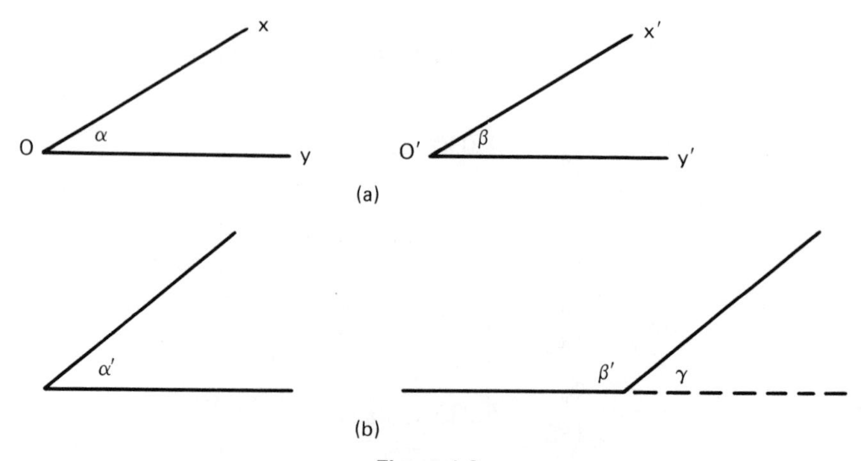


Figure 1.2

*The circumflex, "^", represents an angle. Another symbol, ∠, is also used to represent angles.

also have corresponding sides that are parallel, but now they are unequal, that is: $\alpha' \neq \beta'$ (the symbol, $\neq$, indicates "not equal to"). In this case, since $\alpha'$ is less than 90° and $\beta'$ is greater than 90°, they obviously cannot be equal. The fact is that $\alpha'$ is equal to its *supplement*, that is: $\alpha' = \gamma$. (When two angles subtend an arc of 180°, they are called *supplementary angles*. Similarly, when they subtend an arc of 90°, they are called *complementary angles*.)

Now consider two angles whose sides are perpendicular to each other as shown in Fig. 1.3(a). In this case, x′O′ is perpendicular to xO and y′O′ is perpendicular to yO. The two angles they form are also equal, that is, $\alpha = \beta$. Notice, however, that in Figure 1.3(b) the two angles

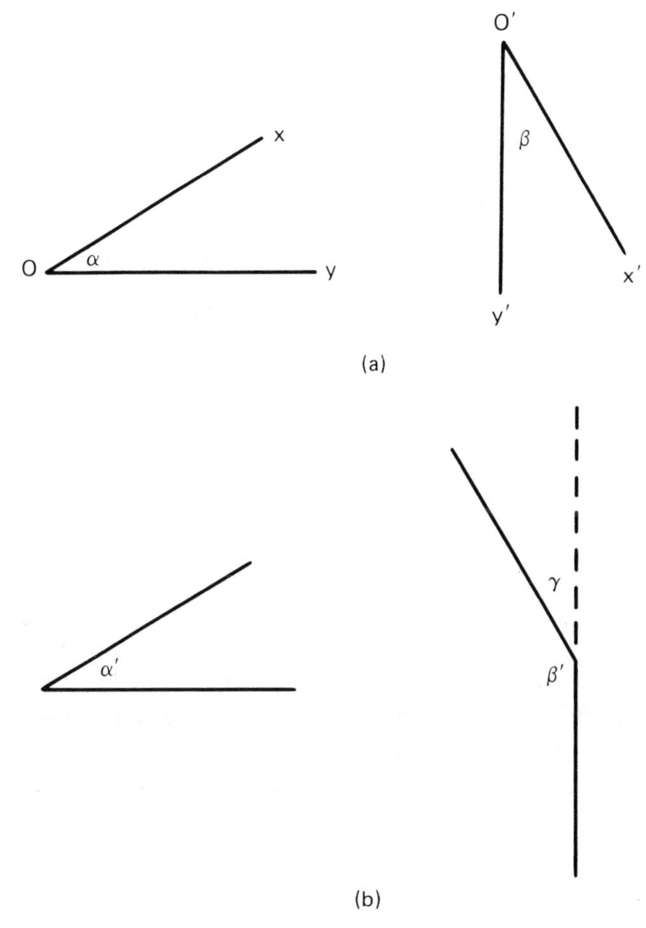


Figure 1.3

again have perpendicular sides, but $\alpha' \neq \beta'$. For the reasons previously discussed, $\alpha' = \gamma$.

## 1.2 TRIANGLES

Consider three points on a plane that do not lie on the same line. By connecting these three points, a *triangle* is constructed, as shown in Fig. 1.4. Certain relationships exist between the lengths of the sides and the values of the angles of triangles that will be reviewed in the following sections. In particular, relationships for right triangles and similar triangles will be discussed in detail. In Fig. 1.4, angles $\alpha$, $\beta$, and $\gamma$ are called *inner angles*, whereas angles $\alpha'$, $\beta'$, and $\gamma'$ are called *outer angles*.

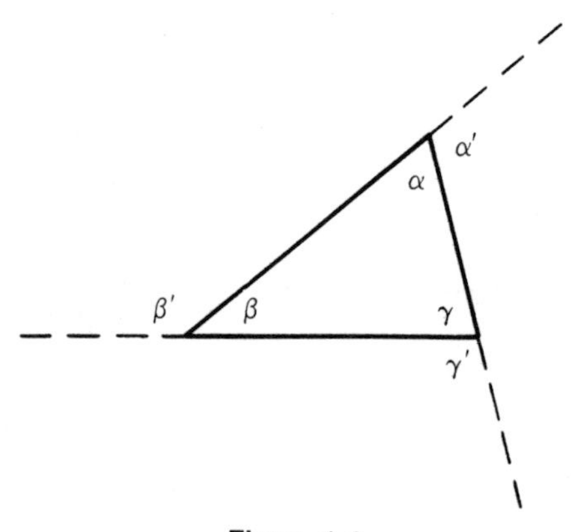


**Figure 1.4**

### Right Triangles

If one of the angles (inner or outer) of a triangle is 90°, that triangle is called a *right triangle*, as shown in Fig. 1.5. Side *AB*, which is opposite to the right angle (indicated by the small box), is called the *hypotenuse*. For angle $\alpha$, sides *AC* and *BC* are called the *opposite* and *adjacent sides*, respectively; for angle $\beta$, sides *BC* and *AC* are so called.

For any triangle (see Fig. 1.4), the sum of the three inner angles is equal to 180°, that is,

$$\alpha + \beta + \gamma = 180° \tag{1.1}$$

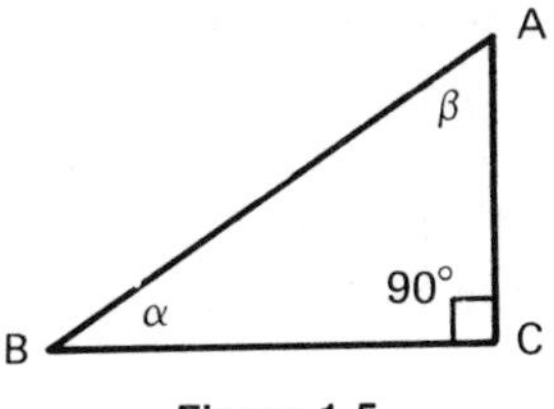


**Figure 1.5**

For a right triangle, $\gamma = 90°$ (see Fig. 1.5), and Eq. 1.1 reduces to: $\alpha + \beta = 90°$. Consequently, if one acute angle of a right triangle, say $\alpha$, is known, the other angle, $\beta$, can be easily found since

$$\beta = 90° - \alpha$$

Also, in any triangle such as the one shown in Fig. 1.4, the following relationship exists between the outer angles:

$$\alpha' + \beta' + \gamma' = 360°$$

and the outer angle $\alpha'$ is related to the inner angles as follows:

$$\alpha' = 180° - \alpha \quad \text{and} \quad \alpha' = \beta + \gamma$$

The latter equation holds true since $\beta + \gamma = 180° - \alpha$ as well.

In the right triangle of Fig. 1.5, the following relationship holds between the lengths of the sides:

$$AB^2 = AC^2 + BC^2 \tag{1.2}$$

Notice that the longest side of the right triangle is the side that faces the 90° angle, that is, the hypotenuse, $AB$. Of a right triangle it is therefore said that the square of the hypotenuse is equal to the sum of the squares of the other two sides.

## Similar Triangles

If two triangles have two of their angles mutually equal (consequently, the third angles will also be equal since the sum of the three angles must equal 180° in both cases), they are called *similar triangles*. There are certain relationships among the lengths of the sides of two similar triangles that are extremely useful.

Consider the two triangles, ABC and A′B′C′, shown in Fig. 1.6(a). If $\hat{A} = \hat{A}'$ and $\hat{B} = \hat{B}'$ and therefore $\hat{C} = \hat{C}'$, the relationships between the lengths of the sides of these two triangles can be written as:

$$\frac{AB}{A'B'} = \frac{AC}{A'C'} = \frac{BC}{B'C'}$$

It is customary to denote the length of the side opposite an angle $\hat{A}$ by the lower-case letter $a$ and those opposite $\hat{B}$ and $\hat{C}$ by lower-case letters

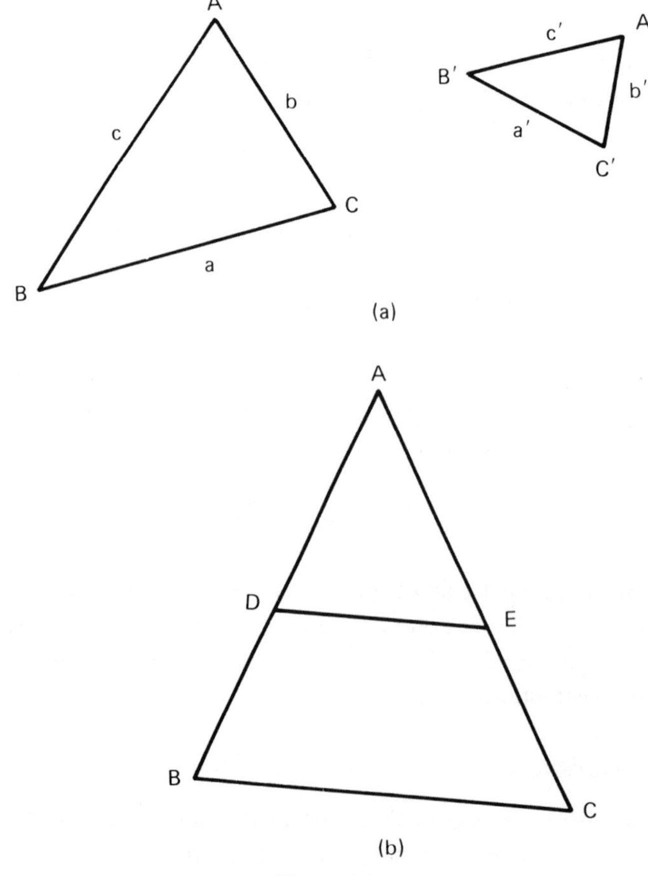


**Figure 1.6**

$b$ and $c$, respectively. Accordingly, the above relationship will look as follows:

$$\frac{c}{c'} = \frac{b}{b'} = \frac{a}{a'} \tag{1.3}$$

In Eq. 1.3, we really have two independent equations, for example:

$$\frac{a}{a'} = \frac{b}{b'} \qquad \text{and} \qquad \frac{a}{a'} = \frac{c}{c'}$$

Since all other relationships can be easily obtained from Eq. 1.3, this is the only formula for similar triangles that needs to be memorized. Note that all the denominators in Eq. 1.3 belong to one of the triangles and that all the numerators belong to the other and, furthermore, that the two elements of each fraction in Eq. 1.3 are the sides opposite the equal angles of the two triangles. Once the original relationship has been established, it can be manipulated in various ways; for example, making use of the algebraic principles discussed in Sec. 1.4,

$$\frac{a}{a'} = \frac{b}{b'}$$

can be converted to

$$\frac{a}{b} = \frac{a'}{b'} \qquad ab' = a'b \qquad a = a'\left(\frac{b}{b'}\right)$$

and so forth. These are simple but extremely useful and frequently used relationships. Students, therefore, should become fluent in handling them.

Now consider triangle ABC in Fig. 1.6(b). If a line is drawn from $D$ to $E$ parallel to $BC$, the two triangles ABC and ADE are similar triangles, and the above relationships can now be written as

$$\frac{AD}{AB} = \frac{AE}{AC} = \frac{DE}{BC}$$

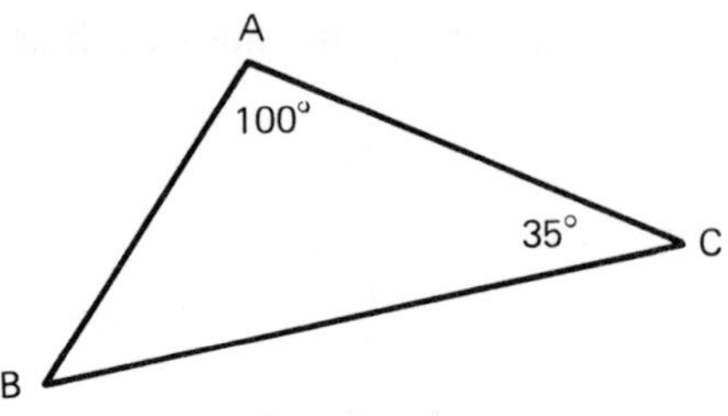


Figure 1.7

and as

$$\frac{AD}{DE} = \frac{AB}{BC}$$

and so forth.

EXAMPLE 3: In triangle ABC in Fig. 1.7, what is the value of angle $B$?

SOLUTION:

$$\hat{A} + \hat{B} + \hat{C} = 180°$$
$$100° + \hat{B} + 35° = 180°$$

Therefore, $\hat{B} = 180° - 100° - 35°$, or

$$\hat{B} = 45°$$

EXAMPLE 4: In the right triangle ABC in Fig. 1.8, what is the length of side $AC$?

SOLUTION: Using Eq. 1.2,

$$BC^2 = AB^2 + AC^2$$
$$AC^2 = BC^2 - AB^2 = (5)^2 - (3)^2 = 25 - 9 = 16$$

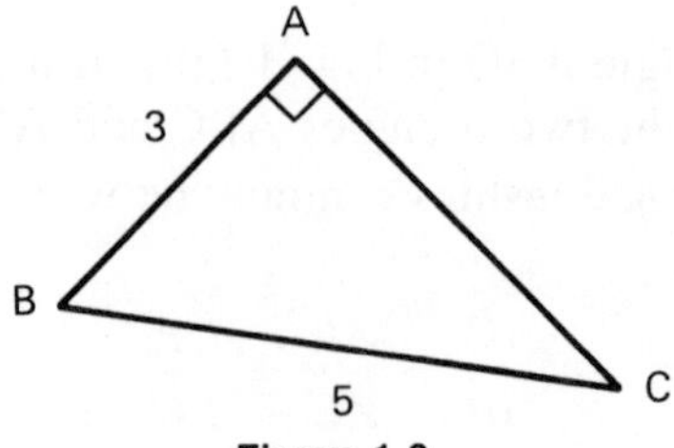


Figure 1.8

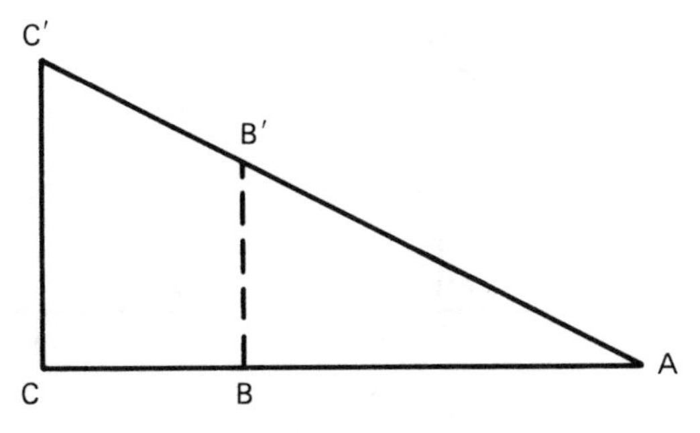


Figure 1.9

Therefore, since $AC^2 = 16$,

$$AC = 4$$

EXAMPLE 5: In Fig. 1.9, $BB'$ is parallel to $CC'$ and $AB = 4''$, $BC = 2''$, and $CC = 3''$. Calculate $BB'$.

SOLUTION: Since the two triangles ABB′ and ACC′ are similar triangles, we use Eq. 1.3 as follows:

$$\frac{AB}{AC} = \frac{BB'}{CC'}$$

$$\frac{4}{4 + 2} = \frac{BB'}{3}$$

Solving for $BB'$, we get $6BB' = 12''$, or

$$BB' = 2''$$

## 1.3 TRIGONOMETRY

### Definition of Sine, Cosine, and Tangent

Consider the right triangle ABC shown in Fig. 1.10(a) with angle $C$ being 90°. The ratios of any two of the three sides of this triangle are called *sine* (abbreviated "sin"), *cosine* (cos), and *tangent* (tan) and are defined in the following way;

$$\sin \alpha = \frac{AC}{AB}$$

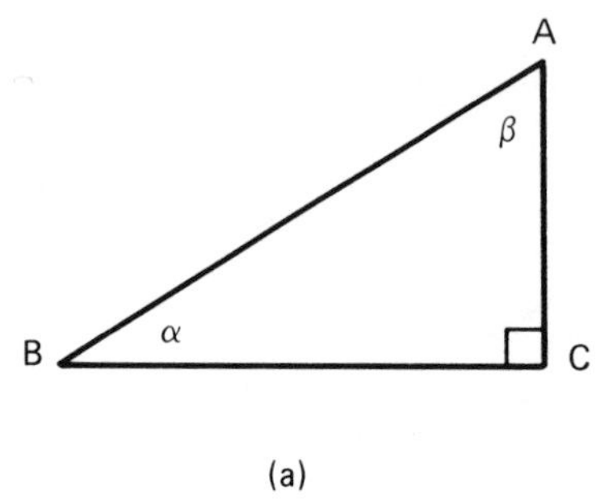


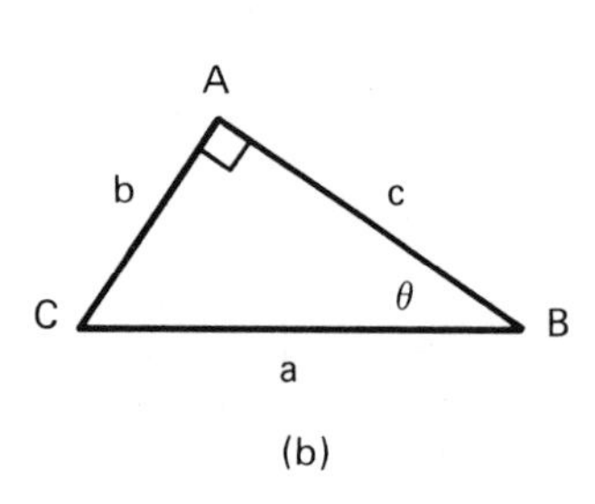


**Figure 1.10**

$$\cos \alpha = \frac{BC}{AB}$$

$$\tan \alpha = \frac{AC}{BC}$$

Notice that in both sin $\alpha$ and cos $\alpha$, the denominator of the ratio is the hypotenuse. For sin $\alpha$, however, the numerator is the *opposite* side; for cos $\alpha$, the *adjacent* side. Tan $\alpha$ is constructed by dividing the opposite side by the adjacent side and really represents the *slope* (or grade) of the line *AB*. (The inverse of the tangent function, or *cotangent*, is seldom used for our purposes; it is represented by *BC/AC*.)

It is extremely important for the student to be able to construct these ratios regardless of the orientation of triangle ABC. For example, these ratios for angle $\beta$ in Fig. 1.10(a) would be as follows:

$$\sin \beta = \frac{BC}{AB}$$

$$\cos \beta = \frac{AC}{AB}$$

$$\tan \beta = \frac{BC}{AC}$$

and, similarly, for angle $\theta$ of Fig. 1.10(b), we have

$$\sin \theta = \frac{b}{a} \tag{1.4a}$$

$$\cos \theta = \frac{c}{a} \tag{1.4b}$$

$$\tan \theta = \frac{b}{c} \tag{1.4c}$$

Now consider these equations. If Eq. 1.4(a) is divided through by Eq. 1.4(b), we get

$$\frac{\sin \theta}{\cos \theta} = \frac{b/a}{c/a} = \frac{ba}{ca} = \frac{b}{c}$$

But $b/c$ is equal to $\tan \theta$ as in Eq. 1.4(c); therefore, the tangent of an angle is equal to the sine divided by the cosine of the same angle, that is,

$$\tan \theta = \frac{\sin \theta}{\cos \theta} \tag{1.4d}$$

Now, let's construct the sum of the squares of Eqs. 1.4(a) and 1.4(b) as follows. First,

$$\sin^2 \theta = \frac{b^2}{a^2}$$

$$\cos^2 \theta = \frac{c^2}{a^2}$$

Adding these two equations results in

$$\sin^2 \theta + \cos^2 \theta = \frac{b^2}{a^2} + \frac{c^2}{a^2} = \frac{b^2 + c^2}{a^2}$$

Referring back to Eq. 1.2, the sum of the squares of the two sides of a right triangle is equal to the square of the hypotenuse, or $b^2 + c^2 = a^2$. Thus, substituting $a^2$ for $b^2 + c^2$ in the above equation yields

$$\sin^2 \theta + \cos^2 \theta = 1 \tag{1.5}$$

The above equation is one of the most important trigonometric identities there is. It simply says that the sine and cosine of a given angle must satisfy this equation, that the sum of their squares must equal unity. Once the sine of an angle is known, its cosine can obviously be calculated using this equation, and vice versa.

EXAMPLE 6: In the triangle shown in Fig. 1.11, calculate sin $A$, cos $A$, tan $A$, sin $B$, cos $B$, and tan $B$. Also, what are the values of the angles $A$ and $B$ in degrees?

SOLUTION:

$$AB^2 = AC^2 + BC^2 = 4^2 + 8^2 = 80$$

$$AB = 8.94$$

$$\sin A = \frac{BC}{AB} = \frac{8}{8.94} = 0.89$$

$$\cos A = \frac{AC}{AB} = \frac{4}{8.94} = 0.45$$

$$\tan A = \frac{BC}{AC} = \frac{8}{4} = 2$$

$$\sin B = \frac{AC}{AB} = \frac{4}{8.94} = 0.45$$

$$\cos B = \frac{BC}{AB} = \frac{8}{8.94} = 0.89$$

$$\tan B = \frac{AC}{BC} = \frac{4}{8} = 0.5$$

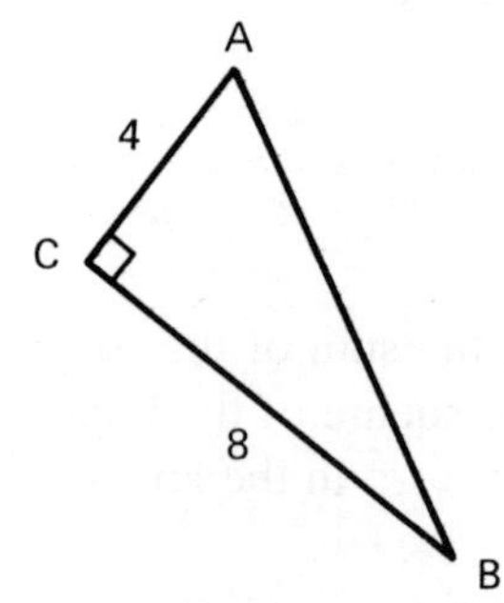


Figure 1.11

The values of angles $A$ and $B$ can be calculated by using any one of the trigonometric functions evaluated above (after a function has been evaluated, merely push the INV trigonometric function of a calculator).* For example, if $\tan A = 2$, then INV tan 2 (or arctan 2 or $\tan^{-1} 2$) = 63.4°. If $\cos B = 0.89$, then INV cos 0.89 (or arccos 0.89 or $\cos^{-1}$ 0.89) = 27.1°.

Notice that $\hat{A} + \hat{B} = 63.4° + 27.1° = 90.5°$, but we know that the sum of angles $A$ and $B$ in a right triangle should be equal to 90°. The discrepancy is due to the fact that in calculating angle $B$ we have used the value, $\cos B = 0.89$, which is merely a two-digit approximation. The accuracy can be improved if we enter the values 8 and 8.94 and obtain the ratio, 8/8.94. Leaving all its decimal places in the calculator, push $\cos^{-1}$ key. This results in a value for angle $B$ of 26.51°, not 27.1°. The thing to remember here is that one should not truncate the intermediate results; that is, all the calculations should be performed with the highest accuracy (the greatest number of decimal places in the calculator) and only the very last result truncated—in this case, to one decimal place.

It is strongly recommended that to improve the accuracy of calculations, students take full advantage of their calculators and avoid recording intermediate operations. If some value is to be used in later calculations, this value should be entered (without truncating it) into the memory of the calculator and recalled later. This procedure assures that all values will have high accuracy, and only the final result will have to be rounded off.

## Trigonometric Circle

Trigonometric functions can alternatively be studied by using the *trigonometric circle*, which is defined as a circle with unit radius.

Consider horizontal and vertical axes with their origin coinciding with the center of such a circle, as shown in Fig. 1.12. A point $P$ on the trigonometric circle can be represented by an angle $\alpha$, which can be determined by connecting the origin, $O$, to point $P$ as shown in the figure. Point $P$ can now be projected onto the horizontal and vertical axes by dropping two perpendicular lines to these axes from $P$. In triangle $OPx$, $Ox$ is the projection of $OP$ on the horizontal axis, and, since

*More on this subject in Sec. 1.4.

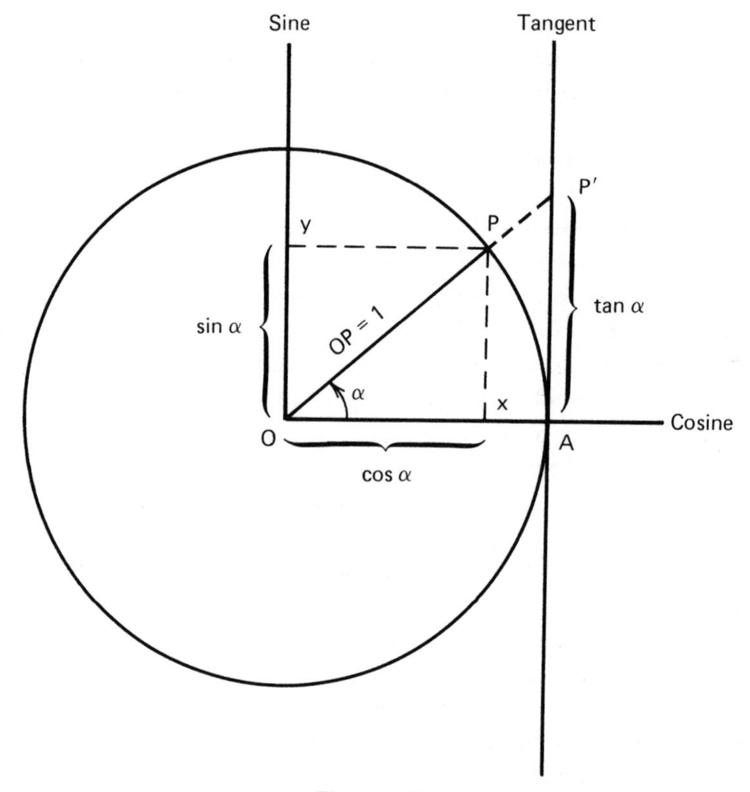


**Figure 1.12**

$\cos \alpha = Ox/OP$, its value can be calculated by the equation, $Ox = OP \cos \alpha$. Since, however, the radius of this circle is equal to unity (that is, $OP = 1$), therefore

$$Ox = \cos \alpha$$

Similarly, $Oy = Px = OP \sin \alpha$, or

$$Oy = \sin \alpha$$

These relations indicate that the projections of $OP$ on the horizontal and vertical axes are actually the cosine and sine of angle $\alpha$, which is the angle between $OP$ and the horizontal axis. The horizontal and vertical axes are thus assigned cosine and sine values, respectively.

Now, if we draw an axis from point A that is parallel to the *sine axis*, which would then be perpendicular to the *cosine axis*, of course, and then extend radius $OP$ to intersect this axis at $P'$, in triangle $OAP'$ we would have

$$\tan \alpha = \frac{AP'}{OA}$$

and since $OA = 1$,

$$AP' = \tan \alpha$$

which says that $AP'$ is nothing other than the tangent of angle $\alpha$. The axis on which it lies is called the *tangent axis*.

Notice that when a circle is divided into four quarters (called *quadrants*), the values of cosine to the right of the perpendicular axis at $O$ are considered to be positive and those to the left of it, negative. The values of sine and tangent above the horizontal axis are considered to be positive and those below it, negative. Therefore, a point in the second quadrant (quadrants are numbered counterclockwise from upper right to left) will have negative cosine and tangent but positive sine values. Figure 1.13 shows the signs of sine, cosine, and tangent in all four quadrants.

The following relationships are also readily evident from Figs. 1.12 and 1.13:

$$\begin{array}{ll} \cos 0° = 1 & \cos 90° = 0 \\ \sin 0° = 0 & \sin 90° = 1 \\ \tan 0° = 0 & \tan 90° = +\infty \end{array}$$

$$\begin{array}{lll} \cos 180° = -1 & \cos 270° = 0 & \cos 360° = 1 \\ \sin 180° = 0 & \sin 270° = -1 & \sin 360° = 0 \\ \tan 180° = 0 & \tan 270° = -\infty & \tan 360° = 0 \end{array}$$

Now consider two triangles, $OPx$ and $OPy$, as shown in the trigonometric circle of Fig. 1.14. These two triangles are equal since they both have $OP$ in common, each has one 90° angle, and the other angles are equal ($P\hat{O}x = y\hat{P}O = \alpha$). Let's denote $y\hat{O}P$ by $\beta$, which is equal to $90° - \alpha$.

In triangle $OPx$, $Ox = \cos \alpha$; in triangle $OPy$, $Py = \sin \beta$ (remember that the radius of the trigonometric circle is one unit); but, since $Py =$

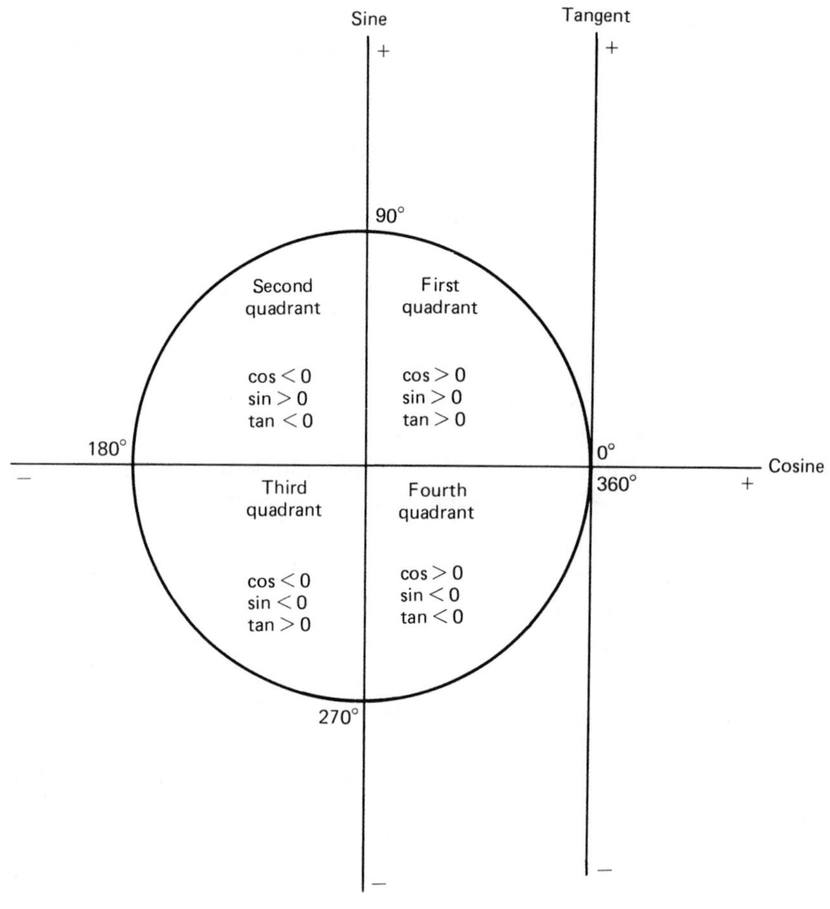


**Figure 1.13**

$Ox$, we have

$$\cos \alpha = \sin \beta = \sin (90° - \alpha)$$

Thus, if the sum of two angles is 90°, the sine of one is equal to the cosine of the other.

Now if $Py$ is extended to the second quadrant, a mirror image of the points and lines of the first quadrant is produced in the second quadrant. Since similar arguments can be made about triangle $OP'x'$, we have the following:

$$\cos \alpha = -\cos (180° - \alpha)$$

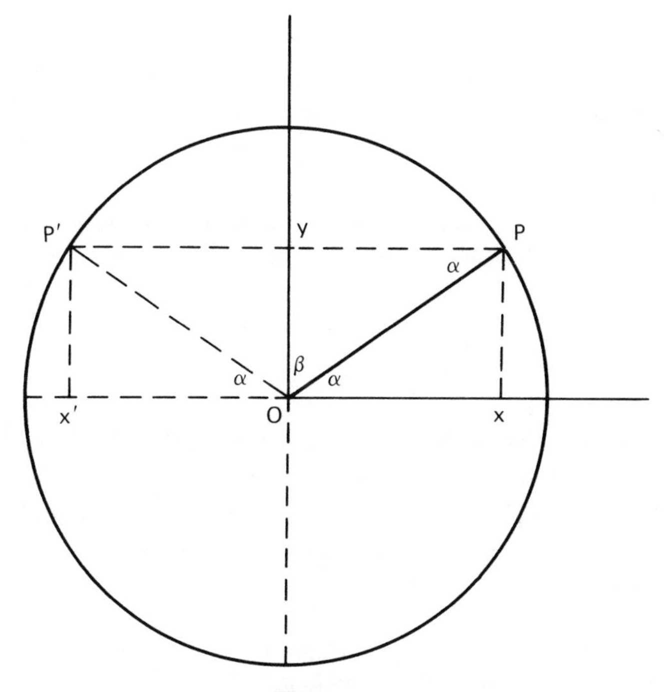


**Figure 1.14**

Considering trigonometric functions of angle $\alpha$ and their relationships with trigonometric functions of angles $-\alpha$, $180 \pm \alpha$, $360 \pm \alpha$, and $90 \pm \alpha$, results similar to those demonstrated above can be obtained. These relationships, known as *trigonometric reduction formulas*, are listed in Table 1.1. It is important to point out that students should not merely memorize this table, but each time such a relationship is needed, they should think of Fig. 1.14 instead and work out the appropriate formula.

**TABLE 1.1. Trigonometric Reduction Formulas.**

| | | |
|---|---|---|
| $\sin(-\alpha) = -\sin\alpha$ | $\cos(-\alpha) = \cos\alpha$ | $\tan(-\alpha) = -\tan\alpha$ |
| $\sin(180° + \alpha) = -\sin\alpha$ | $\cos(180° + \alpha) = -\cos\alpha$ | $\tan(180° + \alpha) = \tan\alpha$ |
| $\sin(180° - \alpha) = \sin\alpha$ | $\cos(180° - \alpha) = -\cos\alpha$ | $\tan(180° - \alpha) = -\tan\alpha$ |
| $\sin(90° + \alpha) = \cos\alpha$ | $\cos(90° + \alpha) = -\sin\alpha$ | $\tan(90° + \alpha) = -\dfrac{1}{\tan\alpha}$ |
| $\sin(90° - \alpha) = \cos\alpha$ | $\cos(90° - \alpha) = \sin\alpha$ | $\tan(90° - \alpha) = \dfrac{1}{\tan\alpha}$ |
| $\sin(360° + \alpha) = \sin\alpha$ | $\cos(360° + \alpha) = \cos\alpha$ | $\tan(360° + \alpha) = \tan\alpha$ |
| $\sin(360° - \alpha) = -\sin\alpha$ | $\cos(360° - \alpha) = \cos\alpha$ | $\tan(360° - \alpha) = -\tan\alpha$ |

## Sine and Cosine Laws

Consider triangle ABC in Fig. 1.15. There exist two sets of relationships between the angles and the lengths of the sides of this triangle. In the first set, sines of angles are involved and, in the second set, cosines.

*Sine Law.* Each side of triangle ABC in Fig. 1.15 is proportional to the sine of the angle opposite to that side, that is,

$$\frac{\sin A}{a} = \frac{\sin B}{b} = \frac{\sin C}{c} \tag{1.6}$$

This relationship, called the *sine law*, is frequently used in solving triangles. For example, if two sides and one angle (other than the included angle) of a triangle are known, the other angles and remaining side can be calculated using Eq. 1.6, or, if two angles and one side are known, the other angle and sides can be easily calculated.

*Cosine Law.* If two sides (in this case, $a$ and $c$) and an *included* angle of a triangle (in this case, $B$) are known, one can easily find the third side ($b$) by using the *cosine law*, as follows:

$$b^2 = a^2 + c^2 - 2ac \cos B \tag{1.7}$$

Similarly, if any pairs of two sides and their included angle are known, the following similar formulas can be used:

$$a^2 = b^2 + c^2 - 2bc \cos A$$

$$c^2 = a^2 + b^2 - 2ab \cos C$$

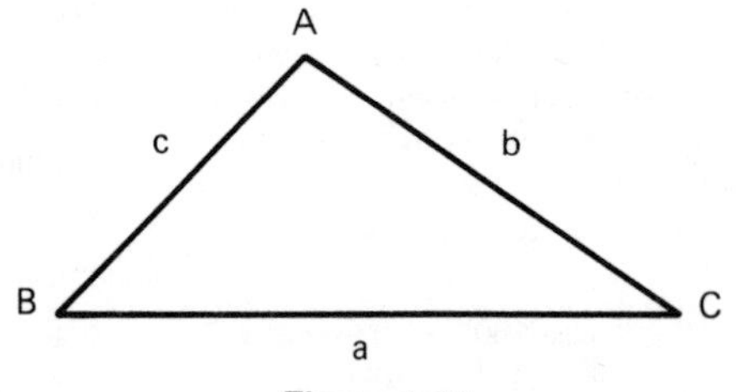


**Figure 1.15**

When the included angle is 90°, cos 90° = 0, and the cosine law thus reduces to the Pythagorean theorem (see Eq. 1.2).

EXAMPLE 7: Find the missing angle and sides of the triangle shown in Fig. 1.16.

SOLUTION: In this problem, two angles and one side are known. The third angle can be found as follows:

$$A + B + C = 180°$$

$$A + 70° + 30° = 180°$$

$$A = 80°$$

The two sides can be found by using the sine law twice:

$$\frac{\sin A}{a} = \frac{\sin B}{b}$$

$$\frac{\sin 80°}{10} = \frac{\sin 30°}{b}$$

$$b = \frac{10 \sin 30°}{\sin 80°} = 5.08$$

and

$$\frac{\sin A}{a} = \frac{\sin C}{c}$$

$$\frac{\sin 80°}{10} = \frac{\sin 70°}{c}$$

$$c = \frac{10 \sin 70°}{\sin 80°} = 9.54$$

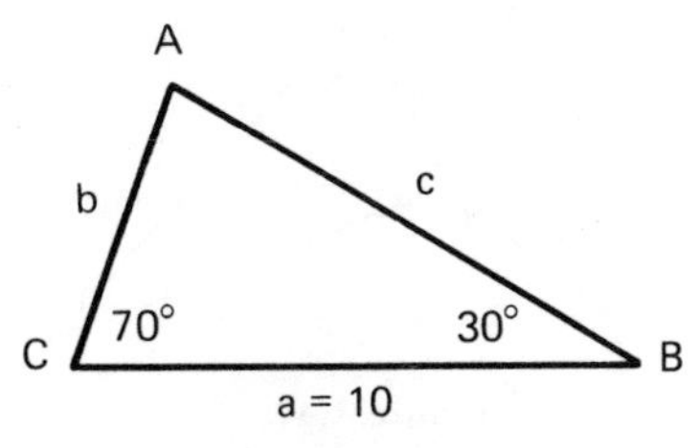


**Figure 1.16**

(Prior to the availability of calculators, students were encouraged to memorize the exact trigonometric values of certain round angles such as 30°, 45°, 60°, etc., and use trigonometric tables to obtain the sine, cosine, and tangent of other angles. These days, life is much simpler; the entire trigonometric table with very high accuracy is stored in your calculator. To obtain sin 65°, say, just punch in 65 and push the SIN key; the value of sin 65° will appear instantly.)

EXAMPLE 8: Find the missing angles and side of the triangle shown in Fig. 1.17.

SOLUTION: In this problem, two sides and an angle other than the included angle are given. Using the sine law, the unknowns can be found as follows:

$$\frac{\sin B}{b} = \frac{\sin A}{a}$$

$$\frac{\sin 25^\circ}{18} = \frac{\sin A}{12}$$

$$\sin A = 0.28$$

$$A = 16.4^\circ$$

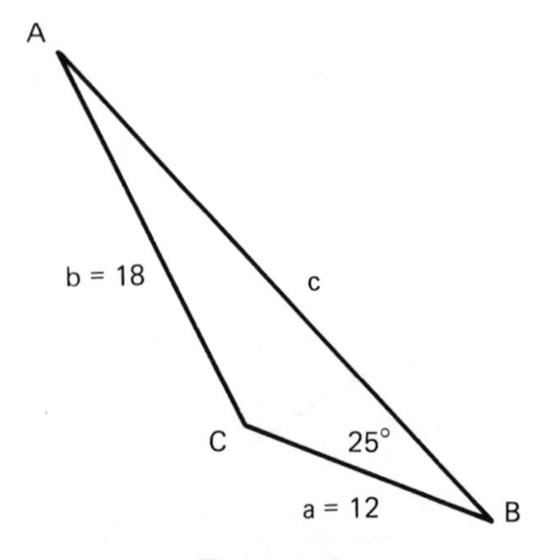


Figure 1.17

and

$$A + B + C = 180°$$

$$16.4° + 25° + C = 180°$$

$$C = 138.6°$$

and

$$\frac{\sin B}{b} = \frac{\sin C}{c}$$

$$\frac{\sin 25°}{18} = \frac{\sin 138.6°}{c}$$

$$c = 28.17$$

EXAMPLE 9: Find the missing side and angles of the triangle shown in Fig. 1.18.

SOLUTION: In this problem, two sides and the included angle are known. The missing side can be found by using the cosine law and the second angle can be found by using the sine law, as follows:

$$c^2 = a^2 + b^2 - 2ab \cos C$$

$$= 25^2 + 30^2 - 2 \times 25 \times 30 \cos 60°$$

Solving for $c$

$$c = 27.84$$

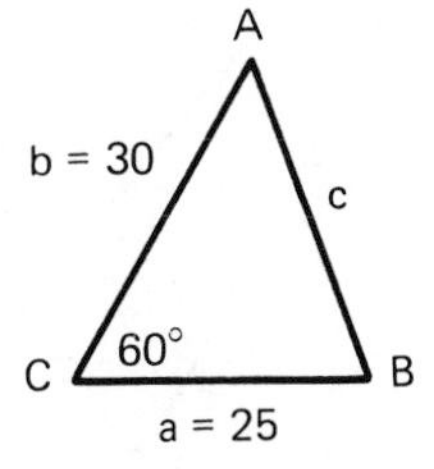


**Figure 1.18**

Then,

$$\frac{\sin A}{a} = \frac{\sin C}{c}$$

$$\frac{\sin A}{25} = \frac{\sin 60°}{27.84}$$

$$\sin A = \frac{25 \sin 60°}{27.84} = 0.78$$

$$A = 51°$$

and

$$A + B + C = 180°$$

$$51 + B + 60° = 180°$$

$$B = 69°$$

EXAMPLE 10: Find the missing side and angles of the triangle shown in Fig. 1.19.

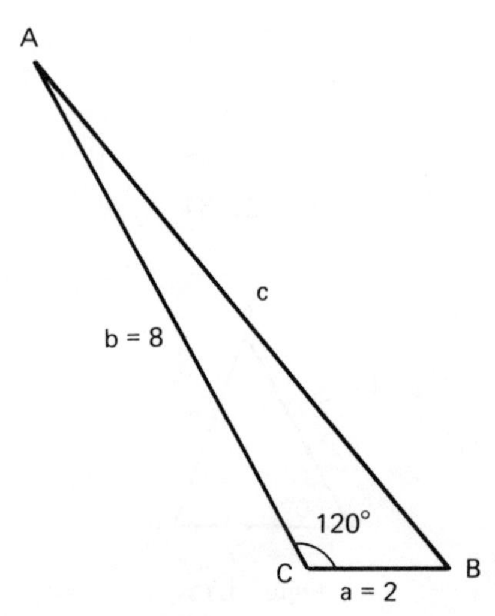


Figure 1.19

SOLUTION: This is similar to the problem solved above except that the included angle is greater than 90°. By the cosine law,

$$c^2 = a^2 + b^2 - 2ab \cos C = 2^2 + 8^2 - 2 \times 2 \times 8 \cos 120°$$

Solving for $c$,

$$c = 9.17$$

and

$$\frac{\sin A}{a} = \frac{\sin C}{c}$$

$$\frac{\sin A}{2} = \frac{\sin 120°}{9.17}$$

$$\sin A = \frac{2 \sin 120°}{9.17} = 0.19$$

$$A = 10.9°$$

$$A + B + C = 180°$$

$$10.9° + B + 120° = 180°$$

$$B = 49.1°$$

EXAMPLE 11: Find the missing angles in the triangle in Fig. 1.20.

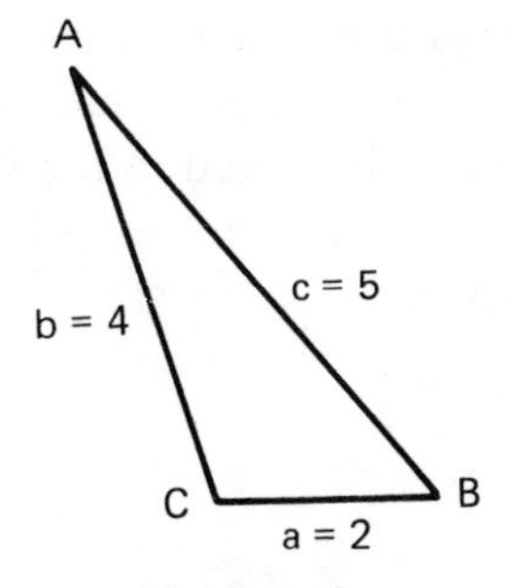


**Figure 1.20**

SOLUTION: Use the cosine law first and then the sine law, as follows:

$$a^2 = b^2 + c^2 - 2ab \cos A$$
$$2^2 = 4^2 + 5^2 - 2 \times 4 \times 5 \cos A$$

Solving for $A$,

$$\cos A = 0.93$$
$$A = 22.3°$$

Then,

$$\frac{\sin A}{a} = \frac{\sin B}{b}$$
$$\frac{\sin 22.3°}{2} = \frac{\sin B}{4}$$
$$\sin B = \frac{4 \sin 22.3°}{2} = 0.76$$
$$B = 49.4°$$

$$A + B + C = 180°$$
$$22.3° + 49.4° + C = 180°$$
$$C = 108.3°$$

## 1.4 ARITHMETICS AND ALGEBRA

Solving mechanics problems often involves dealing with fractions, exponents or powers, coordinate systems, equations, and derivatives, for which the governing rules are discussed in the following sections.

### Manipulation of Fractions

Consider the following equality:

$$\frac{a}{b} = \frac{c}{d} \tag{1.8}$$

Equation 1.8 can be manipulated to advantage in many ways. Some useful options are listed below:

$$\frac{b}{a} = \frac{d}{c}$$ (Both sides inverted)

$$\frac{a}{b} + k = \frac{c}{d} + k$$ (Constant $k$ added to both sides)

$$k\left(\frac{a}{b}\right) = k\left(\frac{c}{d}\right)$$ (Both sides multiplied by constant $k$)

$$a = b\left(\frac{c}{d}\right)$$ (Solved for $a$ by multiplying both sides by $b$)

$$\sqrt{\frac{a}{b}} = \sqrt{\frac{c}{d}}$$ (Square root of both sides taken)

$$\left(\frac{a}{b}\right)^2 = \left(\frac{c}{d}\right)^2$$ (Both sides squared, that is, raised to the second power)

$$\frac{a^2}{b^2} = \frac{c^2}{d^2}$$ (Alternative form of both sides squared)

$$ad = bc$$ (This equation equivalent to Eq. 1.8, in which fraction form is eliminated)

Note also that the expression

$$\frac{a + b}{c} = \frac{d - e}{f}$$

can be rewritten, among other ways, as follows:

$$\frac{a}{c} + \frac{b}{c} = \frac{d}{f} - \frac{e}{f}$$

$$(a + b)f = (d - e)c$$

$$af + bf = dc - ec$$

## Manipulations Involving Powers

In this section, a few simple and useful operations involving exponents or powers are listed, as follows:

*Meaning of Exponents.* $a^2$ means $a \times a$; $a^3$ means $a \times a \times a$; and $a^n$ ("$a$" raised to power "$n$") means $a$ is multiplied by itself $n$ times. If $a$ is raised to a negative power, it can be written as follows:

$$a^{-n} = \frac{1}{a^n}$$

In other words, negative exponents provide the inverse of an original quantity.

*Multiplication of Exponents.* $a^n$ and $a^m$ are multiplied according to the following rule:

$$a^n \times a^m = a^{n+m}$$

EXAMPLE 12: Evaluate $x^2 \times x^5$.

SOLUTION:

$$x^2 \times x^5 = x^{2+5} = x^7$$

*Division of Exponents.* $a^n$ and $a^m$ are divided according to the following rule:

$$\frac{a^n}{a^m} = a^{n-m}$$

EXAMPLE 13: Evaluate $\frac{x^5}{x^3}$.

SOLUTION:

$$\frac{x^5}{x^3} = x^{5-3} = x^2$$

*Raising of Exponents to a Power.* $a^n$ is raised to the $m$'th power according to the following rule:

$$(a^n)^m = a^{nm}$$

*Roots.* $\sqrt{a}$ means the square root of "$a$"; it can also be written as $a^{1/2}$. $\sqrt[3]{a}$ means the cubic root of "$a$"; it can also be written as $a^{1/3}$. $\sqrt[n]{a}$ means the $n$'th root of "$a$"; it can also be written as $a^{1/n}$.

EXAMPLE 14: $\sqrt{4} = \pm 2$ and $\sqrt[3]{27} = 27^{1/3} = 3$.

*Important Identities.* The following identities are also very important and frequently used:

$$(a \times b)^m = a^m \times b^m$$

$$(a + b)^2 = a^2 + b^2 + 2ab$$

$$(a - b)^2 = a^2 + b^2 - 2ab$$

## Coordinate Systems

Solving most problems in mathematics (or mechanics) will be much easier if an appropriate coordinate system is used. Different types of coordinate systems offer different advantages for different applications. The simplest type is the rectangular coordinate system, the only coordinate system used in this book. Its principal purpose, like that of all coordinate systems, is to provide a single frame to which all measurements can be referred.

The simplest coordinate system involves nothing more than a one-dimensional coordinate axis. Suppose several cars are moving along a straight line, as shown in Fig. 1.21. To keep track of their positions

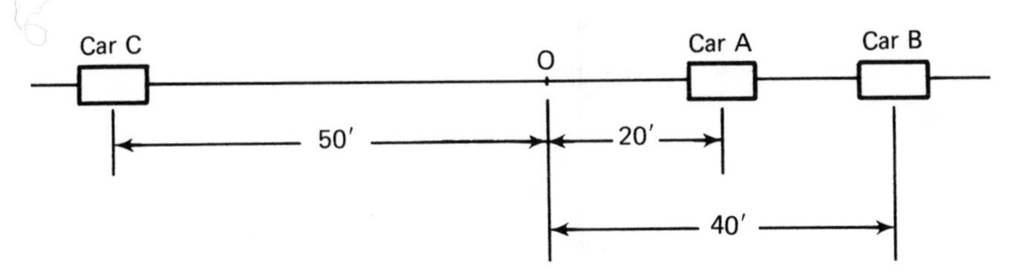


**Figure 1.21**

(their locations), a common point "$O$" is used as a reference. This point is called the *origin*, and the location of each car is measured from this origin.

The distance of a car from origin "$O$" is called its *coordinate*. To indicate that a car is located to the right of the origin, a positive sign is assigned to its coordinate value; the coordinate for car A, therefore, is (+) 20. The coordinate of car C is −50, which indicates that it is located to the left of the origin at a distance of 50′.

Since not many problems are one-dimensional like this one, there is a need for a two-dimensional coordinate system. The two-dimensional rectangular coordinate system consists of two straight lines perpendicular to one another. Each of these straight lines is called a *coordinate axis*, and their point of intersection is called the *origin*. These two axes divide the plane into four quadrants, as shown in Fig. 1.22.

Consider point $A$ in the first quadrant. Its location can be identified by knowing its perpendicular distances from the two coordinate axes, $x$ and $y$. In this example, the perpendicular distance from the $y$ axis is equal to 2″, and the perpendicular distance from the $x$ axis is equal to 3″. For

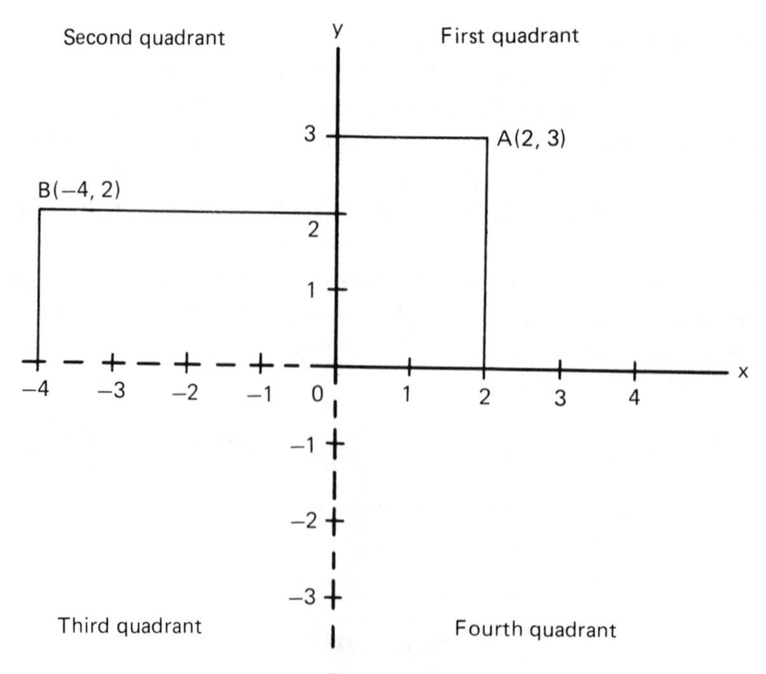


**Figure 1.22**

convenience, the two coordinates of point $A$ are collectively shown in shorthand form as $A(2,3)$ or $A|_3^2$ or $A_x = 2''$, $A_y = 3''$. Both these coordinates are positive. Notice that points in the second quadrant have a negative $x$ coordinate and a positive $y$ coordinate, for example, $B(-4,2)$. Points in the third quadrant have both coordinates negative, whereas points in the fourth quadrant have positive $x$ and negative $y$ coordinates.

Now consider point $A$ in Fig. 1.23 with coordinates $A_x = 5''$ and $A_y = 2''$. Since triangle $OAA'$ is a right triangle ($\hat{A}' = 90°$), we can write

$$OA^2 = OA'^2 + AA'^2 = 5^2 + 2^2 = 29$$

or

$$OA = \sqrt{29} = 5.39''$$

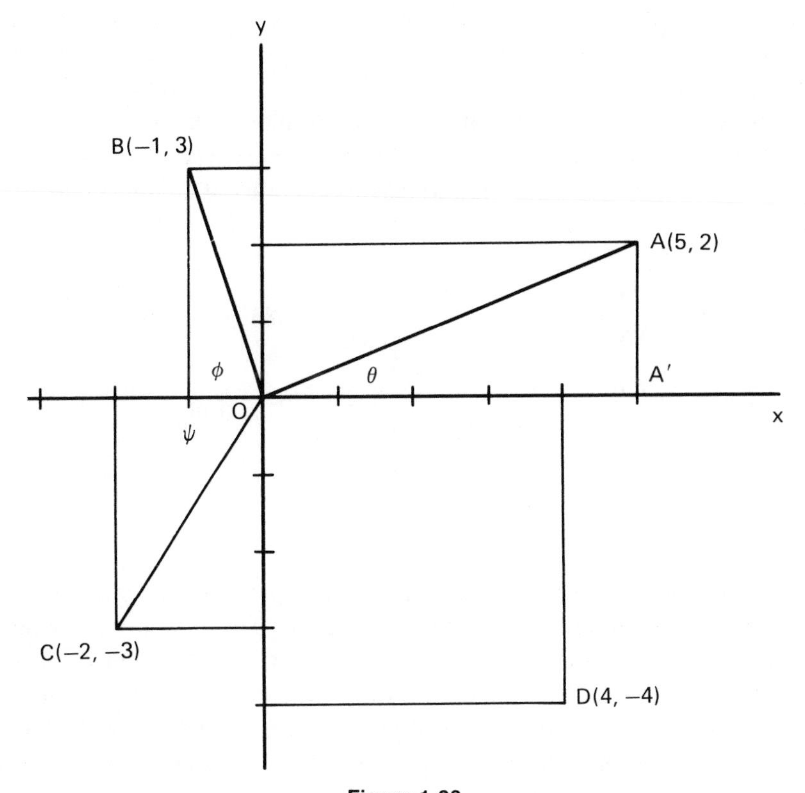


Figure 1.23

Using the trigonometric functions defined earlier, we can write

$$\sin\,\theta = \frac{AA'}{OA} = \frac{2}{5.39} = 0.371''$$

$$\cos\,\theta = \frac{OA'}{OA} = \frac{5}{5.39} = 0.928''$$

$$\tan\,\theta = \frac{AA'}{OA'} = \frac{2}{5} = 0.4''$$

where angle $\theta$ is the angle between $OA$ and the $x$ axis. The trigonometric functions of angle $\theta$ associated with point $A$ are defined as the ratios of the $x$ or $y$ coordinate of $A$ to length $OA$ (positive value) for the sine and cosine and the ratio of the $y$ and $x$ coordinates for the tangent.

Because the signs of both coordinates in the first quadrant are positive, all trigonometric functions of this angle are positive. For related reasons, angle $\phi$ in the second quadrant has positive sine but negative cosine and tangent functions. Note points $C$ and $D$ and their coordinates, all negative in the third and positive and negative in the fourth quadrants. Their trigonometric functions carry signs that reflect these values. For example, let's evaluate signs of trigonometric functions of a point in the third quadrant such as point $C$:

$$\sin\,\Psi = \frac{CC'}{OC'} = \frac{-3}{(-2)^2 + (-3)^2} = -.083, \text{ a negative value}$$

$$\cos\,\Psi = \frac{OC'}{OC} = \frac{-2}{(-2)^2 + (-3)^2} = -.055, \text{ a negative value}$$

$$\tan\,\Psi = \frac{CC'}{OC'} = \frac{-3}{-2} = 1.5, \text{ a positive value}$$

Notice that angle $\Psi$ is an *acute angle* (any angle less than 90°), and its trigonometric functions should all be positive. The reason we obtained negative values for sin $\Psi$ and cos $\Psi$ is that to identify point $C$ we use an angle whose actual value is $\Psi + 180°$ (in other words, the angle that $OC$ makes with the positive $x$ axis). It is customary to use acute angles such as $\phi$ and $\Psi$ and add 90°, 180°, or 270° to them as the situation warrants (that is, to extend them into the second, third, or fourth quadrants). If coordinates of a point are entered in a calculator with their proper signs, the calculator will automatically take care of such additions.

Now consider the line segment *AB* in relation to the *x-y* coordinate system in Fig. 1.24. By drawing two lines from *A* and *B* perpendicular to the *x* axis, we simply project line segment *AB* onto the *x* axis. Similarly, by drawing two more lines from *A* and *B* perpendicular to the *y* axis, we project line segment *AB* onto the *y* axis. The length, *A'B'*, is called the *x* component of line *AB*, and, similarly, the length, *A"B"*, is called its *y* component. These components are indicated by using subscripts *x* and *y* as follows:

$$AB_x \text{ and } AB_y$$

If *A"A* is extended to the right and its angle with line segment *AB* is denoted by $\alpha$, one can easily find the *x* and *y* components of *AB* if one knows the length of *AB* and the magnitude of angle $\alpha$, as follows:

$$\begin{aligned} AB_x &= A'B' = AB \cos \alpha \\ AB_y &= A''B'' = AB \sin \alpha \end{aligned} \tag{1.9}$$

Note that angle $\alpha$ is actually the same as the angle that line segment *AB* makes with the *x* axis.

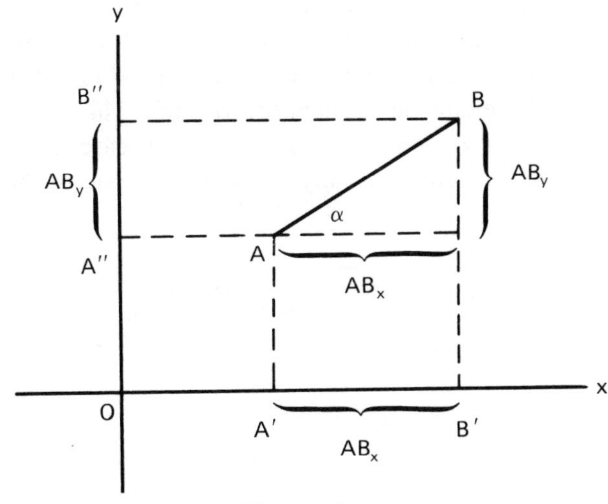


**Figure 1.24**

Suppose now that the $x$ and $y$ components of $AB$ are given and the length of $AB$ and the value of angle $\alpha$ are to be found. To solve this problem, we square both sides of Eq. 1.9 and then take their sum, as follows:

$$AB_x^2 + AB_y^2 = AB^2 \cos^2 \alpha + AB^2 \sin^2 \alpha = AB^2 (\cos^2 \alpha + \sin^2 \alpha)$$

but since $\cos^2 \alpha + \sin^2 \alpha = 1$, therefore

$$AB^2 = AB_x^2 + AB_y^2$$

$$AB = \sqrt{AB_x^2 + AB_y^2} \qquad (1.10)$$

In taking the square root of both sides of the above equation, we have taken only the positive value due to the fact that length is always positive.

Now, by dividing the second equation in 1.9 by the first equation term by term, we get

$$\frac{AB_y}{AB_x} = \frac{AB \sin \alpha}{AB \cos \alpha} = \frac{\sin \alpha}{\cos \alpha} = \tan \alpha$$

In other words, to find the tangent of angle $\alpha$, one only has to divide the value of the $y$ component of $AB$ by its $x$ component. Doing so provides only the tangent of the angle, however, not the angle itself. To arrive at angle $\alpha$, the preceding equation has to be inverted. We must find the angle whose tangent is equal to $AB_y/AB_x$. This process requires the use of *inverse trigonometric functions*—in this case, the inverse tangent. The inverse tangent is usually expressed by one of the following forms:

$$\alpha = \text{arc tan} \frac{AB_y}{AB_x}$$

or

$$\alpha = \tan^{-1} \frac{AB_y}{AB_x}$$

which means the angle whose tangent is equal to $AB_y/AB_x$. Note that in the second notation, $-1$ does not represent a power of the tan function but rather its inverse, that is, the angle whose tangent is equal to $AB_y/AB_x$.

Similarly, the inverse trigonometric functions of sine and cosine are written as follows:

$$\alpha = \text{arc sin}\,\frac{AB_y}{AB} \qquad \text{or} \qquad \alpha = \sin^{-1}\frac{AB_y}{AB}$$

$$\alpha = \text{arc cos}\,\frac{AB_x}{AB} \qquad \text{or} \qquad \alpha = \cos^{-1}\frac{AB_x}{AB}$$

Calculators have three keys for SIN, COS, and TAN and three other keys for $\text{SIN}^{-1}$, $\text{COS}^{-1}$, and $\text{TAN}^{-1}$. To obtain an inverse trigonometric function, some calculators require one to activate the INV (inverse) or 2ndF (second function) key first and then SIN, etc., whereas in using calculators having the inverse trigonometric function keys, simply pressing these keys will give the value of the angle. Note that to obtain angle $\alpha$ using a calculator, the value of the angle will be displayed in degrees or radians, depending on which mode the calculator is set for. In all of the applications in this book, we will use degrees; therefore, keep your calculator set to the degree mode so that all angles will be given in degrees.

## Equations

In solving practically any technical problem, one finally has to solve an equation of some kind in order to obtain a numerical answer. In this section, we will review only those equations that are germane to the subject of this book.

*Equations with One Unknown.* Equations with one unknown have the following general form:

$$f(x) = 0 \tag{1.11}$$

where the left-hand side—which, in general, is denoted by $f(x)$, meaning function of $x$—is an algebraic expression involving the unknown $x$ and some constants. Some examples of the forms this type of equation may take are the following:

$$2x - 1 = 0$$

$$3x^2 + 2x - 5 = 0$$

$$9 \sin x + 17/2 = 0$$

*Linear Equations.* One general form that Eq. 1.11 can take is the *polynomial form.* The simplest form of polynomial is the *linear equation* (the highest power of the unknown $x$ is unity), which is expressed as

$$ax + b = 0 \tag{1.12}$$

where $a$ and $b$ are known constants.

To solve Eq. 1.12, first transfer $b$ to the other side of the equal sign (in this process, the sign of $b$ must be changed), and then divide both sides of this new equation by $a$ (the coefficient of $x$). This process is shown step-by-step in the following two examples:

EXAMPLE 15: Solve the equation: $2x - 1 = 0$ (find the $x$ that will satisfy this equation).

SOLUTION:

Step 1. Transpose $-1$ to the other side, as follows:

$$2x = 1$$

Step 2. Divide both sides of the above equation by 2, as follows:

$$x = \frac{1}{2} = 0.5$$

EXAMPLE 16: Solve the equation $-3x + 9 = 0$

SOLUTION:

Step 1. $-3x = -9$

Step 2. $x = \dfrac{-9}{-3} = 3$

*Quadratic Equations.* The second type of polynomial used in this book is the *second-order polynomial*, which has the following general form:

$$ax^2 + bx + c = 0 \tag{1.13}$$

where $a$, $b$, and $c$ are constants, with $a \neq 0$. (If $a$ is equal to zero, then this equation reduces to the linear equation of Eq. 1.12.) Equation 1.13 is called a second-order polynomial because the highest power of $x$ is 2; it is also called a *parabolic* or *quadratic equation*.

To solve Eq. 1.13, use the following formula:

$$x = \frac{-b \pm \sqrt{b^2 - 4ac}}{2a} \tag{1.14}$$

In general, therefore, any quadratic equation will have two solutions (note the plus-or-minus sign preceding the radical):

$$x_1 = \frac{-b + \sqrt{b^2 - 4ac}}{2a}$$

and

$$x_2 = \frac{-b - \sqrt{b^2 - 4\,ac}}{2a}$$

For applications in this book, the expression, $b^2 - 4ac$, will always have *nonnegative* values.

EXAMPLE 17: Solve $x^2 - 1 = 0$

SOLUTION: In this case, $a = 1$, $b = 0$, and $c = -1$. Consequently, one can use Eq. 1.14 to solve this equation, but since it has two terms only (the second-order term, $x^2$, and the constant term, $-1$), it is much simpler merely to transfer the constant term to the other side of the equal sign and then take the square root of both sides. First,

$$x^2 = 1$$

Then, taking the square root of both sides,

$$x = \pm 1$$

which means that both $+1$ and $-1$ are solutions to the equation, $x^2 - 1 = 0$.

EXAMPLE 18: Solve $3x^2 - 5x = 2$

SOLUTION:

Step 1. Put the equation in the general form (all terms on one side), as follows:

$$3x^2 - 5x - 2 = 0$$

In this case, therefore, $a = 3$, $b = -5$, and $c = -2$:

Step 2. Substitute these values in Eq. 1.14 as follows:

$$x = \frac{-(-5) \pm \sqrt{(-5)^2 - 4(3)(-2)}}{2(3)}$$

Since the $\pm$ sign in front of the square root sign means that once the value of the square root has been evaluated, it is to be added to the first term to get $x_1$ and then subtracted from the first term to get $x_2$, we obtain:

$$x_1 = \frac{-(-5) + \sqrt{(-5)^2 - 4(3)(-2)}}{2(3)} = \frac{5 + \sqrt{49}}{6}$$

$$= \frac{5 + 7}{6} = 2$$

and

$$x_2 = \frac{-(-5) - \sqrt{(-5)^2 - 4(3)(-2)}}{2(3)} = \frac{5 - \sqrt{49}}{6}$$

$$= \frac{5 - 7}{6} = -\frac{1}{3} = -0.33$$

*Simultaneous Equations:* A third type of equation used in this book is the system of *simultaneous equations*. A linear system of two equations with two unknowns will look, in general, as follows:

$$\begin{cases} ax + by = c \\ a'x + b'y = c' \end{cases}$$

To solve this system, one can obtain one of the unknowns, say $x$ from the first equation, in terms of the other unknown and then substitute it in the second equation. This substitution will result in a single equation with but one unknown that can be solved as shown in Examples 15 and 16.

EXAMPLE 19: Solve the following system of simultaneous equations.

$$\begin{cases} 2x + 3y = 8 \\ 4x - y = 2 \end{cases} \tag{1a}$$

SOLUTION: Consider the first equation, $2x + 3y = 8$, assume that $y$ in this equation is not unknown, and solve the equation for $x$. By transposing $3y$ to the other side, we get $2x = 8 - 3y$, and then, by dividing both sides by the coefficient of $x$, we get

$$x = 4 - \frac{3}{2}y \tag{1b}$$

In this way we have solved the first equation for $x$ in terms of $y$. Next, we substitute this value of $x$ in the second equation as follows:

$$4\left(4 - \frac{3}{2}y\right) - y = 2$$

or

$$16 - 6y - y = 2$$

or

$$-7y = -14$$

Therefore

$$y = \frac{-14}{-7} = 2$$

Substituting this value of $y$ is either Eq. 1a or 1b will give the value of the unknown $x$. Since we solved the first equation for $x$, it will be easier to substitute $y = 2$ in Eq. 1b as follows:

$$x = 4 - \frac{3}{2}(2) = 1$$

Therefore the answer to the system of equations in this problem is ($x$ = 1, $y$ = 2). To check these answers, one can substitute these values in the first or second equation of Eq. 1a. If the values obtained are correct, they will satisfy this equation.

EXAMPLE 20: Solve the following system of three equations with three unknowns.

$$\begin{cases} x + y - z = 0 \\ 2x - 4y + z = -3 \\ 5x - y + 3z = 12 \end{cases}$$

SOLUTION: Find $x$ from the first equation in terms of $y$ and $z$ as follows: $x = -y + z$. Substitute this value for $x$ in the second and third equations:

$$\begin{cases} 2(-y + z) - 4y + z = -3 \\ 5(-y + z) - y + 3z = 12 \end{cases}$$

After simplification, we have

$$\begin{cases} -2y + z = -1 \\ -3y + 4z = 6 \end{cases}$$

Notice that in this process the unknown $x$ has been entirely eliminated, and we have two equations and two unknowns that can be solved as shown in the previous example. Find one of the unknowns, say $z$, in terms of $y$ from the first equation in the system:

$$z = 2y - 1$$

Substituting this in the second equation, we obtain

$$-3y + 4(2y - 1) = 6$$

or, after simplification,

$$5y = 10, \text{ or } y = 2$$

Substituting $y = 2$ into $z = 2y - 1$ results in $z = 3$, and finally substituting $y = 2$ and $z = 3$ into any one of the three given equations will result in the value of $x$. Using the first equation, $x + y - z = 0$:

$$x + 2 - 3 = 0, \text{ or } x = 1$$

Therefore, the solution of the given system is ($x = 1$, $y = 2$, $z = 3$).

## Derivatives

Consider the function, $y = f(x)$. If $x$ changes by a very small amount, $\Delta x$, this will cause the value of $y$ to change by a small amount, $\Delta y$. By definition, the *derivative* of $y$ with respect to $x$ is the rate of change of $y$ with respect to $x$ when $\Delta x$ is very small. Since the derivative of $y$ is denoted by $y'$, we therefore have

$$y' = \frac{\Delta y}{\Delta x} \quad \text{or} \quad \frac{f(x + \Delta x) - f(x)}{\Delta x}$$

The derivative has various applications. In this book we will use it only to obtain the slope and find the maximum or minimum value of a function—in particular, simple functions. (The most complicated function used in this book is the polynomial.) Methods used to obtain the derivatives will now be summarized as follows:

1. If the function $y$ has a constant value, its derivative is equal to zero, or

$$\text{If } y = a \ (a \text{ is a constant value}), \text{ then } y' = 0.$$

2. The derivative of $y = x$ is 1, or

$$\text{If } y = x, \text{ then } y' = 1.$$

3. If a function is multiplied by a constant, its derivative will also be multiplied by that constant, or

$$\text{If } y = ax, \text{ then } y' = a.$$

4. The derivative of $n$th powers of $x$, that is, when $y = x^n$, is obtained by multiplying the power $n$ by $x$ to the power $n - 1$, for example:

$$\text{If } y = x^2, \text{ then } y' = 2x^{2-1} = 2x$$

$$\text{If } y = x^3, \text{ then } y' = 3x^{3-1} = 3x^2$$

$$\text{If } y = x^7, \text{ then } y' = 7x^{7-1} = 7x^6$$

5. If a function is the sum of several terms, its derivative is equal to the sum of the derivatives of each individual term, for example:

$$\text{If } y = x^7 + x^3 + x^2 + 9, \text{ then } y' = 7x^6 + 3x^2 + 2x + 0.$$

In the following examples, the derivative of polynomials will be obtained by using a combination of rules outlined above.

EXAMPLE 21: Find the derivative of $y = 2x^3 - 5x^2 + \frac{1}{2}x - 2.$

SOLUTION:

$$y' = (2 \times 3x^2) - (5 \times 2x) + \frac{1}{2} - 0 = 6x^2 - 10x + \frac{1}{2}$$

EXAMPLE 22: Find the derivative of $y = 3x^6 - 15x^4 + 0.12x^3 - x + 5.$

SOLUTION:

$$
\begin{aligned}
y' &= (3 \times 6x^5) - (15 \times 4x^3) + (0.12 \times 3x^2) - 1 + 0 \\
&= 18x^5 - 60x^3 + 0.36x^2 - 1
\end{aligned}
$$

## Determination of the Maximum or Minimum of a Function

In determining the value of the maximum or minimum of a function we will consider only simple functions such as the polynomial.

A function is maximum or minimum when its derivative is equal to zero (except in some exceptional cases not applicable here). To determine the maximum or minimum of a given function, the following steps are required:

1. Find the derivative of the function.
2. Set the derivative equal to zero and solve it for $x$.
3. Substitute the value or values of $x$ obtained in step 2 into the original function to find the value of its maximum or minimum.

EXAMPLE 23: Find the maximum or minimum of the function, $y = 2x^2 + 3x - 5$.

SOLUTION:

Step 1: $y' = 4x + 3$

Step 2: When $y' = 0$, $4x + 3 = 0$, or $x = -\frac{3}{4}$.

Step 3: Substituting $x = -\frac{3}{4}$ in the original function obtains

$$
y\left(x = -\frac{3}{4}\right) = 2\left(-\frac{3}{4}\right)^2 + 3\left(-\frac{3}{4}\right) - 5 = -6\frac{1}{8}
$$

In Step 3 of Example 23, $y(x = -3/4)$ means that the value of function $y$ is evaluated at point $x = -3/4$. Figure 1.25 shows the curve for the given function. To plot this function, its values were evaluated for different values of $x$, and the corresponding pairs of $x$ and $y$ were plotted and the smooth curve then passed through these points. The minimum value of the function as revealed by Fig. 1.25 is equal to $-6.125$ at the point $x = -3/4$.

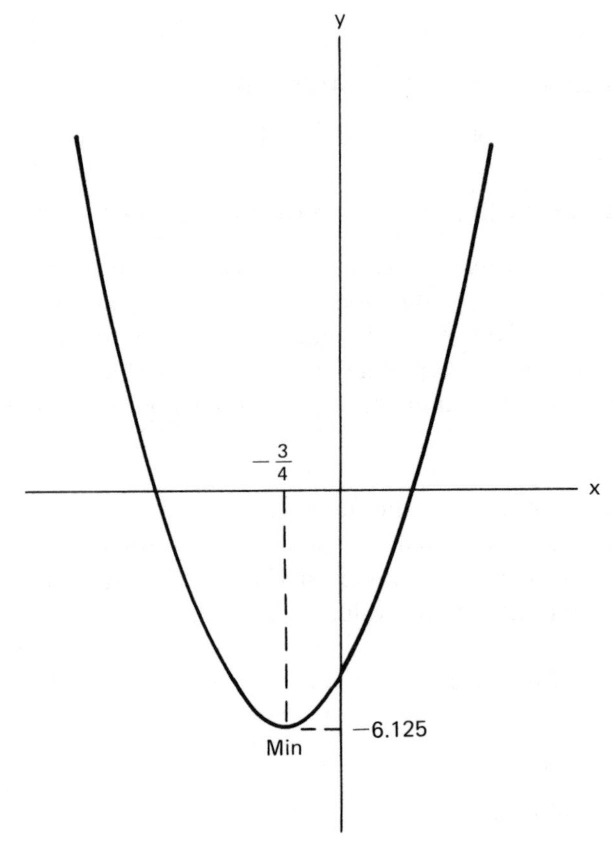


**Figure 1.25**

EXAMPLE 24: Find the maximum or minimum of the function, $y = -x^2 + 7x + 10$.

SOLUTION:

Step 1: $y' = -2x + 7$
Step 2: When $y' = 0$, $-2x + 7 = 0$, or $x = 3.5$.
Step 3: $y(x = 3.5) = -(3.5)^2 + 7(3.5) + 10 = 22.25$.

As shown in Fig. 1.26, the function in Example 24 assumes a maximum value, $y_{max}$, of 22.25 at $x = 3.5$.

Although there are methods to distinguish whether the point at which $y' = 0$ corresponds to the maximum or minimum of a function, we will use the plot of the function to make such distinctions. More examples

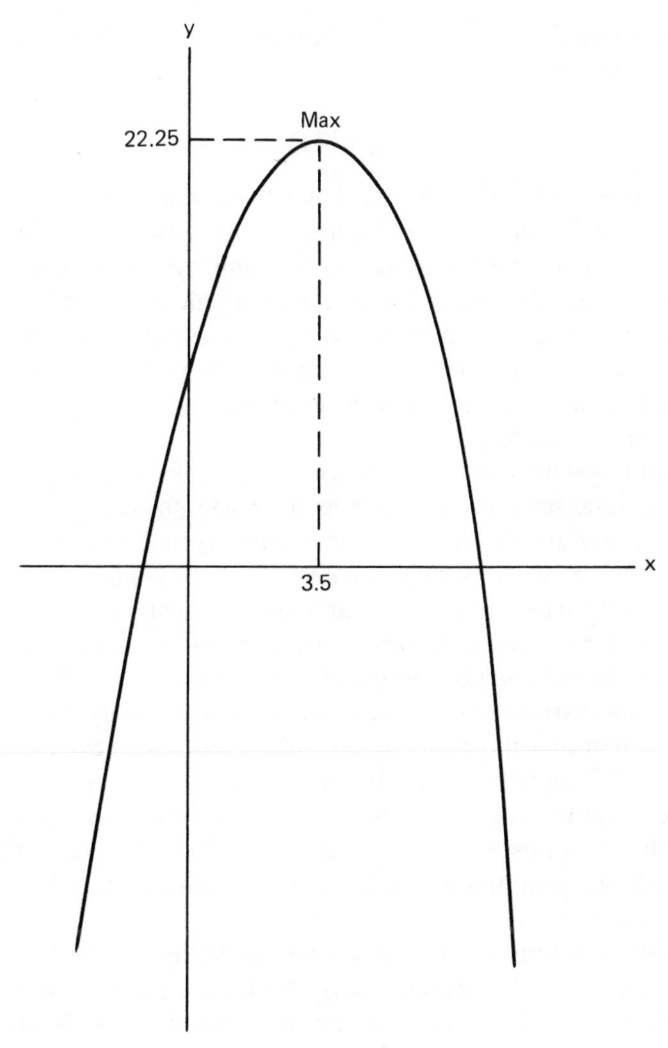


**Figure 1.26**

will be demonstrated in Chap. 7 when maximum shear and bending moment for beams are discussed. In that chapter, we will be interested only in quadratic forms; these, in general, have a shape similar to that of Fig. 1.25 or Fig. 1.26. To plot such functions after evaluating the location of its maximum or minimum, simply evaluate the function at two additional points, one to the right and one to the left of $x_{max}$ or $x_{min}$ and then pass a smooth curve similar to those shown in Figs. 1.25 and 1.26. If the curve looks like that in Fig. 1.25, it reveals a minimum; otherwise, it reveals a maximum.

## UNITS

Although units will not be discussed in detail in this book, it is important to notice the difference between *basic units* and *derived units*. Basic units are the minimum number of units each unit system requires for the derivation of all other units. For example, a unit of length is chosen to be one of the basic units in all systems, but quantities such as area and volume are not because they are derived units. The unit for area is merely the square of the unit for length, and the unit for volume is merely the cube of the unit for length.

For most systems, either (1) length, mass, and time, or (2) length, force, and time have been chosen as the basic units, and units for all other quantities are derived from them. Primary standards for measurement of these basic units have been established by international agreement. The accuracy of these standards for most engineering works and specifically for solving problems in structures is far beyond our present needs and will not even be discussed.

Among the various systems of units, we will use only the U.S. customary system and the "SI" system (the French abbreviation for the International System of units). In the U.S. customary system of units, the basic quantities are length, force, and time, which leaves mass (not used in this book) as a derived quantity or unit. In the "SI" system of units, the basic quantities are length, mass, and time, which leaves force as a derived quantity.

The unit of length in the U.S. customary system is the "foot" (ft), which is equal to 12 "inches" (in.). A larger unit, the "yard" (yd), equal to 3 feet, is also used. The unit for force in the U.S. customary system is the "pound" (abbreviated as "lb" or "#"). For larger values of force, the "ton," which is equal to 2000 lb, is introduced. In structural load applications, the term "kip" (kilo pound) is used as a unit of force; it is equal to 1000 lb.

In the SI system, the unit of length is the "meter" (m). For larger units we have the "kilometer" (1000 m = 1 km). For smaller units we have the millimeter (1 m = 1000 mm) and the centimeter (1 m = 100 cm). In technical calculations, the centimeter is seldom used. The unit for mass in the SI system is the "kilogram" (kg). The unit for force is a derived unit called the "newton" (N). To convert kilograms to newtons, one must multiply the former by 9.81.

Some important and frequently used conversion factors for the two systems are as follows:

1 lb = 4.45 N
1 N = 0.22 lb
1 in. = 25.4 mm
1 mm = 0.039 in.

As noted before, this brief discussion of units is sufficient for our present needs. It is important for students to be consistent in the use of units in solving problems and in converting units from one system to another. Some simple examples of unit conversion will now be demonstrated.

EXAMPLE 25: Express 45 in. in (a) ft and (b) mm.

SOLUTION:
(a) Since 1 ft = 12 in., or 1 in. = $\frac{1}{12}$ ft, 45 in. = 45 × $\frac{1}{12}$ ft, or 45 in. = 3.75 ft. (b) Since 1 in. = 25.4 mm, 45 in. = 45 × 25.4 mm, or 45 in. = 1143 mm.

EXAMPLE 26: Express 12 sq in. in $mm^2$.

SOLUTION: Substitute 25.4 mm for each inch as follows:

$$\begin{aligned} 12 \text{ sq in.} &= 12 \times (25.4 \text{ mm})^2 \\ &= 12 \times 25.4^2 \text{ mm}^2 = 7742 \text{ mm}^2 \end{aligned}$$

EXAMPLE 27: Express 1250 lb ft in lb in.

SOLUTION:
1250 lb ft = 1250 lb × 12 in. = 15000 lb in.

EXAMPLE 28: Express 5000 lb ft in N m.

SOLUTION: The force unit, lb, has to be converted to N by using 1 lb = 4.45 N, and the length unit, ft, has to be converted to m using 1 ft = 12 in. = 12 × 25.4 mm. Since 1 m = 1000 mm, therefore 1 mm = .001 m. Thus, 5000 lb ft = 5000 × 4.45 N × 12 × 25.4 × .001 m = 6782 N m.

## ACCURACY

Although we like to perform all our calculations as accurately as possible, we should take care to approximate our numerical results to match the accuracy of our numerical data. As an example, suppose one has to divide the length of a 20-ft beam by 3. Using a calculator, we get 6.6666667 ft, or, on some calculators, 6.666666667 ft. Obviously, these answers, though apparently very accurate, are ridiculous because in routine engineering practice these lengths cannot be measured so accurately. Even when using a machinist's micrometer, the closest we can measure is down to 0.0001 in. (Remember, moreover, that no one uses a micrometer to measure the length of a 20-ft beam.) If the smallest graduation of the scale used is 1/32 in., the best estimate we can make is one-half of that smallest graduation, or 1/64 in.

It has to be recalled that the accuracy of the results in solving problems depends on both the accuracy of the data provided as well as the accuracy of our calculations. No matter how accurately we perform the latter, the results should be no more accurate than the accuracy of the data given. For example, if a 100,000-lb load is applied to a structure, the best we can say is that this load is accurate within $\pm$ 100 lb, which is an accuracy of 100/100000 = .001, or 0.1 percent. If the given load is accurate only up to 0.1 percent, our calculations should not provide results with an accuracy exceeding 0.1 percent. For example, if this load has to be divided by 3, although our calculator gives an answer of 33333.333, this has to be modified to match the accuracy of the given data, or 0.1 percent. Since $(0.1/100) \times 33333.333$ is approximately 33 lb, our answer is correct only within $\pm 33$ lb, that is, 33,333 $\pm$ 33, and therefore a meaningful answer might be 33,300 lb.

Although the answers to problems should be given with a proper level of accuracy, experience in teaching architectural students reveals that once a student solves a problem and his answer is, say, 1 lb off the book value, he often gets discouraged, failing to realize, for example, that in the above example 33333 lb is the equivalent of 33300 lb but the latter is more meaningful. This fine point, which takes a long time for students to digest, must be continuously emphasized by instructors. With this in mind, this book will give all numerical values to the problems to the last whole number so that students will feel reassured when their results match those on the printed page.

## PROBLEMS

**1.1.** Lines $l_1$, $l_2$, and $l_3$ in Fig. P1.1 are parallel, and all three are crossed by yet another line, *AB*. Indicate which angles are equal angles.

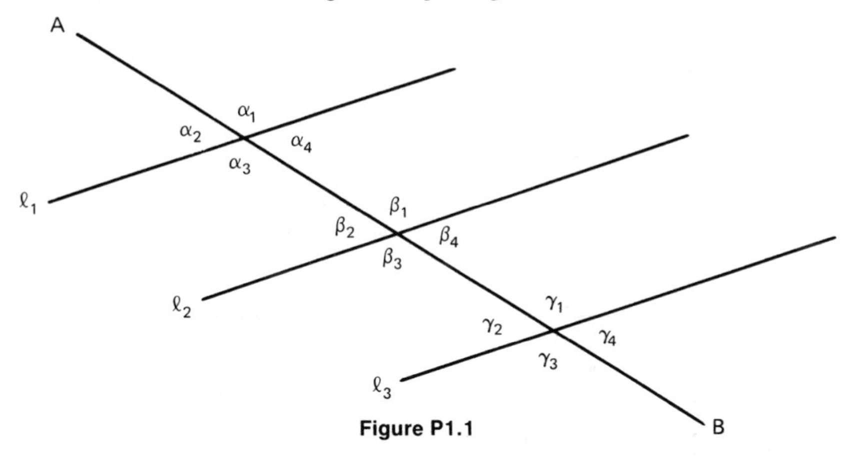


**Figure P1.1**

**1.2.** In each of the triangles shown in Fig. P1.2, what are the values of angle *A*?

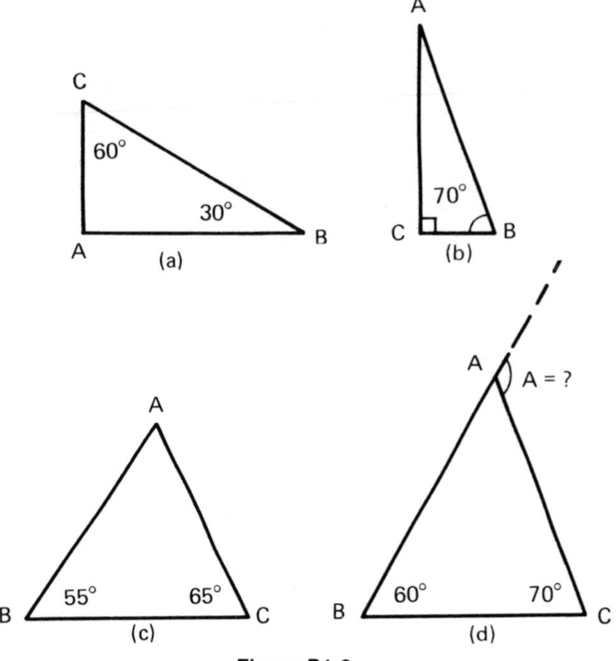


**Figure P1.2**

**1.3.** In Fig. P1.3, *AD* is ⊥ to *AB* (*AD* is perpendicular to *AB*), *DE* ⊥ *AC*, and *DF* ⊥ *BC*. What are the values of angles $\alpha_1$ and $\alpha_2$?

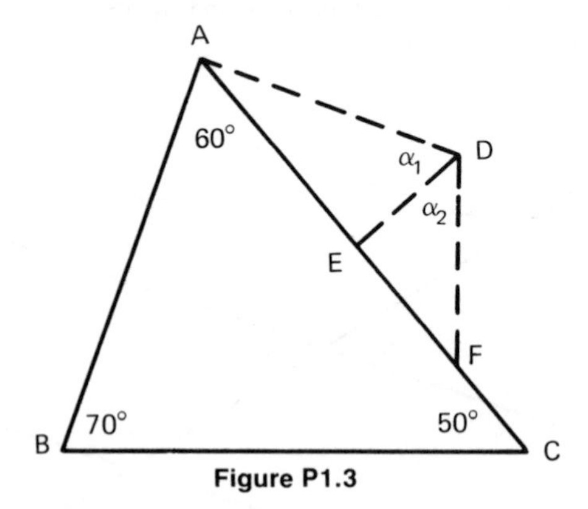


**Figure P1.3**

**1.4.** In triangle ABC in Fig. P1.4, three normals are drawn from point D, which is located inside the triangle, to the three sides of the triangle. Calculate angles $\alpha_1$, $\alpha_2$, and $\alpha_3$.

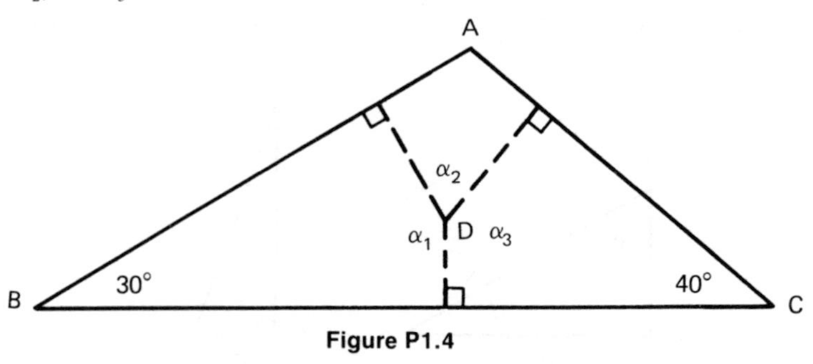


**Figure P1.4**

**1.5.** Calculate angles $\alpha_1$ and $\alpha_2$ in Fig. P1.5. *DE* is ∥ to *BC* (*DE* is parallel to *BC*).

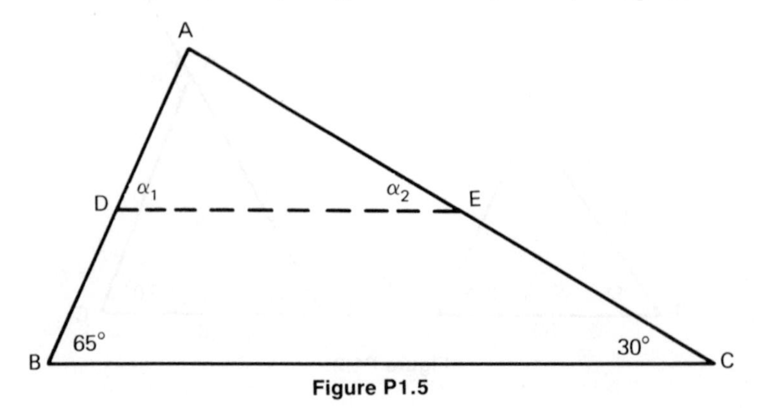


**Figure P1.5**

**1.6.** From point D inside the triangle in Fig. P1.6, two lines are drawn parallel to $AB$ and $AC$. What is the value of angle $\alpha$?

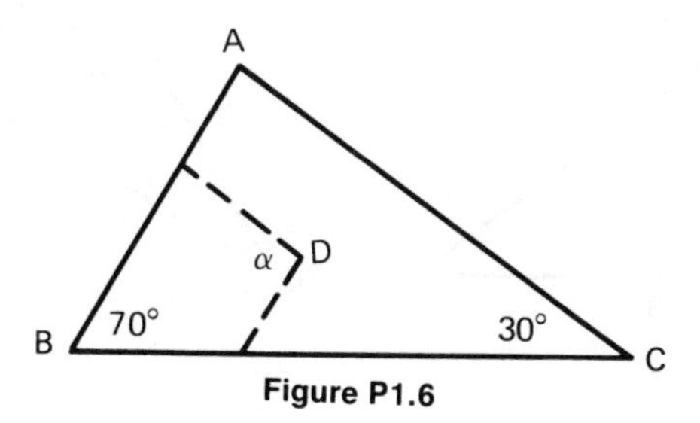


**Figure P1.6**

**1.7.** Calculate the lengths of $DE$ and $EF$ in Fig. P1.7 if $DE \parallel CB$, $EF \parallel AB$, $CB = 20'$, $AD = 15'$, and $AB = 55'$.

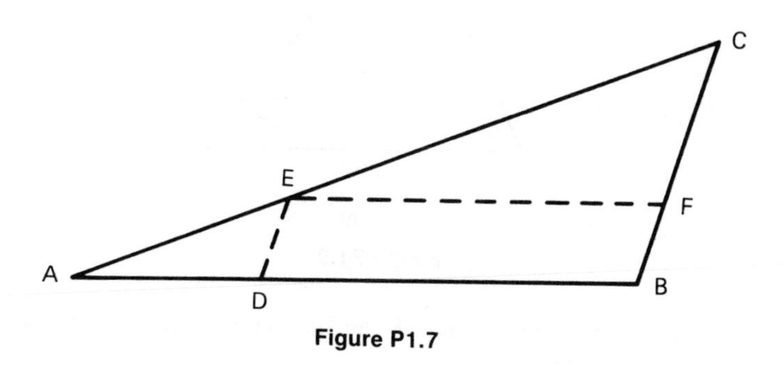


**Figure P1.7**

**1.8.** Calculate the sine, cosine, and tangent of angles $A$ and $B$ in Fig. P1.8.

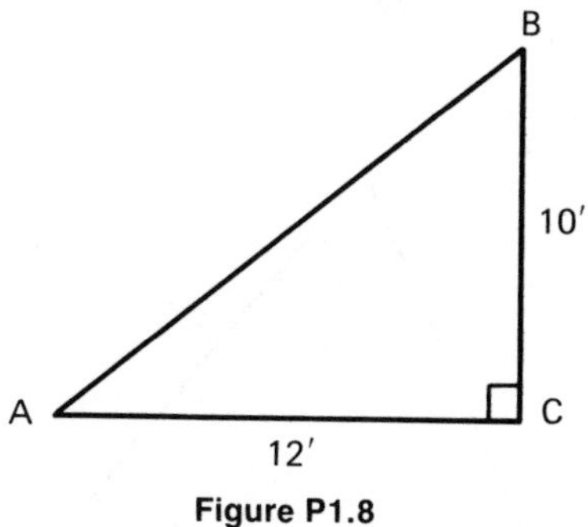


**Figure P1.8**

**1.9.** In the triangles in Fig. P1.9, calculate the missing sides and angles.

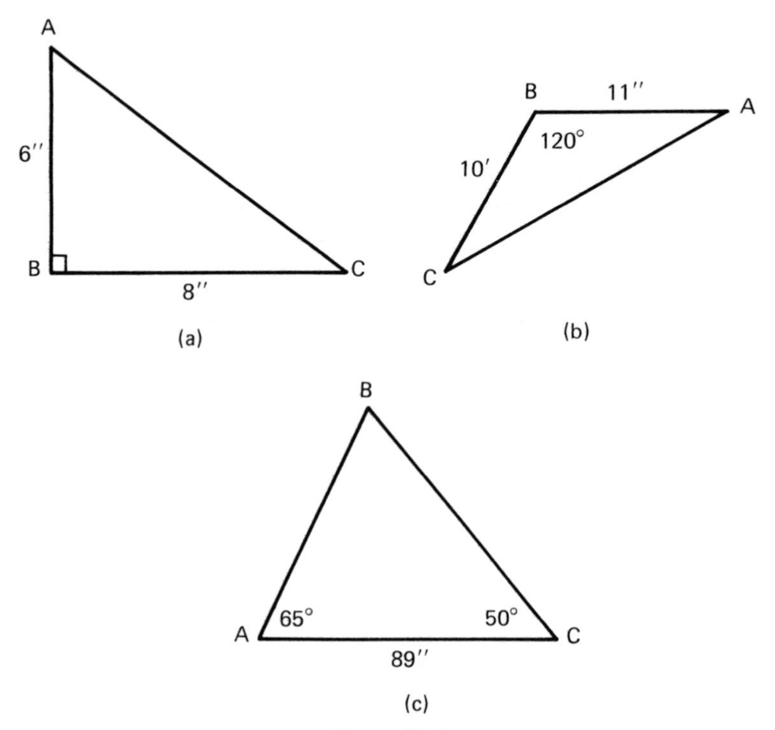


Figure P1.9

**1.10.** In the triangle in Fig. P1.10, find the missing sides and angles under each of the following circumstances:

(a) $a = 10''$; $b = 20''$; $c = 30''$
(b) $a = 10''$; $b = 20''$; $\hat{C} = 30°$
(c) $a = 10''$; $\hat{B} = 85°$; $\hat{C} = 15°$
(d) $\hat{A} = 15°$; $b = 12''$; $\hat{C} = 75°$
(e) $\hat{A} = 45°$; $c = 19''$; $\hat{C} = 55°$

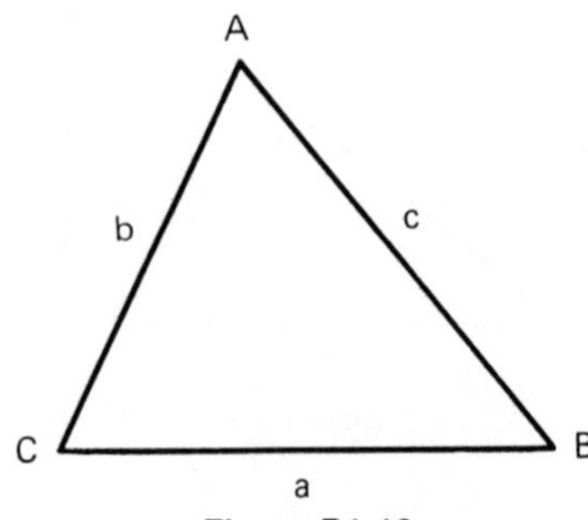


Figure P1.10

**1.11.** Determine the angle between cables $AB$ and $AC$.

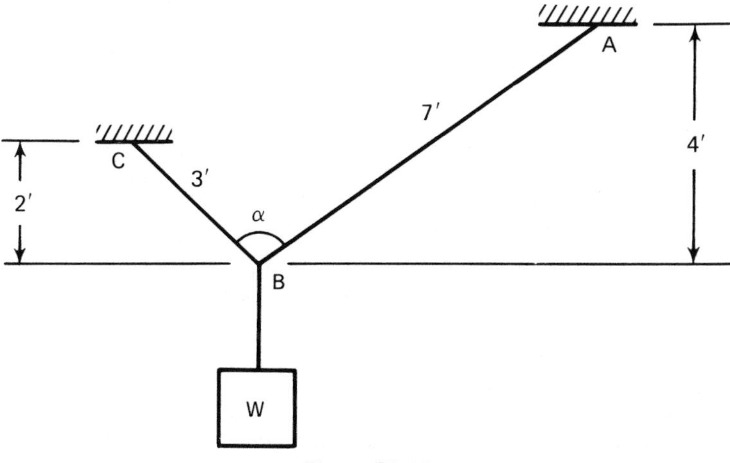


**Figure P1.11**

**1.12.** A surveyor starts from point A and, after going through stations B, C, D, E, F, and G, stops at station H. What is the shortest distance $AH$?

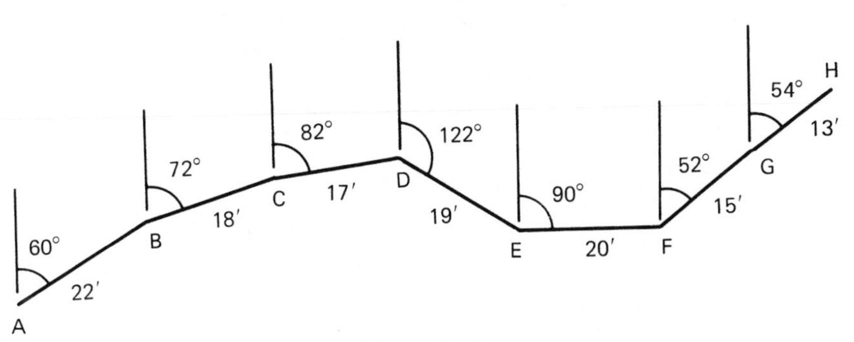


**Figure P1.12**

**1.13.** Solve the following first-order equations:

(a) $5x = 9$
(b) $3x = 0$
(c) $1/3x - 5/8 = 0$
(d) $1000x - 25 = 0$

**1.14.** Solve the following second-order equations:

(a) $2x^2 - 3x + 1 = 0$
(b) $3x^2 + 2x = 16$
(c) $1/5\, x^2 - 9x + 20 = 0$
(d) $x^2 - 3/2x + 1/2 = 0$

**1.15.** Solve the following systems of equations:

(a) $\begin{cases} x + 7 = 5 \\ x - y = 1 \end{cases}$

(b) $\begin{cases} 2x - 3y = 5 \\ 3x + y = 35 \end{cases}$

(c) $\begin{cases} x + y + z = 60 \\ -x + y + 3z = 100 \\ 2x + 3y - 2/3z = 60 \end{cases}$

(d) $\begin{cases} 2x + 3y + 4z = 3 \\ 2x - 30y + 40z = 1 \\ 10x + 15y + z = 10.25 \end{cases}$

**1.16.** Determine the derivative of the following functions:

(a) $y = 2x + 1$
(b) $y = 11x^3 - 21x^2 + 17$
(c) $y = 1/7x^7 + 1/6x^6 + x$
(d) $y = x^{12} - x^6 + x^3 + 1/4x^2 + 1/3$

**1.17.** For the following functions: (a) determine where the maximum or minimum occurs, (b) calculate the values of the maximum or minimum, and (c) plot their curves and indicate the points of maximum or minimum values:

(a) $y = 2x^2 + 3x + 1$
(b) $y = -1/3x^2 + 1/4x + 9$
(c) $y = 1/3x^3 - x + 1$
(d) $y = x^3 + 1/2x^2 - 4x + 15$

# 2
# FORCE

## 2.1 DEFINITION OF FORCE

The simplest way to define a force is by just thinking of pull or push. More generally, *force* is defined as an action that tends to disturb a body. This disturbance can be thought of as rise, fall, translation, rotation, spinning, and so forth.

Consider a block resting on a table. Connect a cable to the block and start pulling the block. If the right amount of force is applied, the block will start to move. The amount of movement obviously depends on the amount and duration of the applied force. The force can be applied directly to a body, such as pulling a car using a towrope, or indirectly, such as gravitational force. *Gravitational force* is the force the earth exerts on objects that pulls them toward its center. This force is known as the *weight* of the object and can be obtained by using a weighing scale.

## 2.2 VECTORS

To describe a force completely, one needs to specify its magnitude and line of action as well as its direction. Therefore, three distinct pieces of information are necessary for a complete description. This requirement is contrary to that for some other quantities such as temperature, length, area, etc., which need only one value (with a proper unit) to be completely defined. You are familiar with these kinds of quantities, all of which indicate "size" or "magnitude." They are known as *scalar quantities*. On the other hand, quantities such as force, displacement, and velocity are called *vector quantities*. Magnitude, line of action, and direction must be completely specified for all vector quantities.

A *vector* is a quantity that has magnitude, line of action, and direction. Graphically, a vector is shown by an arrow (Fig. 2.1). The length of this arrow is drawn so as to be proportional to the magnitude of the

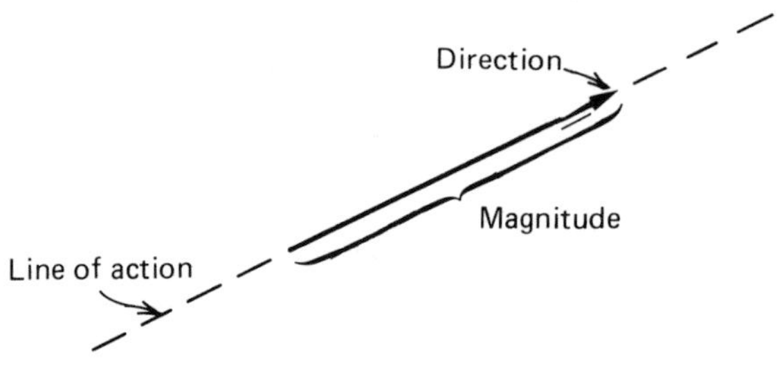


**Figure 2.1**

vector using an arbitrary scale. For example, if a vector represents a force of 2000 lb and we arbitrarily say that a length of 1 inch is equivalent to a 1000-lb force, the length of the vector should be 2 inches. The line of action of the force is along that of the arrow and its direction is indicated by the arrowhead. Notice that two directions are possible for each line of action.

Before beginning to work with vectors, it is necessary to point out the following concepts:

1. To distinguish vectors from scalars, all vector quantities throughout this book will be represented by an appropriate letter with an arrow at its top, whereas scalars will be represented by just plain letters. For example, $\vec{F}$, $\vec{F}_1$, and $\vec{W}$ represent vector quantities, whereas $F$, $F_1$, and $W$ represent scalar quantities.
2. Consider the cart shown in Fig. 2.2(a). A cable is attached to this cart at point A, and a force $\vec{F}$ is applied to this cable. Point A is known as the *point of application* of the force.
3. The action of force $\vec{F}$ on the cart would be the same if the cart were pushed at B or pulled at C with force $\vec{F}$, provided B or C is located at any point along the cable, as shown in Figures 2.2(b) and (c) (in all cases, the magnitude of force $\vec{F}$ and its lines of action are the same). This is known as the *principle of transmissibility*. This principle simply states that the point of application of a force acting on a body can be slid along the line of action of the force without altering its effect on the body.
4. Two vectors are equal if they are parallel and have the same direction and magnitude. In Fig. 2.3, the two vectors $\vec{F}_1$ and $\vec{F}_2$ are equal since their lines of action are parallel, their directions are the same, and their magnitudes are equal.

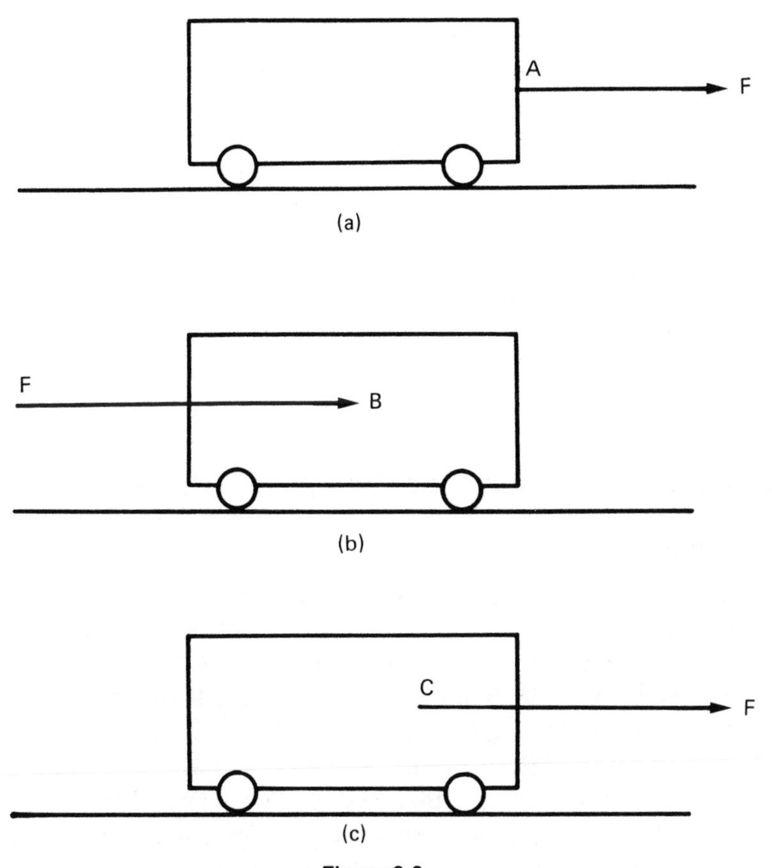

**Figure 2.2**

5. A negative vector is produced by changing the direction of a given vector (while keeping its magnitude and line of action unchanged). If the vector $\vec{F}_1$ shown in Fig. 2.4(a) is considered positive, then the vector $\vec{F}_2$ is negative provided that the lines of action of both of them are parallel and their magnitudes the same. Similarly, the two vectors shown in Fig. 2.4(b) represent positive and negative vectors.

**Figure 2.3**

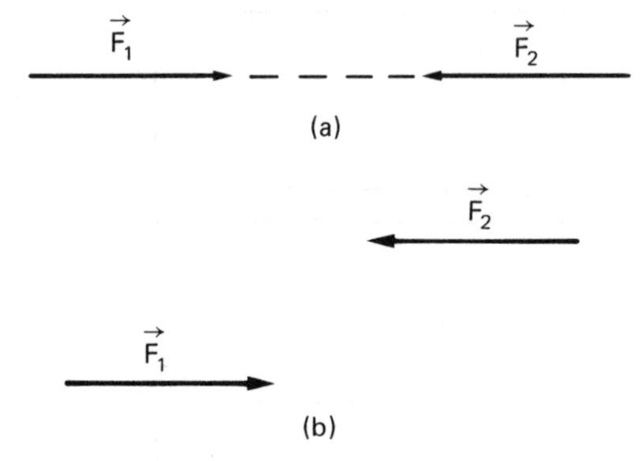


**Figure 2.4**

## 2.3 OPERATIONS WITH VECTORS

### Vector Addition

Vectors cannot be added like numbers. Although 2 plus 3 is equal to 5 in basic arithmetic, vector 2 plus vector 3 is not, in general, equal to vector 5. To illustrate this difference, we shall use a *displacement vector* in the following example. Suppose that the distance between Washington and Philadelphia is about 200 miles and that the distance between Philadelphia and New York is about 150 miles. If one travels first from Washington to Philadelphia (200 miles) and then from Philadelphia to New York (150 miles), it will be wrong to say that if anyone travels from Washington to New York, the distance between these two points is therefore 350 miles (200 + 150 miles). It is correct to say, however, that the displacement vector $\overrightarrow{200}$ plus the displacement vector $\overrightarrow{150}$ is equal to the displacement vector $\overrightarrow{250}$, or

$$\overrightarrow{200} + \overrightarrow{150} = \overrightarrow{250}$$

The reason for this anomaly is that other things besides numbers (miles), such as directions, are involved. The rule for adding two vectors will now be discussed in detail.

Two vectors, $\vec{A}$ and $\vec{B}$, are shown in Fig. 2.5(a). To find the sum of these two vectors, do as follows. First, take a point such as $O$ in space and draw a line parallel to vector $\vec{A}$, choosing length $OA$ to be equal to the length of vector $\vec{A}$. The length of vector $\vec{A}$ can be written as $|\vec{A}|$ or simply $A$ (without the arrow). Therefore,

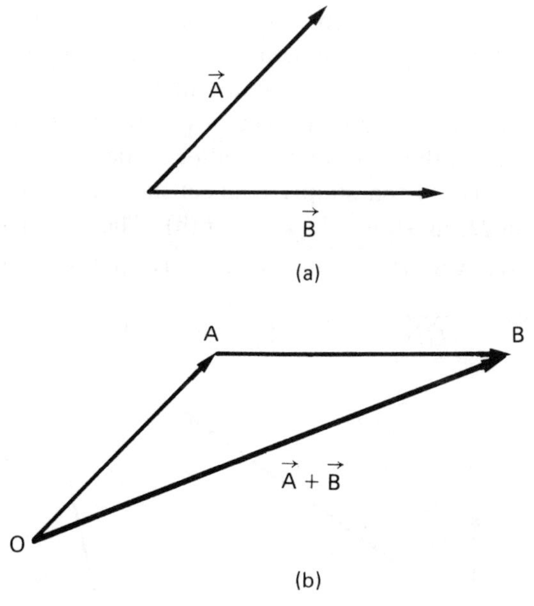


**Figure 2.5**

$$OA = |\vec{A}| = A$$

or, vectorially, $\overrightarrow{OA} = \vec{A}$. Then from point $A$ draw the second line parallel to the line of action of vector $\vec{B}$, and choose the length $AB$ to be equal to the length of vector $\vec{B}$, as shown in Fig. 2.5(b). Then,

$$AB = |\vec{B}| = B$$

or, vectorially, $\overrightarrow{AB} = \vec{B}$.

A vector produced by connecting point $O$ (the starting point) to point B (the terminating point) is the sum of the two vectors $\vec{A}$ and $\vec{B}$, and we can thus write

$$\overrightarrow{OA} + \overrightarrow{AB} = \overrightarrow{OB}$$

It is important to notice that if the length of $OA$ is added to the length of $AB$, the result will be greater than the length of $OB$. This equation is a vectorial equation, and if we remove the arrows from the top of $OA$, $AB$, and $OB$, the equality may no longer hold true.

To obtain the sum of a number of vectors, such as the four vectors shown in Fig. 2.6(a), repeat the same process as that for two vectors. From point $O$ draw a vector parallel to the first vector ($\overrightarrow{OA} = \vec{A}_1$); from the end of this vector draw another vector parallel to the second vector ($\overrightarrow{AB} = \vec{A}_2$); and from this yet others parallel to the third and fourth ($\overrightarrow{BC} = \vec{A}_3$ and $\overrightarrow{CD} = \vec{A}_4$). Then simply connect the starting point $O$ to the terminating point $D$, as shown in Fig. 2.6(b). The vector $\overrightarrow{OD}$ is called the sum of the four vectors $\vec{A}_1$, $\vec{A}_2$, $\vec{A}_3$, and $\vec{A}_4$, and we can write

$$\overrightarrow{OD} = \vec{A}_1 + \vec{A}_2 + \vec{A}_3 + \vec{A}_4$$

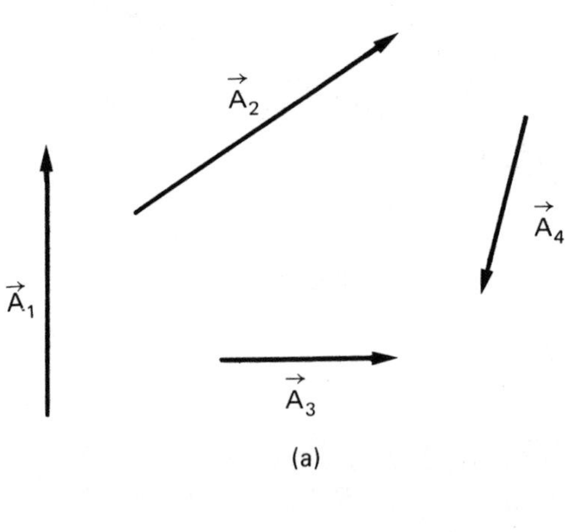


(a)

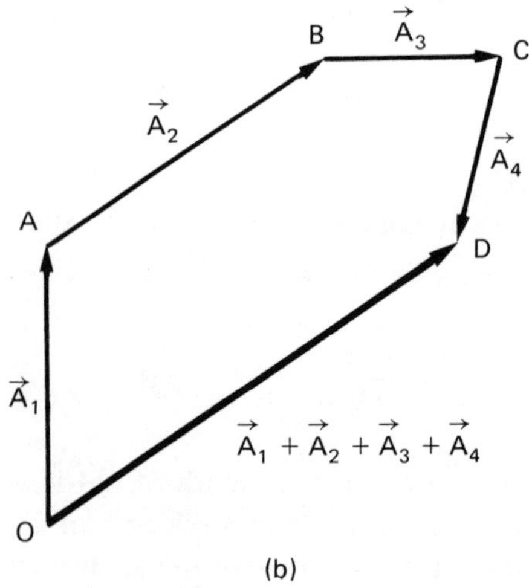


(b)

**Figure 2.6**

The method just described is known as the *string polygon method*, and if vectors $\vec{A}_1, \ldots, \vec{A}_4$ are forces, the polygon OABCD is called a *force polygon*. Notice that in this method, if the terminating point coincides with the starting point, the sum of the vectors is equal to zero.

Alternatively, the sum of two vectors can be determined by using the *parallelogram law*, which is essentially the same as the string polygon method. Its use will now be described.

To find the sum of the two vectors $\vec{A}$ and $\vec{B}$ shown in Fig. 2.7(a), take a point $O$ as origin. From point $O$ draw two vectors $\overrightarrow{OA}$ and $\overrightarrow{OB}$ equal to vectors $\vec{A}$ and $\vec{B}$, as shown in Fig. 2.7(b). Next, from the ends of these vectors draw two lines ($AC$ and $BC$) parallel to $OA$ and $OB$ in order to construct a *parallelogram* (a four-sided figure whose opposite sides are parallel). The diagonal vector $\overrightarrow{OC}$ of this parallelogram is the sum of two vectors $\vec{A}$ and $\vec{B}$. It is obvious to see that triangle OAC in Fig. 2.7(b) is the same as triangle OAB in Fig. 2.5(b); therefore, both methods are essentially the same. It is also evident from Figs. 2.6(b) and 2.7(b) that the way the sum of two vectors is constructed is independent of the order in which the two vectors are added.

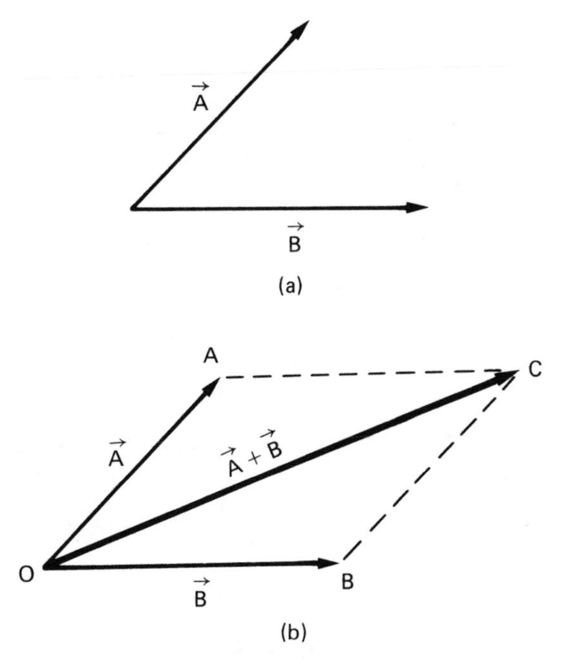


**Figure 2.7**

To find the sum of a number of vectors using this method, just take the first two vectors and find their sum (diagonal of the parallelogram); then add the third vector to this sum to construct a new sum (diagonal of the new parallelogram); and repeat this process to cover all the vectors. Obviously, the string polygon method is the faster method.

If two (or more) vectors are parallel, the process of obtaining the sum will be much simpler. Since the vectors are parallel, their sum will also be parallel to the vectors themselves, and its magnitude (length) will be equal to the algebraic sum of their magnitudes.

EXAMPLE 1: Graphically construct the sum of horizontal force (20 kN) and vertical force (30 kN) as shown in Fig. 2.8(a).

SOLUTION: From point O construct a vector equal to the horizontal 20-kN force [see Fig. 2.8(b)]. Then from the end of this vector construct another vector equal to the vertical 30-kN force [see Fig. 2.8(c)]. Finally, the sum $\vec{R}$ can be obtained by connecting point O to the end of

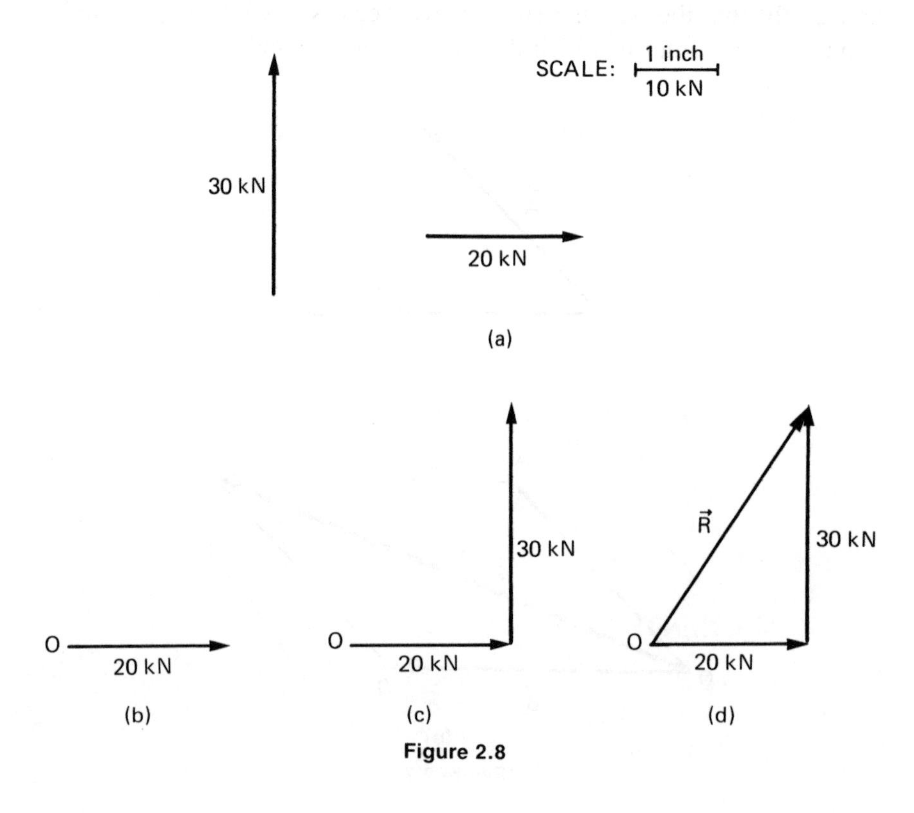


Figure 2.8

the second vector [see Fig. 2.8(d)]. Be sure to use an appropriate scale in this process (we used 1 inch to be equivalent to 10 kN). Using the same scale, measure the length of $\vec{R}$, which gives the magnitude of sum, and, using a protractor, measure the angle the vector $\vec{R}$ makes with the horizontal axis (in this case, the 20-kN force). It will be found that $\vec{R}$ = 36 kN and $\alpha = 56°$.

EXAMPLE 2: Graphically add the three vectors shown in Fig. 2.9(a).

SOLUTION: From point $O$ construct vector $\overrightarrow{OA}$ parallel to the 2000-lb vector [see Fig. 2.9(b)] and equivalent in magnitude by using an arbitrary scale, say 1″ = 1000 lb, and making its length equal to 2″; then from $A$ construct vector $\overrightarrow{AB}$ parallel to the 3000-lb vector and its length equal to 3″; and finally construct $\overrightarrow{BC}$ parallel to the 500-lb vector and its length equal to 1/2″. The vector obtained by connecting point $O$ to point $C$ will give the sum of the three vectors shown in Fig. 2.9(a). To find this sum, simply measure $OC$ by using an inch-scale. The length of $OC$ is found to be equal to 4.9″, which indicates that the sum equals 4900 lb.

### Vector Subtraction

To subtract vector $\vec{B}$ from vector $\vec{A}$, simply change the sign (or direction) of vector $\vec{B}$, and add this new vector to vector $\vec{A}$, as follows:

$$\vec{A} - \vec{B} = \vec{A} + (-\vec{B})$$

### Multiplication of a Vector by a Scalar

To multiply a vector, $\vec{A}$, by a scalar, $m$, simply construct a vector parallel to vector $\vec{A}$ and let it have the magnitude of $m$ times the magnitude of $\vec{A}$ (or $mA$).

In the following sections, the words "vector" and "force" are used interchangeably.

## 2.4 COMPONENTS

Consider a force $\vec{F}$ and a reference axis with positive and negative directions. As shown in Fig. 2.10(a), the direction of the reference axis $Ox$ is positive in the direction of the arrow and negative opposite to it.

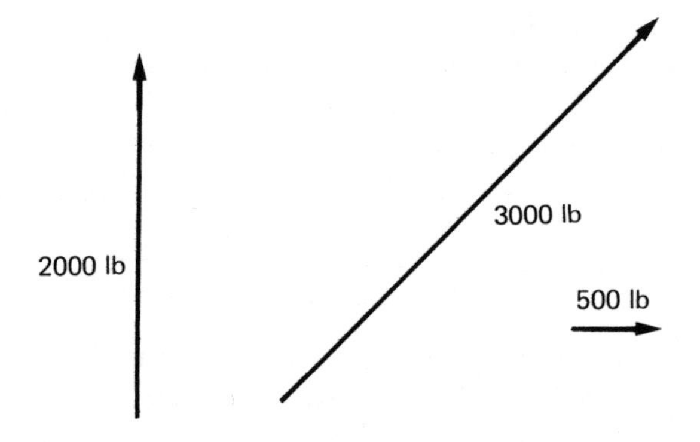


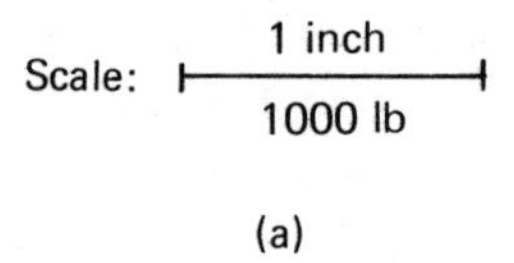


(a)

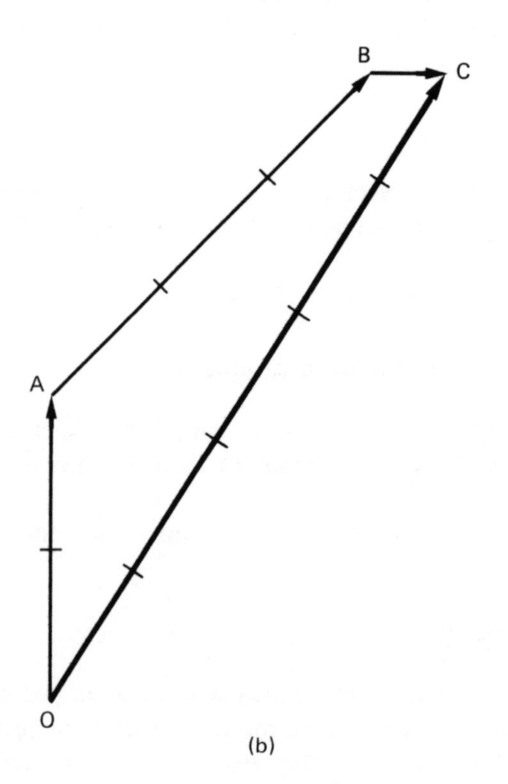


(b)

**Figure 2.9**

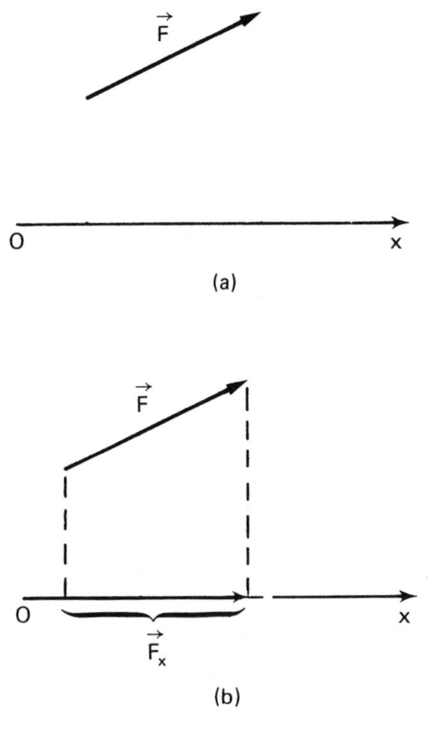


**Figure 2.10**

Now take the end points of vector $\vec{F}$ and draw two perpendicular lines* to the axis $Ox$, as shown in Fig. 2.10(b). By doing so, another vector $\vec{F}_x$ is produced that is called the *component* or *projection* of $\vec{F}$ along $Ox$. Notice that the direction of the original vector is followed, and in this case the component is considered positive with respect to $Ox$ since its direction matches the positive direction of $Ox$. On the other hand, vectors $\vec{Q}$ and $\vec{W}$ in Fig. 2.11 have negative components because they are in opposite direction to the positive $Ox$ axis.

Notice that the "projection" of a vector and the "component" of a vector along a reference axis are used as interchangeable terms. Both have positive or negative values depending on the alignment of the projection or component with the positive or negative direction of the reference axis. Although the terms are used synonymously, projection usually refers to a vector and component to the magnitude of this vector but having

*In this book, we use normal projections only rather than nonperpendicular or oblique projections.

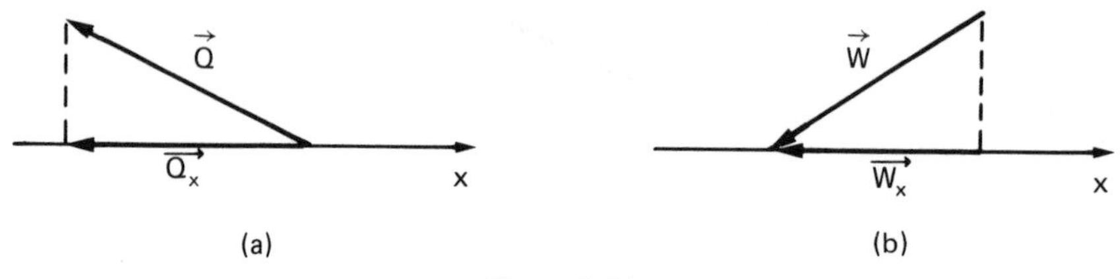


Figure 2.11

a positive or negative sign. The component of force $\vec{F}$ along the axis $Ox$ is shown by $\vec{F}_x$ in Fig. 2.10(b).

A given force has many components along various reference axes (one component for each axis). Consider the force $\vec{F}$ and the rectangular coordinate system $Ox$–$Oy$ shown in Fig. 2.12. The projections of force $\vec{F}$ along these two coordinate axes are called the *rectangular components* of $\vec{F}$. $F_x$ is the $x$ component and $F_y$ is the $y$ component. If the angle between $\vec{F}$ and the $x$ axis is $\alpha$, then the magnitude of $F_x$ is simply the magnitude of $\vec{F}$, or F, multiplied by the cosine of angle $\alpha$, and the magnitude of $F_y$ is F cos $\beta$. Since $\alpha + \beta = 90°$, however, $\cos \beta = \sin \alpha$ and therefore in the two-dimensional rectangular coordinate system only one angle is needed to obtain the components along the two coordinate axes, as follows:

$$
\begin{aligned}
F_x &= F \cos \alpha \\
F_y &= F \sin \alpha
\end{aligned}
\tag{2.1}
$$

$F$ without an arrow on top indicates the magnitude of the vector $\vec{F}$.

If force $\vec{F}$ is applied to a body in a general direction, it can be decom-

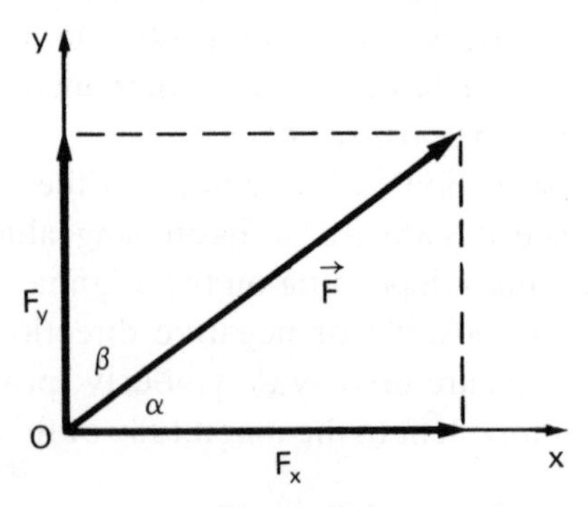


Figure 2.12

posed into two forces with specified directions, one horizontal ($\vec{F}_x$), the other vertical ($\vec{F}_y$). Notice that these are equivalent forces, which means that the action of force $\vec{F}$ on the body will be the same as the actions of $\vec{F}_x$ and $\vec{F}_y$ taken together, but the force is now acting in horizontal and vertical directions. Notice that $\vec{F}_x$ and $\vec{F}_y$ are vectors in $x$ and $y$ directions with magnitudes $F_x$ and $F_y$. Of course, this distinction will prove more useful if many forces are involved, a situation that will be discussed shortly.

Now, consider Eqs. 2.1 again. By squaring both sides of both of them, one gets

$$F_x^2 = F^2 \cos^2 \alpha$$

$$F_y^2 = F^2 \sin^2 \alpha$$

Then, by adding the two,

$$F_x^2 + F_y^2 = F^2 \cos^2 \alpha + F^2 \sin^2 \alpha = F^2(\cos^2 \alpha + \sin^2 \alpha)$$

Since $\cos^2 \alpha + \sin^2 \alpha = 1$, therefore

$$F^2 = F_x^2 + F_y^2$$

By taking the square root of both sides of the above, we get

$$F = \sqrt{F_x^2 + F_y^2} \tag{2.2}$$

Notice that only the positive sign of the square root is considered here since "length $F$" is a positive value. (Recall that the square root of 4 is not only $+2$, it can also be $-2$.)

Again considering Eqs. 2.1, by dividing the second equation by the first on each side, we get

$$\frac{F_y}{F_x} = \frac{F \sin \alpha}{F \cos \alpha} = \frac{\sin \alpha}{\cos \alpha} = \tan \alpha$$

Therefore,

$$\alpha = \arctan \frac{F_y}{F_x} \tag{2.3}$$

Equation 2.2 says that if you have two rectangular components of an unknown vector, you can find the magnitude of that vector by adding the squares of the components and then finding the square root of the result. Equation 2.3 says that if you have two rectangular components of an unknown vector, you can find its direction and line of action by simply dividing the $y$ component by the $x$ component and finding the angle whose tangent is equal to the ratio, $F_y/F_x$. This can be accomplished on your calculator by inputting the $F_y$ and $F_x$ values, finding the $F_y/F_x$ ratio, and then pushing the *arctan key* (or, with some calculators, the *INV tan* or *2ND tan keys* in sequence). Make sure that your calculator is in *degree mode*.

Therefore, if a vector is given, one can find its rectangular components using Eqs. 2.1, and, inversely, if the components of an unknown vector are given, one can identify that vector completely by using Eqs. 2.2 and 2.3.

## 2.5 RESULTANT

The resultant of two or more vectors (forces) has the same effect as all the vectors (forces) combined. Generally, the resultant of forces is used for convenience. When two or more forces are applied to a body, it is more convenient to replace them with a single force that has the same effect on the body as all those forces applied collectively. This equivalent force is called the *resultant*.

The resultant of more than two forces can be found simply by finding the $x$ component of each force separately and then adding these up to get the $x$ component of the system of the forces and thereafter repeating the process for the $y$ components. This will lead to formulations similar to Eqs. 2.1 and 2.2, but generalized for as many forces as desired.

It is customary to represent the resultant by $\vec{R}$. The $x$ and $y$ components of the resultant are as follows:

$$R_x = (F_1)_x + (F_2)_x + \ldots. = \text{sum of } x \text{ components of all forces } (\Sigma F_x)$$

$$R_y = (F_1)_y + (F_2)_y + \ldots. = \text{sum of } y \text{ components of all forces } (\Sigma F_y)$$

or

$$R_x = \Sigma F_x$$
$$R_y = \Sigma F_y \tag{2.4}$$

Using Eq. 2.2,

$$R = \sqrt{R_x^2 + R_y^2} = \sqrt{(\Sigma F_x)^2 + (\Sigma F_y)^2} \tag{2.5}$$

$$\tan \alpha = \frac{R_y}{R_x} = \frac{\Sigma F_y}{\Sigma F_x}$$

or

$$\alpha = \arctan \frac{\Sigma F_y}{\Sigma F_x} \tag{2.6}$$

EXAMPLE 3: Determine the resultant of the two forces $F = 10$ kN and $G = 20$ kN shown in Fig. 2.13.

SOLUTION: Find the $x$ and $y$ components of each force separately by projecting them along the $x$ and $y$ axes using Eqs. 2.1. The x component of the 10 kN force is

$$F_x = 10(\cos 30°) = 8.66 \text{ kN}$$

Similarly,

$$F_y = 10(\sin 30°) = 5 \text{ kN}$$

$$G_x = 20(\cos 15°) = 19.32 \text{ kN}$$

$$G_y = 20(\sin 15°) = 5.18 \text{ kN}$$

Next, algebraically sum the two $x$-component and the two $y$-component forces separately by using Eq. 2.4. "Algebraically" means that if the direction of the $x$-component force matches the positive $x$ direction (to the right), this force is to be considered as positive; otherwise, it is negative. Likewise, if the direction of the $y$-component force matches

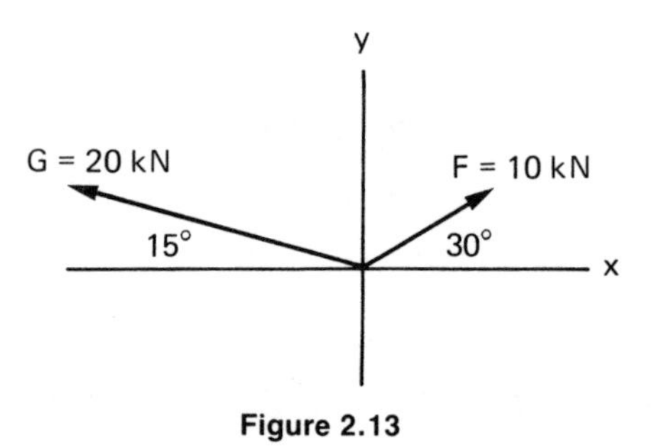


Figure 2.13

the positive $y$ direction (upward), this force is to be considered as positive; otherwise, it is negative.

In this example, $F_x$ is positive, $G_x$ is negative, and $F_y$ and $G_y$ are both positive. Therefore,

$$R_x = \Sigma F_x = F_x - G_x = 8.66 - 19.32 = -10.66 \text{ kN}$$

$$R_y = \Sigma F_y = F_y + G_y = 5 + 5.18 = 10.18 \text{ kN}$$

These are the net components of the resultant. Using Eqs. 2.5 and 2.6, we can also obtain the magnitude of the resultant as well as its orientation (its angle with the $x$ axis):

$$R = \sqrt{(\Sigma F_x)^2 + (\Sigma F_y)^2} = \sqrt{(-10.66)^2 + (10.18)^2} = 14.74 \text{ kN}$$

and

$$\tan \alpha = \frac{R_y}{R_x} = \frac{\Sigma F_y}{\Sigma F_x}$$

Solving for $\alpha$,

$$\alpha = \arctan \frac{R_y}{R_x} = \arctan \frac{10.18}{-10.66} = -43.7° \text{ or } 136.3°$$

Notice that an angle of $-43.7°$ is the same as one of $136.3°$, which is obtained by adding $180°$ to $-43.7°$.

EXAMPLE 4: Determine the resultant of the three forces—25 lb, 50 lb, and 100 lb—shown in Fig. 2.14.

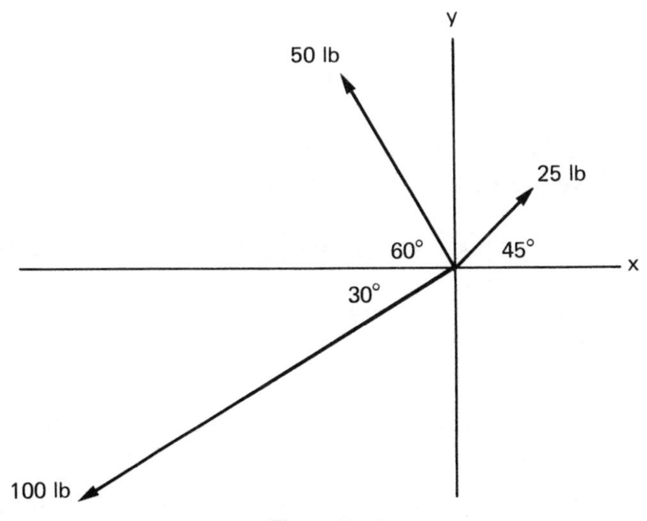


**Figure 2.14**

SOLUTION: The same process used in Example 3 will also be followed here. The $x$ and $y$ components of each force will be indicated by the use of subscripts $x$ and $y$ with the magnitude of that force; i.e., the $x$ component of the 25-lb force, for example, is indicated by $(25)_x$ and is equal to the following:

$$(25)_x = 25 \cos 45° = 17.68 \text{ lb}$$

Similarly,

$$(25)_y = 25 \sin 45° = 17.68 \text{ lb}$$

$$(50)_x = 50 \cos 60° = 25 \text{ lb}$$

$$(50)_y = 50 \sin 60° = 43.30 \text{ lb}$$

$$(100)_x = 100 \cos 30° = 86.60 \text{ lb}$$

$$(100)_y = 100 \sin 30° = 50 \text{ lb}$$

Using Eq. 2.4,

$$R_x = \Sigma F_x = (25)_x + (50)_x + (100)_x$$

$$= 17.68 - 25 - 86.60 = -93.92 \text{ lb}$$

$$R_y = \Sigma F_y = (25)_y + (50)_y + (100)_y$$
$$= 17.68 + 43.30 - 50 = 10.98 \text{ lb}$$

These two values are the $x$ and $y$ components of the resultant. To find the resultant, we use Eqs. 2.5 and 2.6 as follows:

$$R = \sqrt{(\Sigma F_x)^2 + (\Sigma F_y)^2} = \sqrt{(-93.92)^2 + (10.98)^2} = 94.56 \text{ lb}$$

$$\alpha = \arctan \frac{R_y}{R_x} = \arctan \frac{10.98}{-93.92} = -6.7°$$

Notice that $-6.7°$ is the same as $173.3°$, which is obtained by adding $180°$ to $-6.7°$.

EXAMPLE 5: Determine the resultant of the four forces applied to the top of the vertical post shown in Fig. 2.15.

SOLUTION: Find the $x$ and $y$ components of each force separately. Notice that since the 10-kN force is horizontal, its $x$ component is 10 kN and its $y$ component is zero. Similarly, the $x$ component of the 20-kN force is zero and its $y$ component is 20 kN.

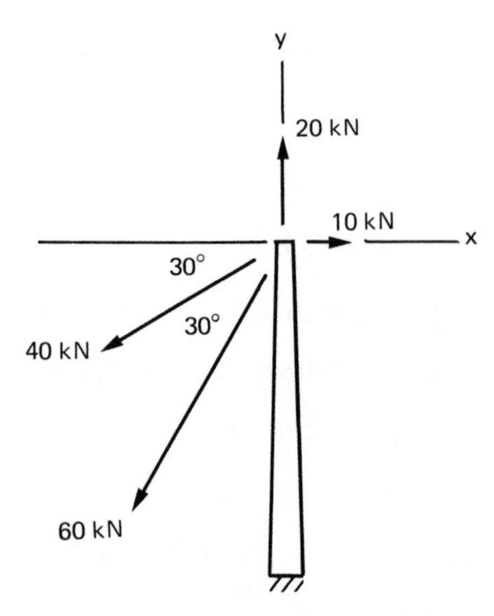


Figure 2.15

$$(10)_x = 10 \text{ kN}$$

$$(10)_y = 0$$

$$(20)_x = 0$$

$$(20)_y = 20 \text{ kN}$$

$$(40)_x = 40 \cos 30° = 34.64 \text{ kN}$$

$$(40)_y = 40 \sin 30° = 20 \text{ kN}$$

$$(60)_x = 60 \cos 60° = 30 \text{ kN}$$

$$(60)_y = 60 \sin 60° = 51.96 \text{ kN}$$

The algebraic sum of the $x$ and $y$ component forces will give the $x$ and $y$ components of the resultant:

$$R_x = \Sigma F_x = (10)_x + (20)_x + (40)_x + (60)_x$$

$$= 10 + 0 - 34.64 - 30 = -54.64 \text{ kN}$$

$$R_y = \Sigma F_y = (10)_y + (20)_y + (40)_y + (60)_y$$

$$= 0 + 20 - 20 - 51.96 = -51.96 \text{ kN}$$

Using Eqs. 2.5 and 2.6, the magnitude and the angle of the resultant can now be calculated:

$$R = \sqrt{(\Sigma F_x)^2 + (\Sigma F_y)^2} = \sqrt{(-54.64)^2 + (-51.96)^2} = 75.40 \text{ kN}$$

$$\alpha = \arctan \frac{R_y}{R_x} = \arctan \frac{-51.96}{-54.64} = 43.6°$$

The method used in obtaining the sum or resultant of vectors (forces) in Examples 1 and 2 is called the *graphical method.* Although basic, quick, and simple, it is not very practical when one has to find the sum or resultant of a large number of forces. In fact, it is not a very accurate method since some error may be involved in drawing parallel lines that will affect the correctness of the result. The larger the number of vectors (forces) involved, the less accurate will be the result. The *components method* used in Examples 3 through 5 is a more practical one, with greater accuracy.

## 2.6 SPACE FORCES

In our previous discussion, all the forces we considered acted in a single plane (a two-dimensional space), and, as we saw, the $x$ and $y$ components ($F_x$, $F_y$) of a force $\vec{F}$ completely identified that force. As a general rule, however, not all the forces applied to a body are two-dimensional, and we must therefore talk about three-dimensional or space forces. A three-dimensional force $\vec{F}$ has three rectangular components ($F_x$, $F_y$, $F_z$) along the $x$-, $y$-, and $z$-coordinate axes as shown in Fig. 2.16.

To construct these three components graphically, pass three planes from point A parallel to each of the coordinate planes ($Ox$–$Oy$; $Ox$–$Oz$;

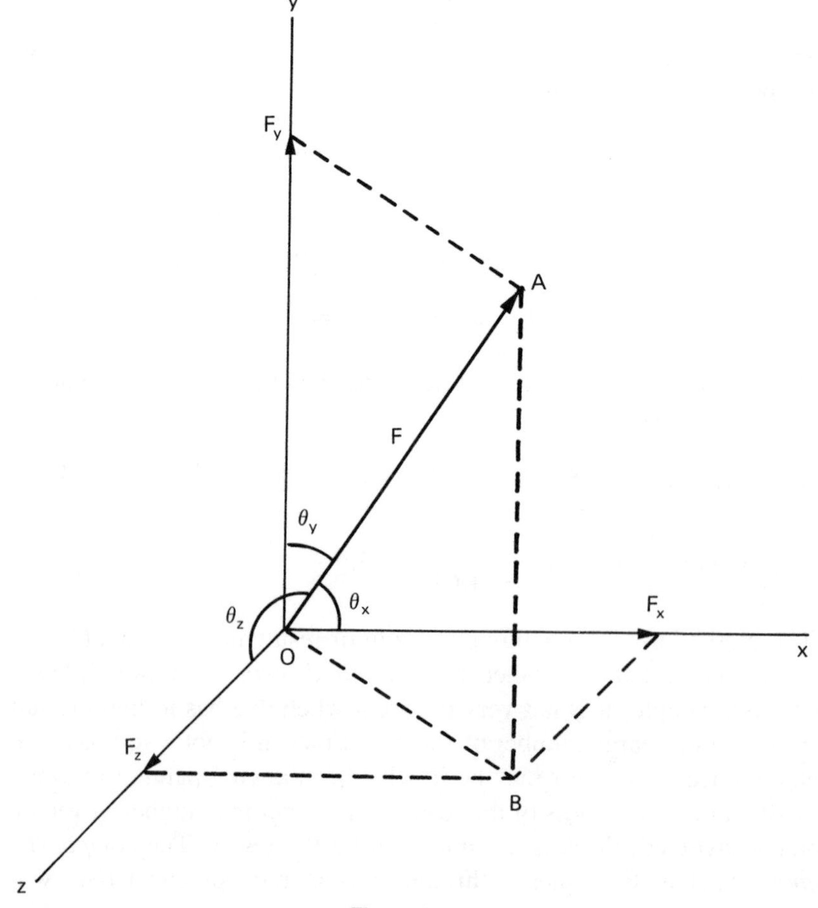


Figure 2.16

and $Oy$–$Oz$). Intersection of these planes (indicated by the broken lines) with the coordinate axes will give the three components.

The force, $\vec{F} = \overrightarrow{OA}$, makes angles $\theta_x$, $\theta_y$, and $\theta_z$ with the $Ox$, $Oy$, and $Oz$ coordinate axes, respectively. $OB$ is the projection of $OA$ in the $Ox$–$Oz$ plane, and in the triangle OAB we have

$$OA^2 = OB^2 + AB^2$$

Since $OB^2 = F_x^2 + F_z^2$ and $AB^2 = F_y^2$, therefore

$$OA^2 = F_x^2 + F_y^2 + F_z^2 = F^2$$

or

$$F = \sqrt{F_x^2 + F_y^2 + F_z^2} \tag{2.7}$$

Notice that this formula reduces to Eq. 2.2 when $F_z = 0$ (that is, when a two-dimensional space is involved.)

Let's denote the angles that $OA$ makes with the $x$, $y$, and $z$ axes by $\theta_x$, $\theta_y$, and $\theta_z$; then we can write:

$$\begin{aligned} F_x &= F \cos \theta_x \\ F_y &= F \cos \theta_y \\ F_z &= F \cos \theta_z \end{aligned} \tag{2.8}$$

or, solving for the cosine of the angles themselves,

$$\begin{aligned} \cos \theta_x &= \frac{F_x}{F} \\ \cos \theta_y &= \frac{F_y}{F} \\ \cos \theta_z &= \frac{F_z}{F} \end{aligned} \tag{2.9}$$

These are called *direction cosines* or *cosine directors*. To identify a vector in space, one needs to know not only its magnitude but also its cosine directors.

EXAMPLE 6: Determine the resultant of the two concurrent space forces shown in Fig. 2.17, whose magnitudes are $F_1 = 150$ lb and $F_2 = 300$ lb. The lines of action of forces $F_1$ and $F_2$ pass through the origin and points $A_1$ and $A_2$, respectively. The coordinates of points $A_1$ and $A_2$ are (2,4,1) and (3,1,2), respectively.

SOLUTION: First, calculate the lengths $OA_1$ and $OA_2$ in order to obtain the cosine directions. For force $F_1$,

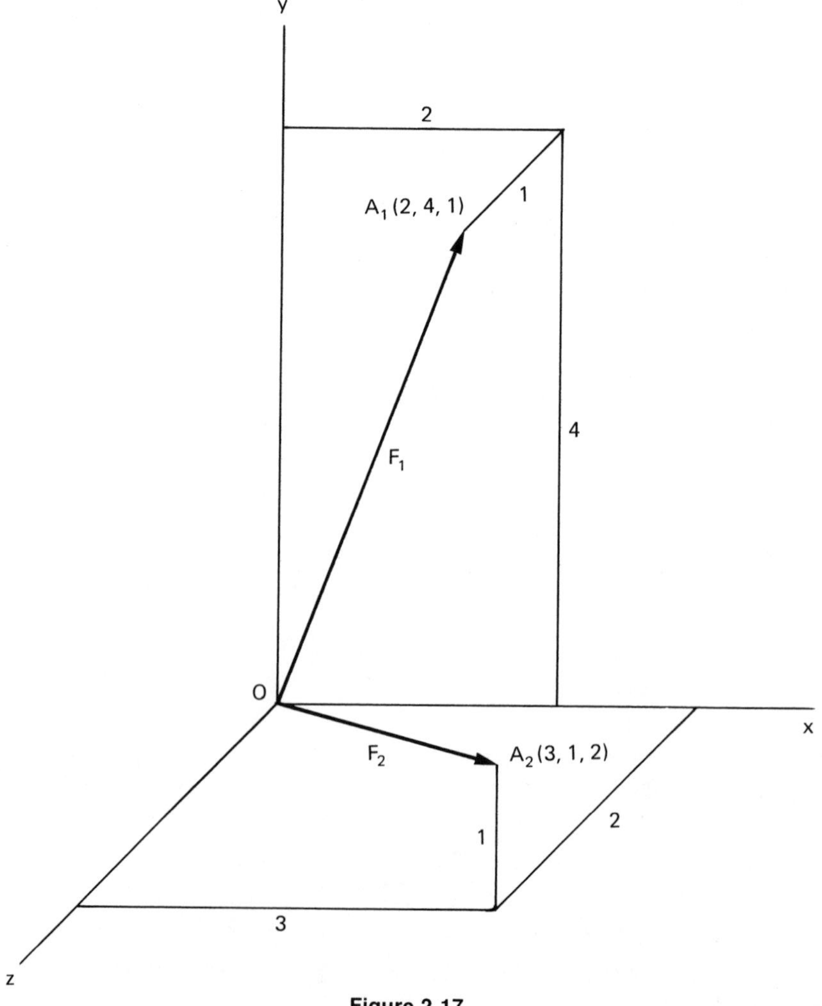


Figure 2.17

$$OA_1 = \sqrt{2^2 + 4^2 + 1^2} = \sqrt{21} = 4.58$$

The cosine directors of $OA_1$ are

$$\cos\theta_x = \frac{2}{4.58} = 0.437$$

$$\cos\theta_y = \frac{4}{4.58} = 0.873$$

$$\cos\theta_z = \frac{1}{4.58} = 0.218$$

Using Eq. 2.8, the components of force $F_1$ can be calculated by substituting the direction cosines obtained above:

$$(F_1)_x = F_1 \cos\theta_x = 150\ (0.437) = 65.6 \text{ lb}$$

$$(F_1)_y = F_1 \cos\theta_y = 150\ (0.873) = 131.0 \text{ lb}$$

$$(F_1)_z = F_1 \cos\theta_z = 150\ (0.218) = 32.8 \text{ lb}$$

Similarly, for force $F_2$

$$OA_2 = \sqrt{3^2 + 1^2 + 2^2} = \sqrt{14} = 3.74$$

$$\cos\theta_x = \frac{3}{3.74} = 0.802$$

$$\cos\theta_y = \frac{1}{3.74} = 0.267$$

$$\cos\theta_z = \frac{2}{3.74} = 0.535$$

and the $x$, $y$, and $z$ components of force $F_2$ are

$$(F_2)_x = 300\ (0.802) = 240.6 \text{ lb}$$

$$(F_2)_y = 300\ (0.267) = 80.1 \text{ lb}$$

$$(F_2)_x = 300\ (0.535) = 160.5 \text{ lb}$$

Once the $x$, $y$, and $z$ components of the two forces are calculated, the $x$, $y$, and $z$ components of the resultant $\vec{R}$ can be calculated by algebraically adding the $x$, $y$, and $z$ components of the two forces:

$$R_x = \Sigma F_x = (F_1)_x + (F_2)_x = 65.6 + 240.6 = 306.2 \text{ lb}$$

$$R_y = \Sigma F_y = (F_1)_y + (F_2)_y = 131.0 + 80.1 = 211.1 \text{ lb}$$

$$R_z = \Sigma F_z = (F_1)_z + (F_2)_z = 32.8 + 160.5 = 193.3 \text{ lb}$$

Finally, using Eq. 2.7, the resultant $R$ can be calculated:

$$R = \sqrt{R_x^2 + R_y^2 + R_z^2} = 419.1 \text{ lb}$$

The direction cosines of the resultant can also be calculated:

$$\cos \theta_x = \frac{306.2}{419.1} = 0.731$$

$$\cos \theta_y = \frac{211.1}{419.1} = 0.504$$

$$\cos \theta_z = \frac{193.3}{419.1} = 0.461$$

Now that the cosine directors of $\vec{R}$ are known, the angles $\vec{R}$ makes with the coordinate axes can be calculated simply by using the INV COS of these values:

$$\theta_x = 43.1°; \ \theta_y = 59.8°; \text{ and } \theta_z = 62.5°$$

## PROBLEMS

**2.1.** Draw each of the following forces to scale (1 inch = 1000 lb); the angular orientation of each is given in parentheses: $F_1$ = 1000 lb (horizontal, to the right); $F_2$ = 550 lb (vertical, upward); $F_3$ = 1 ton (45° NE); and $F_4$ = 1500 lb (30° SW).

**2.2.** Graphically determine the sum of the forces shown in (a) through (f) in Fig. P2.1.

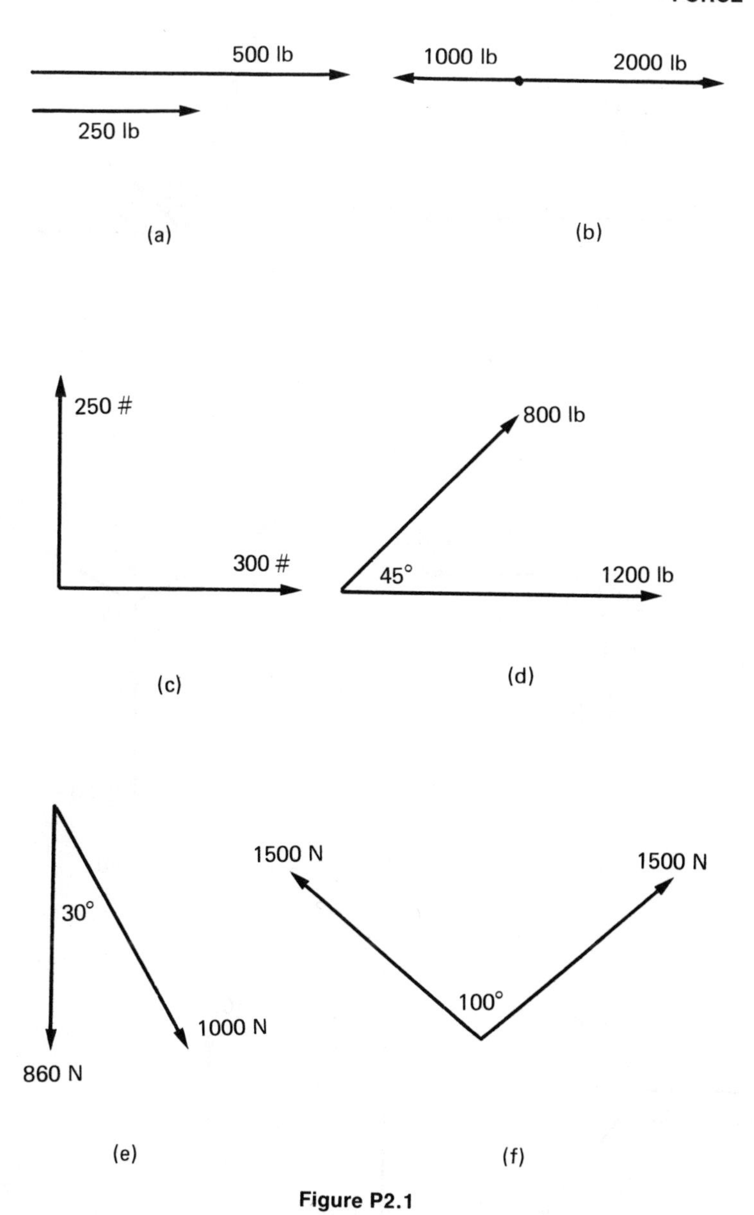


**Figure P2.1**

**2.3.** Graphically determine the sum of the forces in (a) through (f) in Fig. P2.2, and then find the magnitude and the angle that each sum makes with the horizontal axis. Verify your answer by the algebraic method.

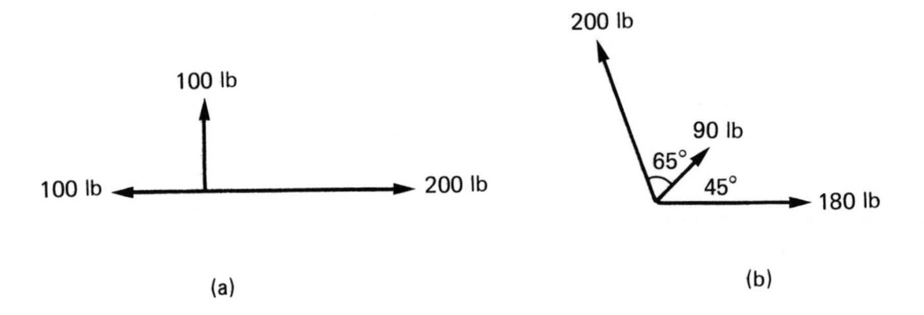
100 lb
100 lb
200 lb
(a)
200 lb
90 lb
65°
45°
180 lb
(b)

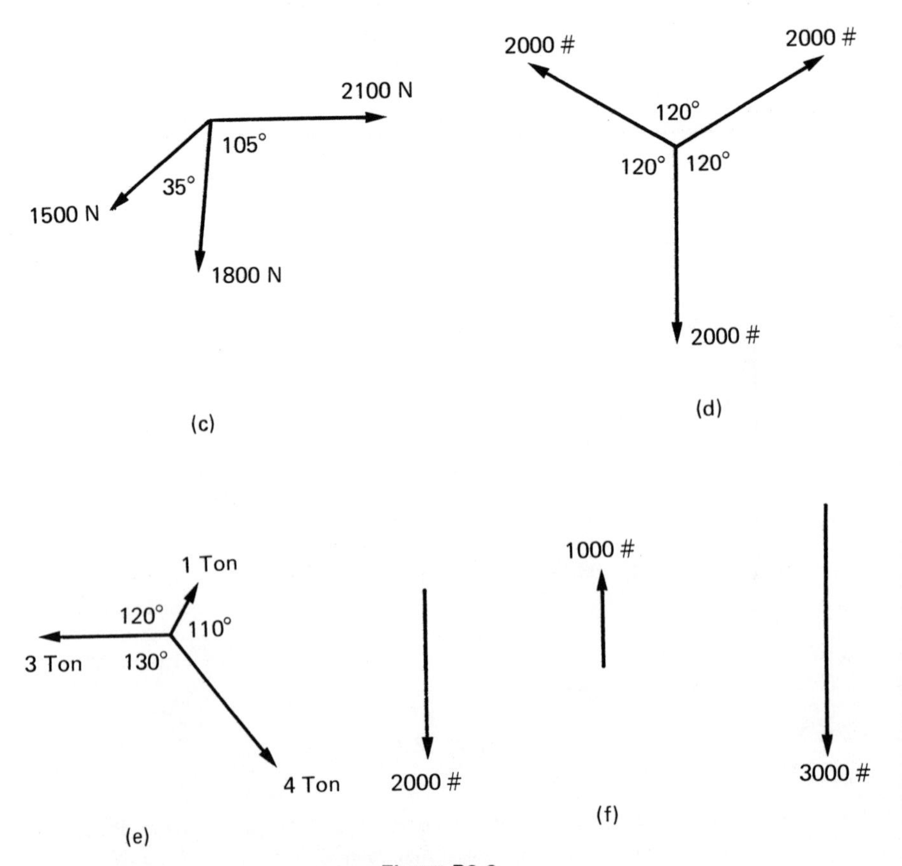
2100 N
105°
35°
1500 N
1800 N
(c)
2000 #
2000 #
120°
120°
120°
2000 #
(d)
1 Ton
120°
110°
3 Ton
130°
4 Ton
(e)
2000 #
1000 #
3000 #
(f)

**Figure P2.2**

**2.4.** Using the polygon of forces, determine the magnitude and direction of the resultant in (a) through (d) of Fig. P2.3. Verify your answer by the algebraic method.

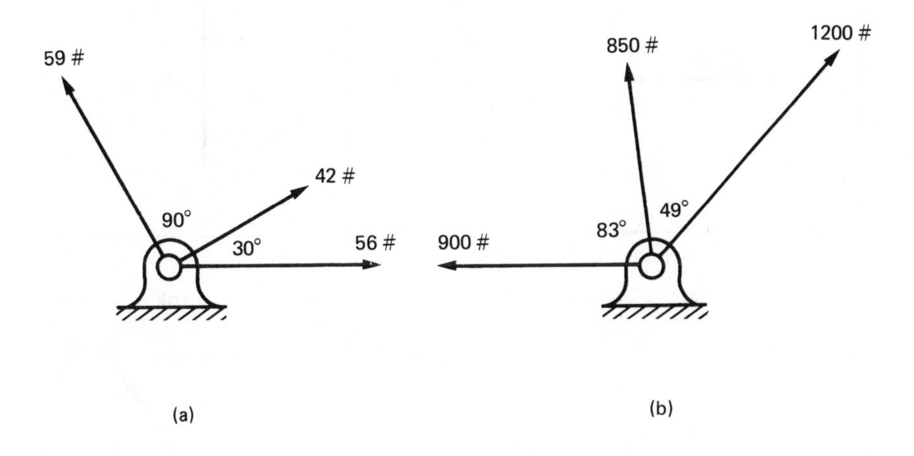


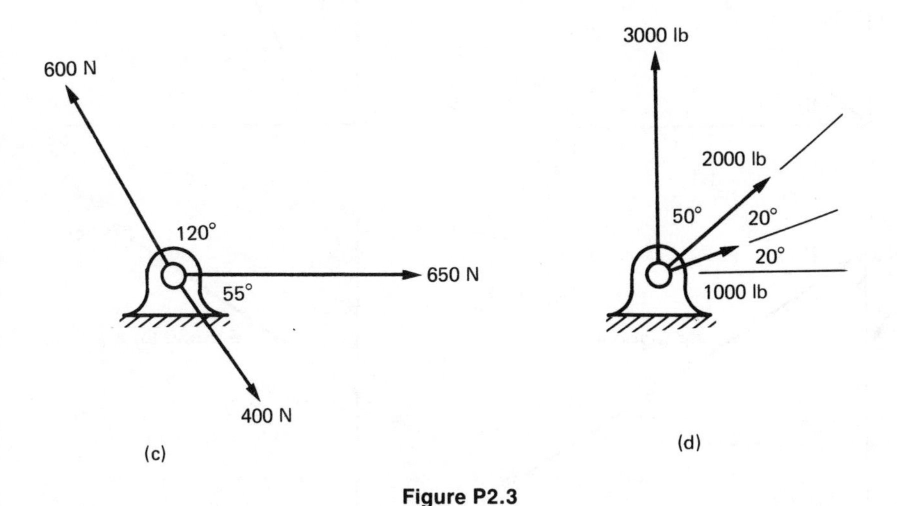


**Figure P2.3**

**2.5.** Find the rectangular components of the forces shown in (a) through (f) of Fig. P2.4.

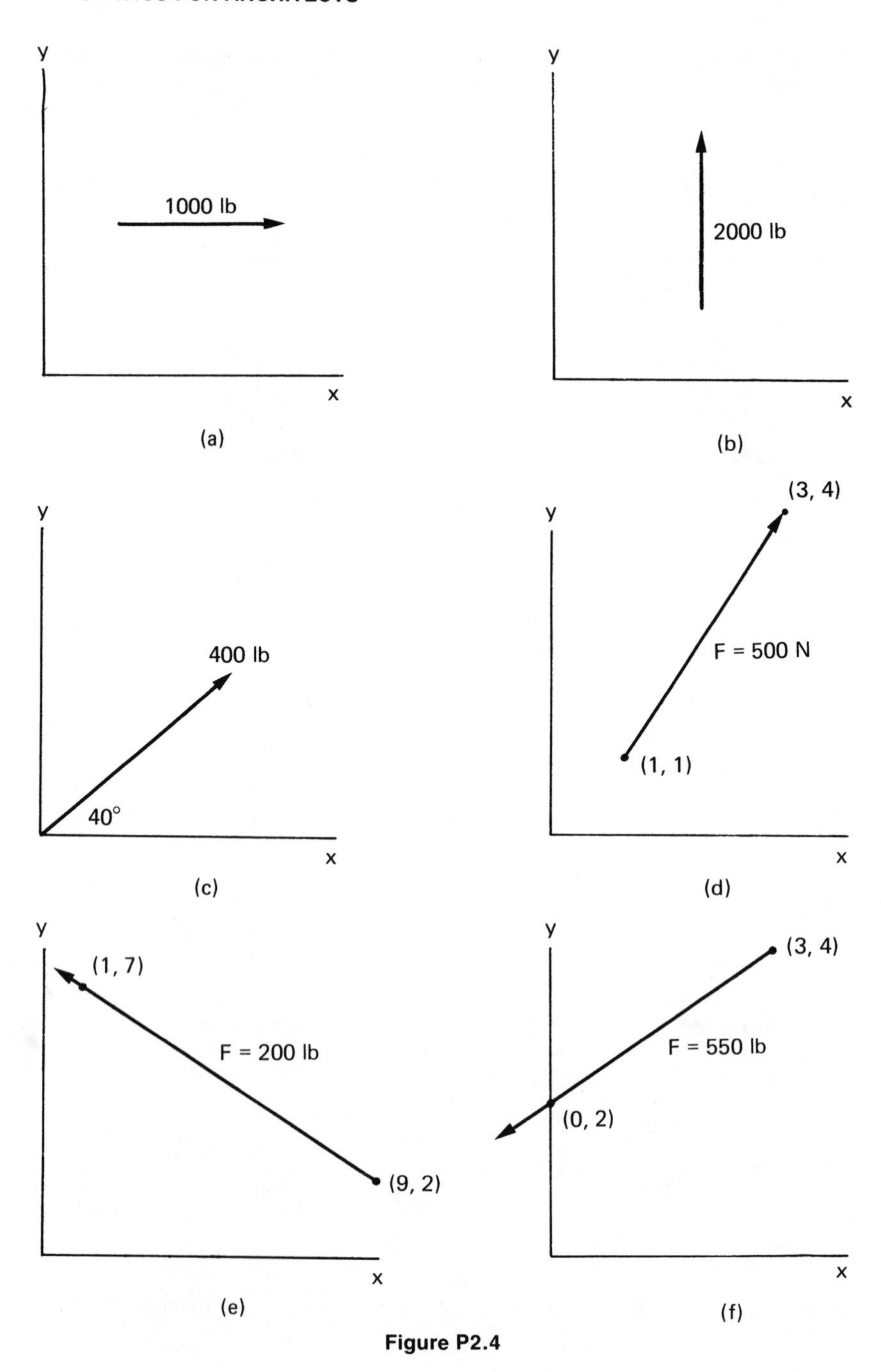
y
x
1000 lb
(a)
y
x
2000 lb
(b)
y
x
400 lb
40°
(c)
y
x
(3, 4)
F = 500 N
(1, 1)
(d)
y
x
(1, 7)
F = 200 lb
(9, 2)
(e)
y
x
(3, 4)
F = 550 lb
(0, 2)
(f)

**Figure P2.4**

**2.6.** Find the rectangular components of the space forces shown in (a) through (d) of Fig. P2.5.

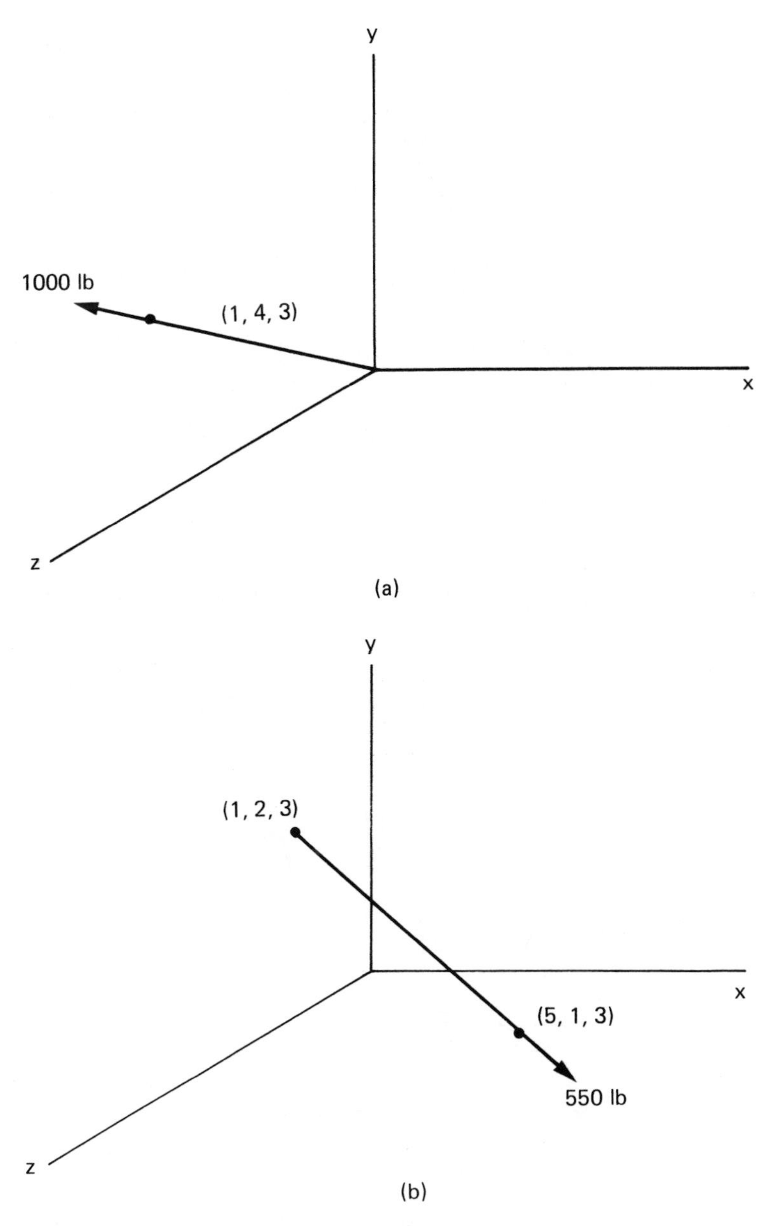


**Figure P2.5**

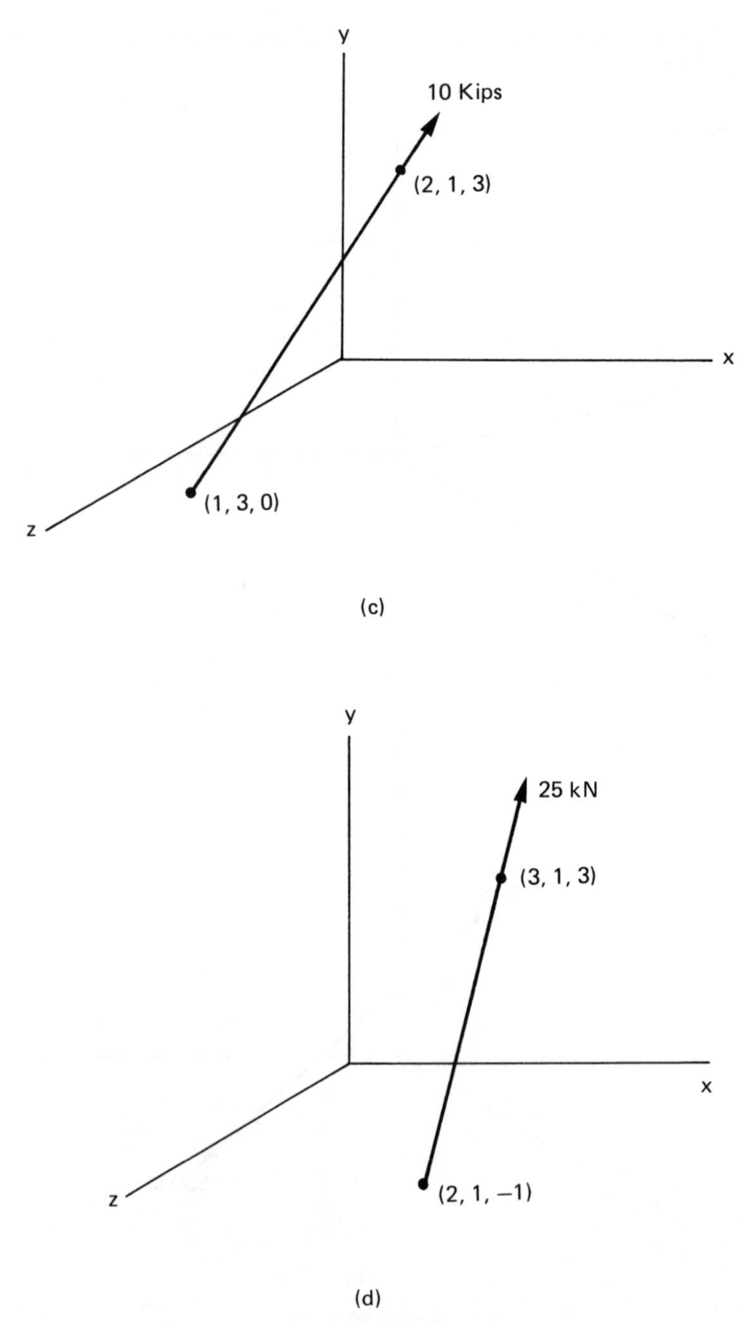


**Figure P2.5.** *(Continued)*

**2.7.** Find the resultant of the forces acting on the rectangular block shown in Fig. P2.6.

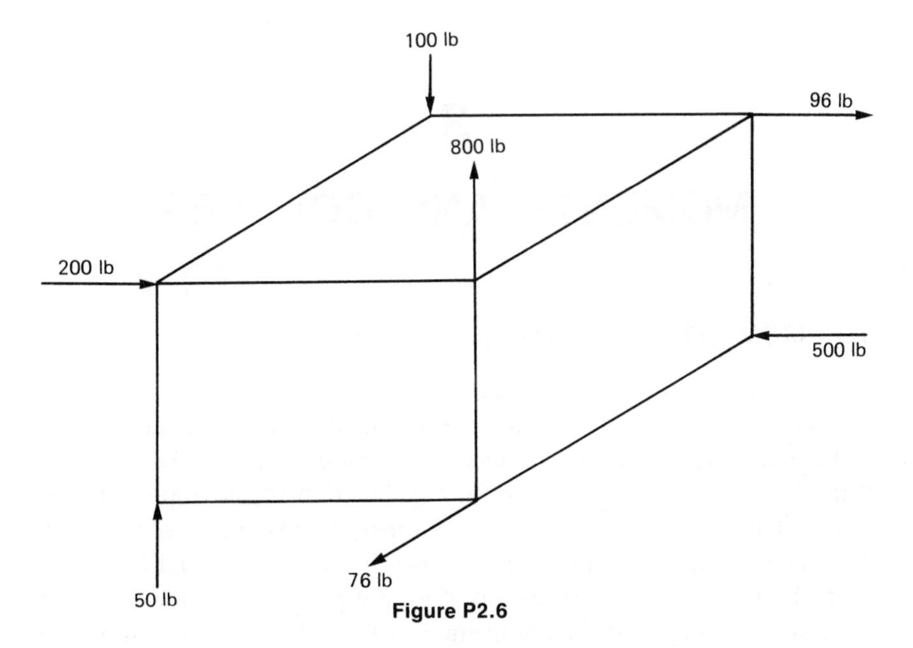

Figure P2.6

**2.8.** Find the resultant of the two space forces, $F_1 = 1000$ lb and $F_2 = 2000$ lb, in Fig. P2.7. The line of action of $F_1$ passes through points (3,1,2) and (2,1,5). The line of action of $F_2$ passes through points (5,4,1) and (6,2,5).

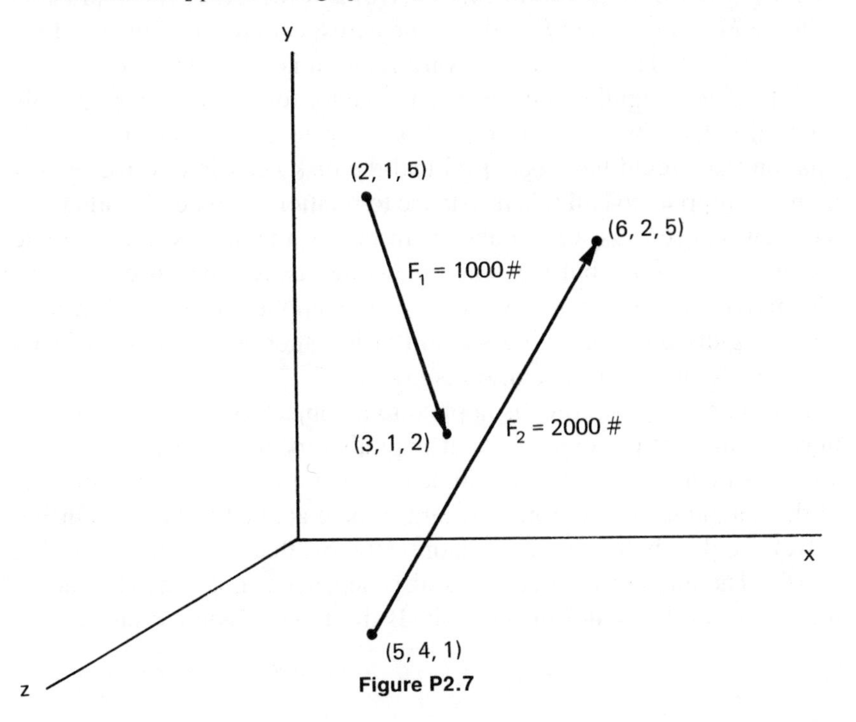

Figure P2.7

# 3

# MOMENTS AND COUPLES

## 3.1 MOMENT OF A SINGLE FORCE

Consider an object (such as a disk) on a table such as that shown in Fig. 3.1. The result of applying a force $\vec{F}$ to this disk in the manner shown in Figures 3.1(a) and (b) is called *translation*, i.e., the force $\vec{F}$ will "translate" or "slide" the disk in the direction of the applied force. Notice that in these cases the force $\vec{F}$ (or its extension) passes through the center of the disk. In the cases shown in Figs. 3.1(c) and (d), however, the force $\vec{F}$ (or its extension) does not pass through the center of the disk. Not only will the application of force $\vec{F}$ move the disk to the right as before (translation), it will also cause the disk to rotate. This result can be experimentally verified by marking two points on the disk such as $A$ and $B$ as shown in Figs. 3.1(c) and (d). After the application of force $\vec{F}$, points $A$ and $B$ will assume new locations, as shown in Figs. 3.1(e) and (f). This tendency toward rotation is called *moment*.

If the disk is rigidly connected to the table, force $\vec{F}$ cannot physically rotate the disk, but a reaction will develop to counteract or resist the rotation that would have occurred had the disk been free to rotate. It is more appropriate to call this resistance to rotation "moment" rather than the rotation itself, as we did above. In fact, if the disk is free to rotate, the moment is zero, but the rotation analogy is used because it is more illustrative and easy to follow. If a force is applied to one end of a rod that is rigidly connected to a support at its other end, a moment exists although no appreciable rotation is noticed.

In general, when a force is applied to an object, not only will it try to move (translate) the object, but it may also try to rotate the object (or apply moment to it). The magnitude of the moment is clearly a function of the magnitude of the force. A larger force applied to the disk in Fig. 3.1(c) would obviously cause more rotation than that shown in Fig. 3.1(f). The amount of rotation is also dependent upon the distance of force $\vec{F}$ from the center of the disk. If the force is applied close to the

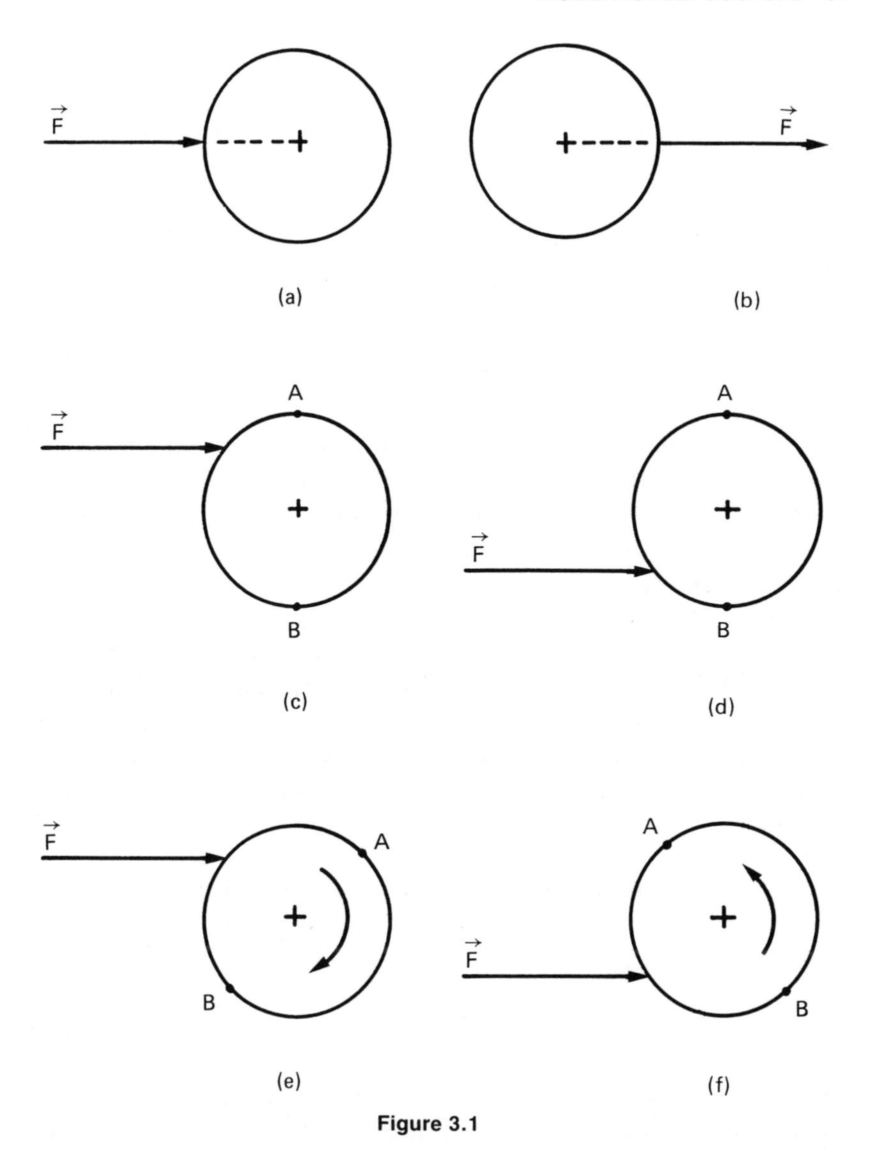


**Figure 3.1**

center of the disk, it will produce less rotation; the same force applied further away from the center of the disk will cause much more rotation. Operating a revolving door is a good example of this principle. If you operate a revolving door by applying a force (pull or push) on the outside edge of the door, you will find it easy to open, but if you try to operate it by applying force in the middle, you may have to pull or push much

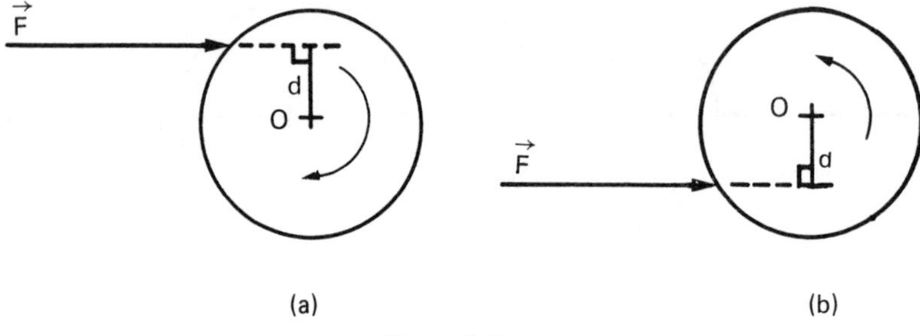


Figure 3.2

harder; finally, if you pull or push the door at a point close to its hinges, you may find yourself not powerful enough to operate it at all.

In quantifying moment (or the amount of rotation), therefore, two factors are involved: force and distance. Hence, *moment is defined as the product of the magnitude of the force and the distance*. This distance is taken as the shortest distance from the point (the center of the disk or the hinge in the door in the previous examples) to the force.

Now consider Fig. 3.2(a). A force $\vec{F}$ is applied to a disk and tends to rotate the disk about, let us say, its center $O$. The amount of rotation or the magnitude of the moment, $M$, will be equal to the product of the magnitude of the force, $F$, and the *perpendicular* distance, $d$, from point $O$ (which is also called the *moment center*) to the force (or its extension), as follows:

$$\text{Moment} = \text{Force} \times \text{Perpendicular distance}$$

or,

$$M = Fd \tag{3.1}$$

In this example, force $\vec{F}$ will try to rotate the disk clockwise. By just changing the direction of the force in Fig. 3.2(a), the rotation (or moment) will be counterclockwise. On the other hand, if force $\vec{F}$ is applied to the disk at distance $d$ below the center of rotation, as shown in Fig. 3.2(b), the same amount of rotation (moment) will be produced but in a counterclockwise direction. Therefore, not only is it important to know how much rotation is involved, but also the direction in which the rotation will take place (clockwise or counterclockwise). From this simplified example, we become aware that to describe moment completely, we must specify not only its magnitude but also its direction. We learned in Chap. 2 that such quantities are vector quantities. Therefore, *moment is not a scalar but a vector quantity.*

To perform calculations involving moments in their vectorial form requires the use of the cross-product of two vectors, an approach that will be altogether avoided in this book because of the math involved. Therefore, in moment calculations, we will calculate not only the magnitude of the moment but also simultaneously its direction of rotation. To do so requires modification of Eq. 3.1 to the following:

$$M = \pm Fd \tag{3.2}$$

We thus have two choices of direction: clockwise and counterclockwise. It does not matter which of these is considered positive and which negative, but it is important that in a given problem (or equation) a sign convention be defined and adhered to throughout the problem. To avoid confusion and for the sake of uniformity and consistency, we will consider the moment of forces producing clockwise rotations as *positive moment* and that producing counterclockwise rotations as *negative moment*.

The unit for moment has to come from Eq. 3.1 or Eq. 3.2. Since moment is determined to be equal to force multiplied by distance, its unit should be the unit used for force multiplied by the unit used for distance. In the following simple examples, the sign convention just decided upon as well as appropriate units in calculating moments of forces are demonstrated.

EXAMPLE 1: Find the moment of force about point $O$ in each of the cases in Fig. 3.3.

SOLUTION:

1. Fig. 3.3(a): The magnitude of the force shown is 100 lb. The perpendicular distance is 2 ft. Since the force will cause the disk to rotate clockwise about point $O$, its sign is considered positive. Its unit is lb ft. Using the subscript $O$ to indicate that the moment of the force is calculated about point $O$, $M_O = +100 \times 2 = 200$ lb ft.

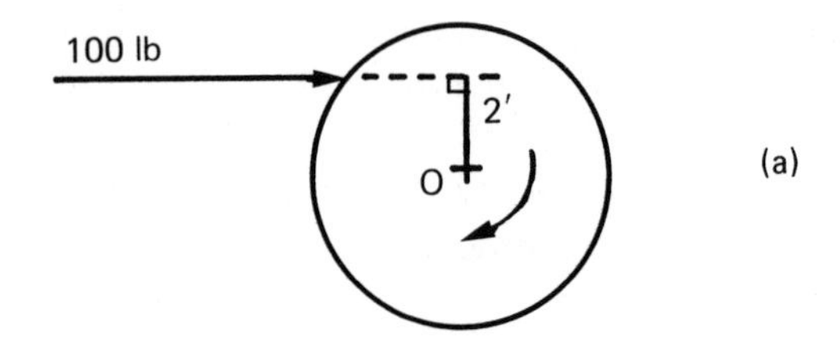


2. Fig. 3.3(b): This problem is identical to the one preceding. The 100-lb force has merely slid along its line of action, and the positive moment it produces is as follows: $M_O = 100 \times 2 = 200$ lb ft.

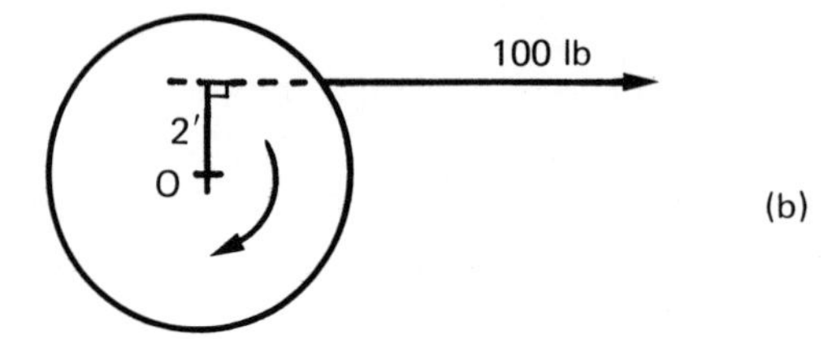


3. Fig. 3.3(c): Since the 200-lb force will cause the disk to rotate counterclockwise about point $O$, the negative sign is needed to indicate counterclockwise moment, as follows: $M_O = -200 \times 50 = -10{,}000$ lb in.

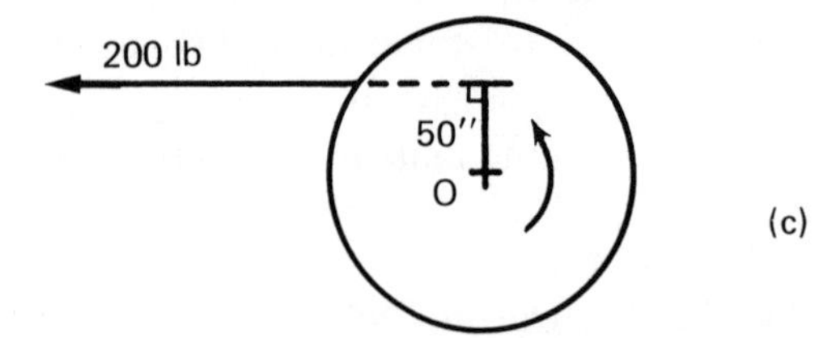


4. Fig. 3.3(d): Since the force is given in newtons (N) and the distance in meters (m), these being the units of force and length in the SI system of units, the unit for moment will be newton × meter, or Nm, as follows: $M_O = +500 \times 2 = 1000$ Nm.

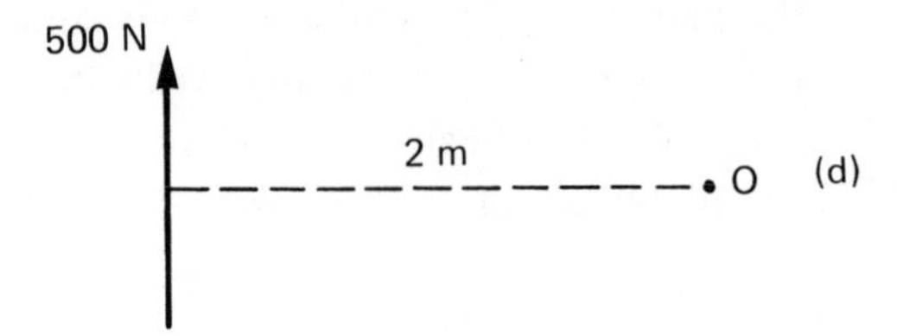

5. Fig. 3.3(e): Since rotation is again clockwise: $M_O = +250 \times 10 = 2500$ Nm.

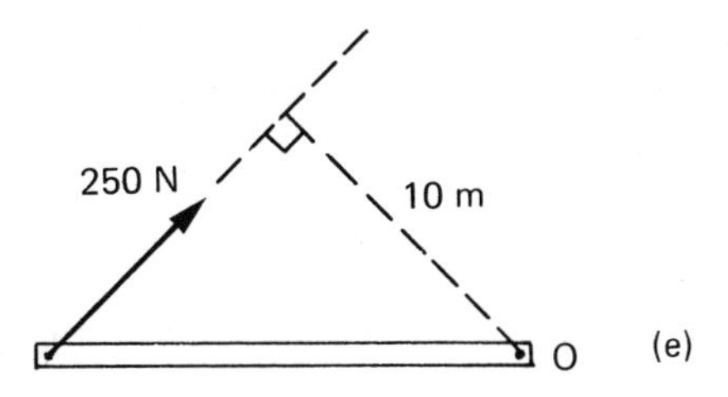


6. Fig. 3.3(f): Since the force 600 N passes through point $O$ (and hence $d = 0$), no rotation is involved. Therefore: $M_O = 600 \times 0 = 0$ Nm.



## 3.2 MOMENT OF MORE THAN ONE FORCE

If more than one force is applied to a body, the moment of all forces about a given point inside or outside that body can be determined by adding the moment of each one of these forces algebraically. This is illustrated in the following example.

EXAMPLE 2: Calculate the moment of all forces applied to the lever in Fig. 3.4 about point $A$.

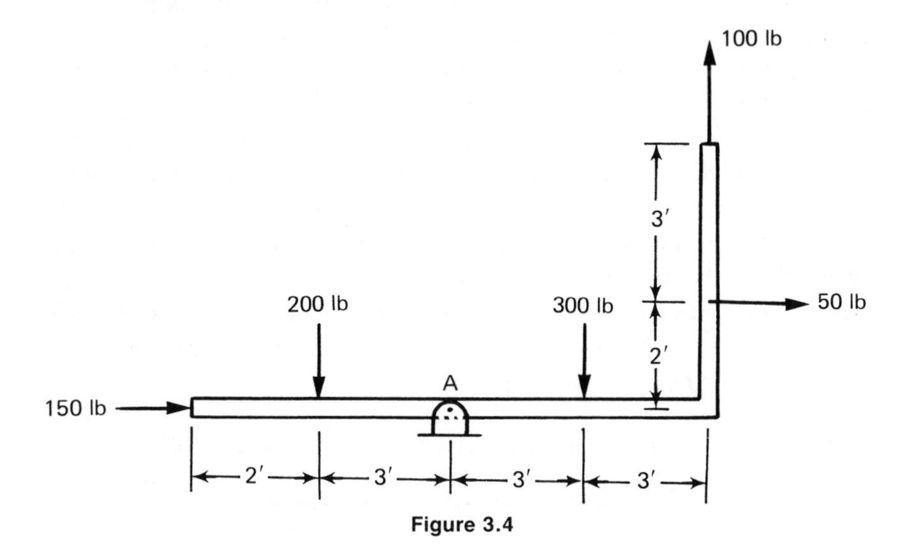


Figure 3.4

SOLUTION: Since five forces are involved, there will be five terms in the moment equation. As usual, we will assume clockwise rotation to indicate positive moment. Since by Eq. 3.2, the moment of each force is equal to that force multiplied by its perpendicular distance from point $A$, we have:

$$\begin{aligned} M_A &= (150 \times 0) - (200 \times 3) \\ &\quad + (300 \times 3) + (50 \times 2) - (100 \times 6) \\ &= -200 \text{ lb ft} \end{aligned}$$

Notice that since the horizontal force at the far left (150 lb) passes directly through the moment center $A$, its perpendicular distance is zero ($d = 0$), and therefore it will not contribute to the total moment.

## 3.3 VARIGNON'S THEOREM

Varignon's theorem states that the moment of a force about any point is equal to the sum of the moments of the individual components of the force about the same point. This very important and useful theorem will be used frequently throughout this book.

EXAMPLE 3: Find the moment of the 100-lb force about point $O$ as shown in Fig. 3.5(a).

SOLUTION: Although the perpendicular distance, $d$, from point $O$ to the line of action of the force is not directly given in this problem, it can be calculated by using the geometric information provided by Fig. 3.5(a). In using Varignon's theorem (which is also called the "moment theorem"), however, it is not necessary to calculate $d$, only to find the horizontal and vertical components of the 100-lb force by projecting it along the $x$ and $y$ axes. As shown in Fig. 3.5(b), the two components will be 100 cos 30°, or approximately 87 lb, and 100 sin 30°, or 50 lb, respectively. Since the 100-lb force is the sum (resultant) of these two components, the above theorem states that the moment of the 100-lb force about point $O$ is equal to the sum of moments of the components 100 cos 30° and 100 sin 30° about point $O$, or

$$\begin{aligned} M_O = -100 \times d &= (100 \cos 30°) \times 10' - (100 \sin 30°) \times 30' \\ &= 1000 \cos 30° - 3000 \sin 30° = -634 \text{ lb ft} \end{aligned}$$

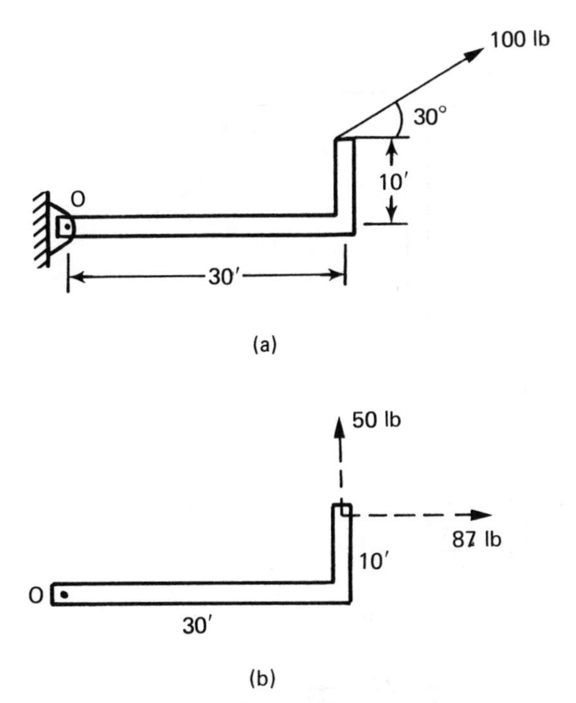


**Figure 3.5**

This is often the more convenient method, especially if more than two forces are involved, since finding the components allows us to deal with forces that are either horizontal or vertical.

EXAMPLE 4: A 1000-N force is applied to the end of a bracket as shown in Fig. 3.6(a). Calculate the moment of this force about base point $A$ in four different ways.

SOLUTION:

METHOD 1. To calculate $d$, draw a perpendicular line from point $A$ to the force, as shown in Fig. 3.6(b). Then,

$$d = 3 \sin 50° + 9 \cos 50° = 8.083 \text{ m}$$

The moment of the 1000-N force about point $A$ is clockwise and therefore positive, and its magnitude is

$$M_A = F \times d = 1000 \times 8.083 = 8083 \text{ Nm}$$

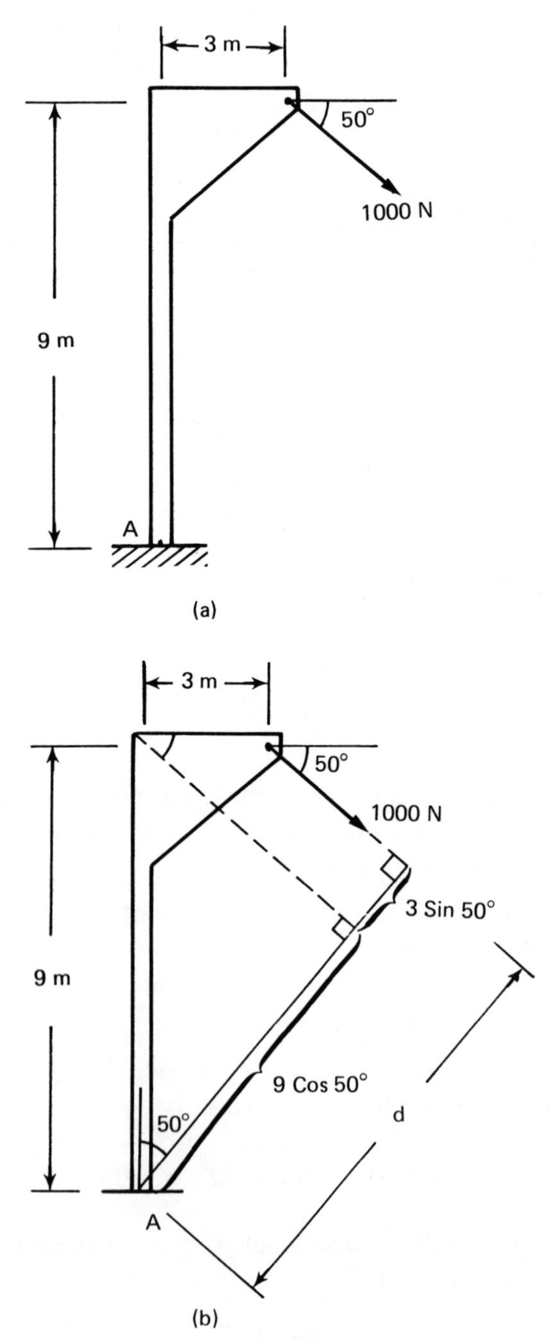


**Figure 3.6**

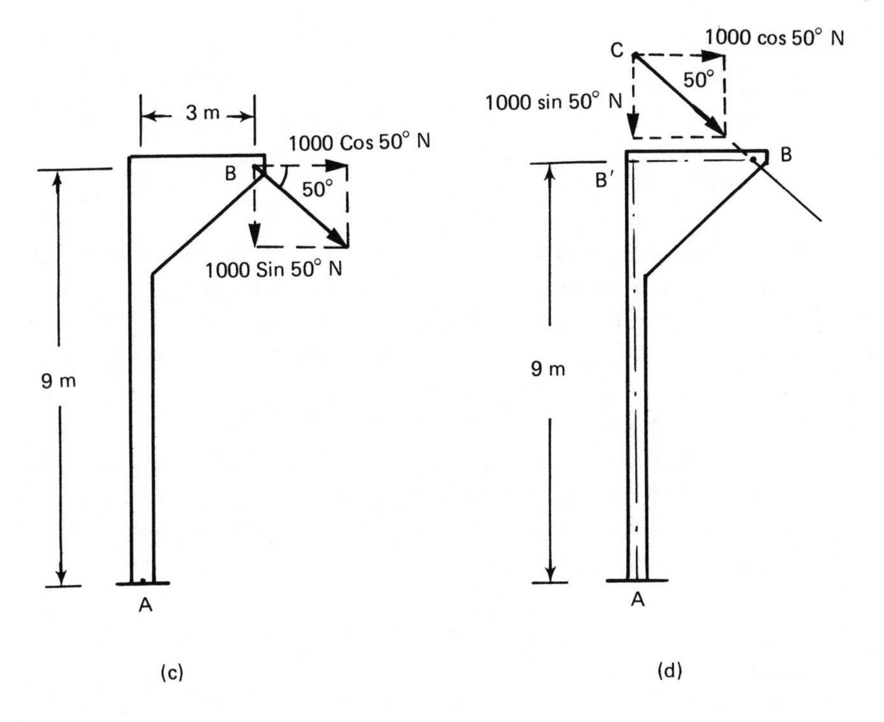


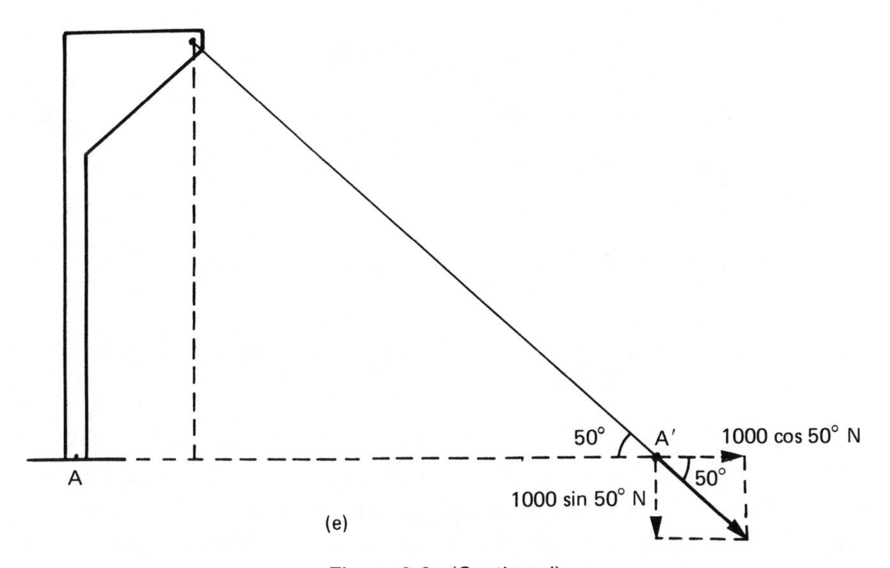


**Figure 3.6.** (*Continued*)

METHOD 2. Using the moment theorem, the moment of the 1000-lb force about point $A$ is equal to the sum of the moments of its rectangular components about point $A$, as shown in Fig. 3.6(c):

$$M_A = (1000 \cos 50°) \times 9 + (1000 \sin 50°) \times 3 = 8083 \text{ Nm}$$

METHOD 3. As we saw in Chap. 2, a force can be slid along its line of action without altering its effect. Using this fact, slide the 1000-N force along its line of action and place it at $C$, which is at the midpoint of the vertical post (Fig. 3.6d). Now, as in Method 2, we must find its rectangular components and use the moment theorem. In this case, since the downward vertical force (1000 sin 50°) passes through the moment center, $A$, it will not contribute to the moment equation (its moment about $A$ is equal to zero), and the only term that remains is the horizontal force (1000 cos 50°), which has to be multiplied by the distance $AC$. To calculate $AC$, we proceed as follows:

$$AC = AB' + B'C$$

Since we know $AB' = 9$ m, we need calculate only $B'C$. Because $B'B = 3$ m and the angle $CBB' = 50°$, we can write

$$\tan 50° = \frac{B'C}{BB'} = \frac{B'C}{3}$$

$$B'C = 3 \tan 50°$$

Therefore,

$$AC = AB' + B'C = 9 + 3 \tan 50° = 12.575 \text{ m}$$

and

$$M_A = (1000 \cos 50°) \times 12.575 = 8083 \text{ Nm}$$

METHOD 4. As in Method 3, slide the force 1000 N to point $A'$ along its line of action (Fig. 3.6e). In this case, the horizontal component (1000 cos 50°) passes through the moment center and will not contribute to the moment equation. The moment will thus be equal to the vertical force (1000 sin 50°) multiplied by the distance $AA'$, which can be calculated as follows:

$$AA' = 3 + \frac{9}{\tan 50°} = 10.552 \text{ m}$$

and, therefore,

$$M_A = (1000 \sin 50°) \times 10.552 = 8083 \text{ Nm}$$

Obviously, as was to be expected, all four methods provide the same answer.

## 3.4 COUPLE

Consider the steering wheel of your car. You usually hold it by two hands at points $A$ and $B$, as shown in Fig. 3.7(a). When you want to turn your car to the right, for example, your right hand exerts a downward force at $B$, and your left hand exerts the same amount of force at $A$, though in an upward direction, as shown in Fig. 3.7(b). Since the two forces are equal and parallel but applied in opposite directions, the net amount of force exerted on the steering wheel is zero. Nevertheless, these two forces will rotate the wheel clockwise. This rotation (or moment) that is produced by applying two equal, parallel, and opposite forces is called a *couple*. The magnitude of the couple is equal to the

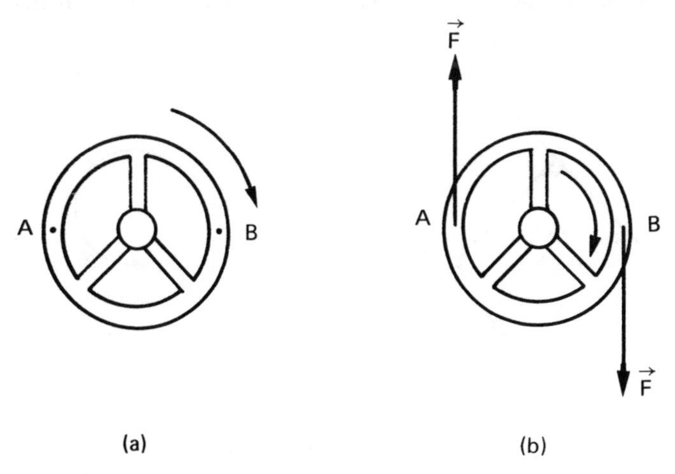


**Figure 3.7**

sum of the two moments that each of the forces produces about the center of the steering wheel. If $D$ is the diameter of the wheel, the force at $A$ will have a moment, $M_1$, equal to $F \times D/2$ and the force at $B$ will have a moment, $M_2$, equal to $F \times D/2$ about the center of the wheel. Therefore, the couple $C$ is equal to

$$C = M_1 + M_2 = F(D/2) + F(D/2) = FD$$

The magnitude of the couple is thus equal to the magnitude of one of the forces multiplied by the perpendicular distance separating the two forces. Positive or negative signs are assigned to couples as they are to moments, i.e., clockwise couples are positive and counterclockwise couples are negative.

To summarize, as shown in Fig. 3.8(a), a couple is produced by the application of two parallel forces of equal magnitudes and opposite directions that are separated by a perpendicular distance $d$. As was mentioned earlier, the net force is zero. Therefore, the two forces can be eliminated, and the action of couple can be shown solely by the direction of the rotation they produce. The magnitude of this couple, $C = Fd$, is usually shown next to the circular arrow, as shown in Fig. 3.8(b).

Moment and couple are really the same thing, except that it is important to know where the moment center is in the case of a moment but not in the case of a couple; in fact, the magnitude of a couple is the same about any point. Consider, for example, the two parallel and opposite forces with magnitudes $F$ and separated by the perpendicular distance $d$

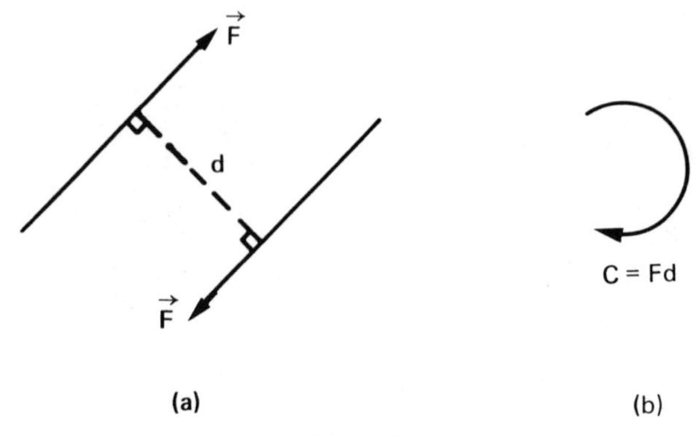


Figure 3.8

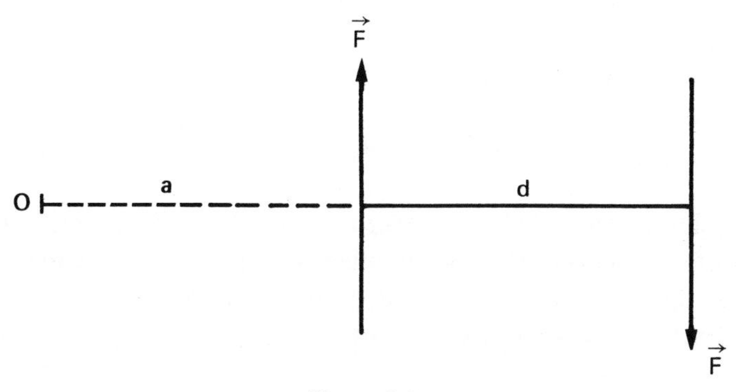


**Figure 3.9**

that are shown in Fig. 3.9. Take an arbitrary point such as $O$, which is located at a distance $a$ from the force on the left. The moment (or couple) of these two forces about the point $O$ can be calculated as

$$M_O = F \times (a + d) - F \times a$$
$$= F \times a + F \times d - F \times a = F \times d$$

which simply means that the magnitude of the moment or couple is $Fd$ and that it does not depend on where the moment center is chosen.

EXAMPLE 5: Determine the magnitude and direction of the couple produced by the two forces applied to the beam shown in Fig. 3.10(a).

SOLUTION: Since the two forces are parallel and equal in magnitude and opposite in direction, they produce a couple; the magnitude of the couple is equal to the magnitude of one of the forces multiplied by their per-

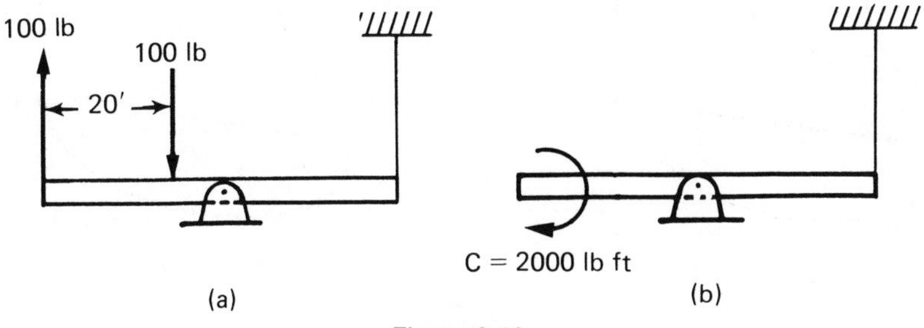


**Figure 3.10**

pendicular distance, or

$$C = 100 \times 20 = 2000 \text{ lb ft}$$

The direction of the couple is clockwise since the parallel forces tend to produce clockwise rotation. Schematically, a clockwise circular arrow, instead of the two forces, is placed on the beam to indicate the couple of 2000 lb ft, as shown in Fig. 3.10(b).

## 3.5 RESOLUTION OF A FORCE INTO A FORCE AND A COUPLE

Often for convenience, it is necessary to move a given force from one point to another while keeping it parallel to itself, but it is not correct to do so because the action of the force will be different when applied at a different point even if its magnitude and direction remain the same. The reason for this is that a force has a tendency to produce moment as well as a pull-push action, and, as we have seen, this moment will depend on the location of the force. We *can* move a force parallel to itself from one point to another, however, provided we add an appropriate moment or couple. This is illustrated in Fig. 3.11(a), which shows a force, $\vec{F}$,

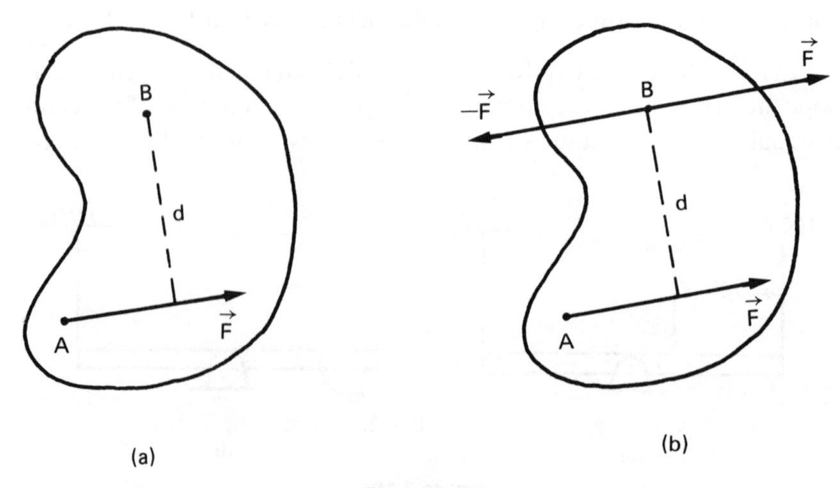


Figure 3.11

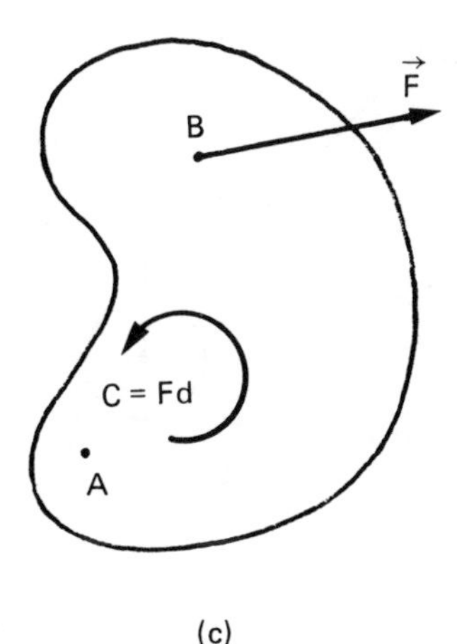


(c)

**Figure 3.11.** *(Continued)*

being applied to a body at point $A$. It is desired to shift this force parallel to itself and place it at point $B$.

To accomplish this, as shown in Fig. 3.11(b), two equal but opposite forces, $\vec{F}$ and $-\vec{F}$, parallel to the original force at $A$ are added at point $B$. Doing so will obviously not introduce any net effect to the body so far as force and moment are concerned. Since the forces $\vec{F}$ at $A$ and $-\vec{F}$ at $B$ are parallel, equal, and opposite, however, their tendency is to produce a couple, which in this case is counterclockwise. Hence these two forces can be replaced by the couple, $C = Fd$. In Fig. 3.11(c), it seems that force $\vec{F}$ has been relocated to point $B$ where it remains parallel to itself, but in the process a couple, $C = Fd$, has been added. The illustration is proof that a force can be relocated to another point parallel to itself provided that a couple is added.

EXAMPLE 6: Replace the horizontal 100-lb force shown in Fig. 3.12(a) by an equivalent force and a couple at point $A$.

SOLUTION: We apply two equal but opposite 100-lb forces at $A$ that are parallel to the given force, as shown in Fig. 3.12(b). Obviously, doing so will not have any effect on lever $AB$. Now consider the given 100-lb force at $B$ and the 100-lb force at $A$ directed to the left. These two equal, parallel, but opposite forces produce a couple. The magnitude of this couple is 100 $d$. Where $d = 10' \cos 45° = 7.07'$, then $C = +100 \times$

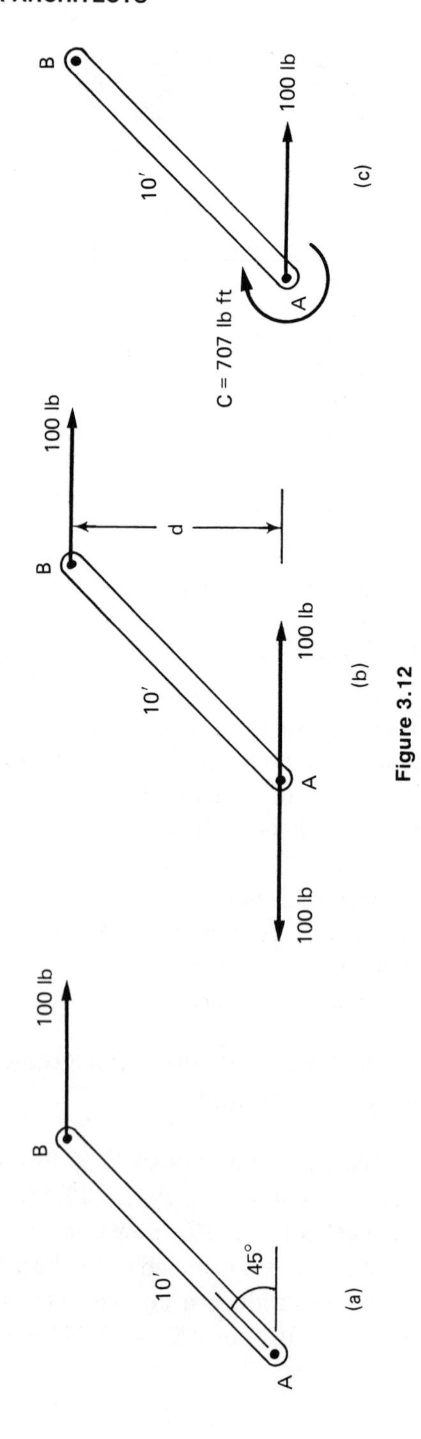
A
B
10′
45°
100 lb
(a)
A
B
10′
100 lb
100 lb
100 lb
d
(b)
A
B
10′
100 lb
C = 707 lb ft
(c)

**Figure 3.12**

7.07 = 707 lb ft. Therefore, the 100-lb force at $B$ can be shifted to $A$, provided that a clockwise couple of magnitude 707 lb ft is added as shown in Fig. 3.12(c). It should be noticed that the two situations shown in Figures 3.12(a) and 3.12(c) are identical.

## PROBLEMS

**3.1.** Calculate the moment of the 150-lb force about point $A$ in Fig. P3.1.

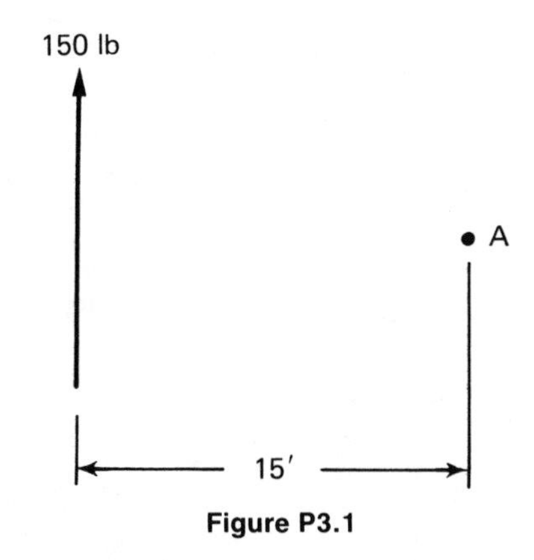


**Figure P3.1**

**3.2.** Calculate the moment of the 10-kN force about point $A$ in Fig. P3.2.

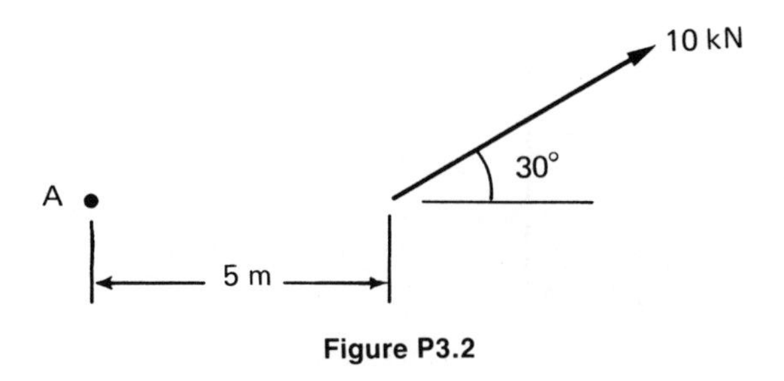


**Figure P3.2**

**3.3.** What is the magnitude and the sign of the moment produced by the 3-ton force about origin $O$ in Fig. P3.3? Use the direct method as well as Varignon's theorem.

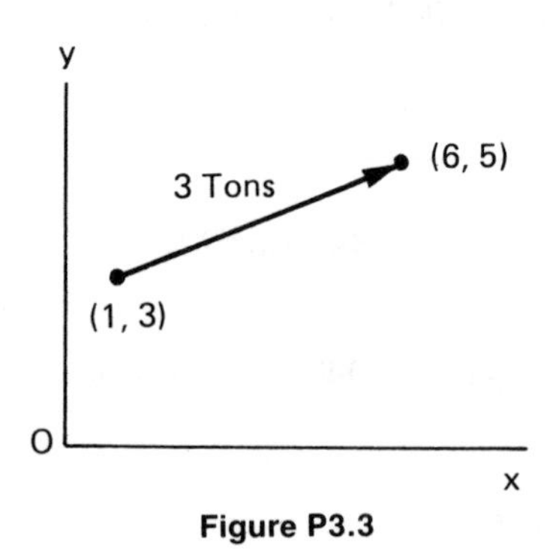


**Figure P3.3**

**3.4.** Calculate the moment of force $\vec{F}$ about point $A$ in Fig. P3.4.

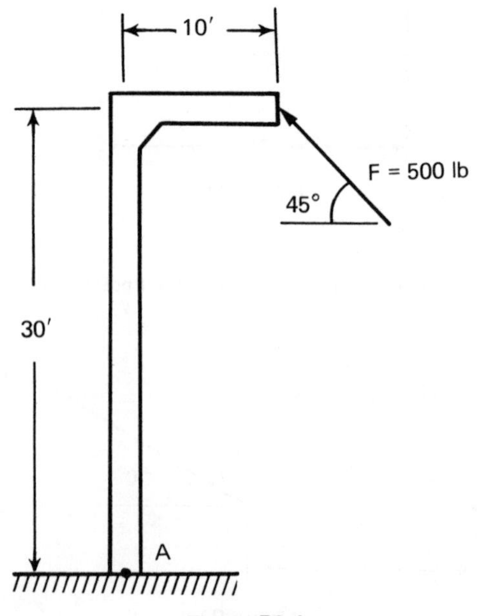


**Figure P3.4**

**3.5.** Calculate the moment produced by three forces $\vec{F}_1$, $\vec{F}_2$, and $\vec{F}_3$ about point $A$ in Fig. P3.5.

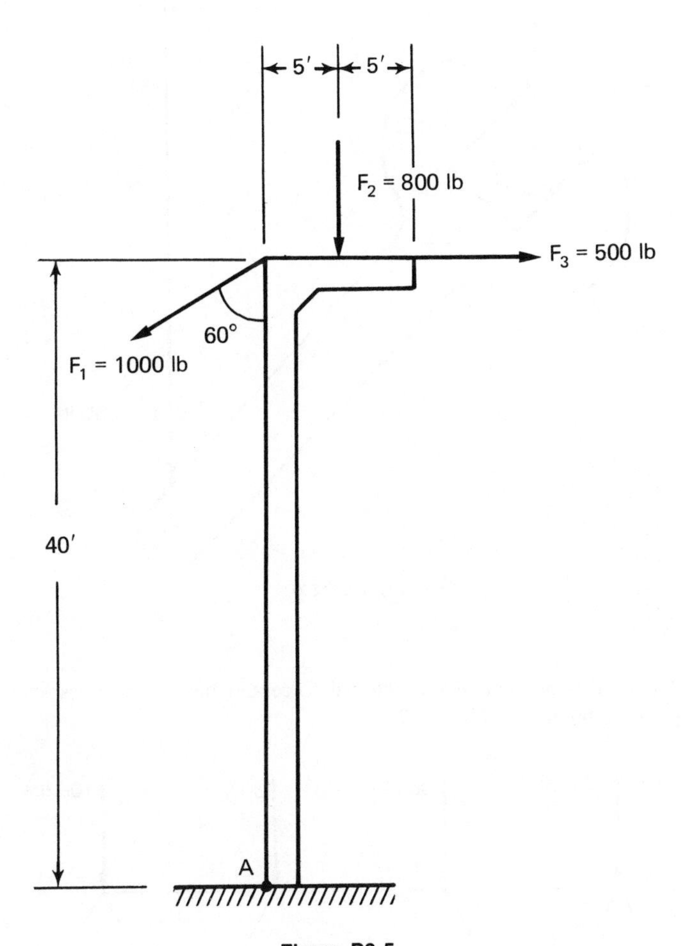


**Figure P3.5**

**3.6.** Calculate the moment of force $\vec{F}$ about the center of the nut in Fig. P3.6.

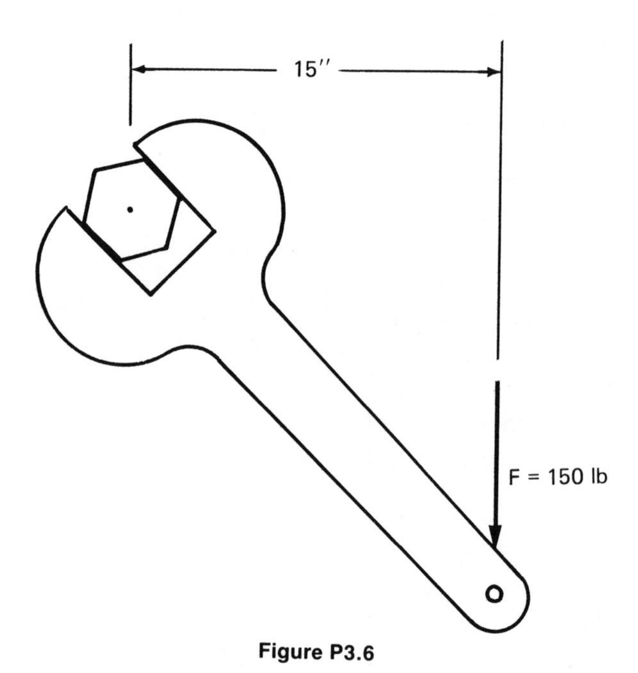


**Figure P3.6**

**3.7.** Calculate the moment of the external forces applied to the truss shown about supporting point $A$ in Fig. P3.7.

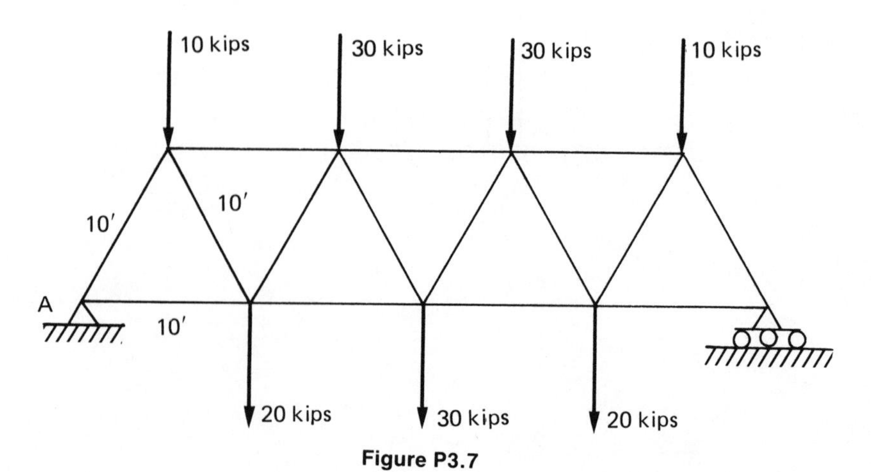


**Figure P3.7**

**3.8.** Calculate the moment of the external forces applied to the truss shown about support $B$ in Fig. P3.8.

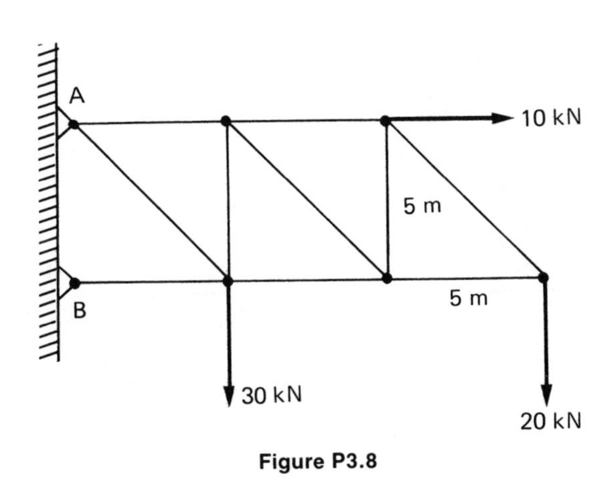


**Figure P3.8**

**3.9.** Find force $\vec{F}$ so that the moment of all forces about point $A$ in Fig. P3.9 is zero.

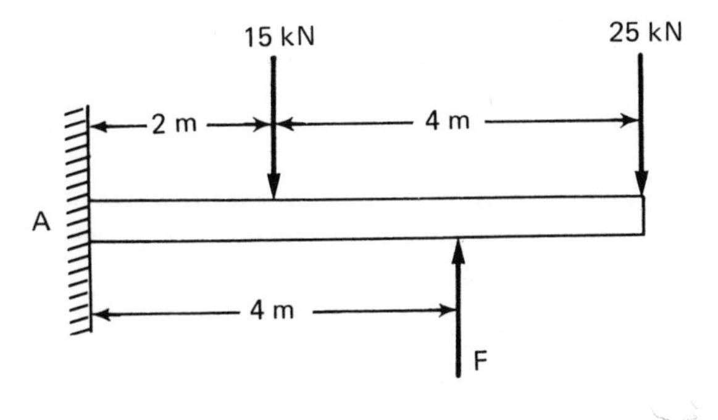


**Figure P3.9**

**3.10.** Calculate the moment of the 100-kN force about points $A$ and $B$ in Fig. P3.10.

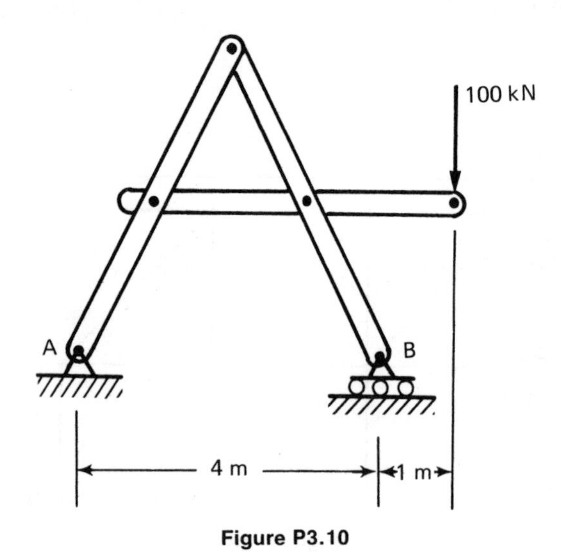


**Figure P3.10**

**3.11.** Calculate the moment of all forces about point $B$ in Fig. P3.11.

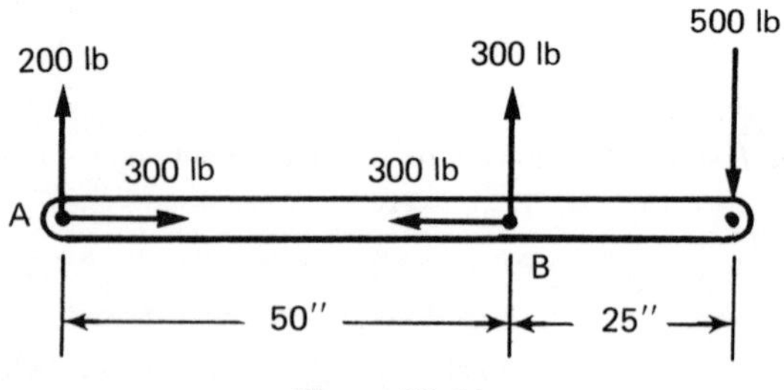


**Figure P3.11**

# 4

# EQUILIBRIUM

## 4.1 INTRODUCTION

Structures and all their individual components are said to be *at rest*. The forces and moments applied to a body are completely counterbalanced by the corresponding reactive forces and moments, a condition called a *state of equilibrium*. The field of statics and, subsequently, of strength of materials and structures deals primarily with the description of the conditions of forces and moments that are needed in order to maintain this state. Equilibrium, therefore, is the most crucial aspect of *statics*, and the subjects covered in subsequent chapters of this book are nothing but an application of the principles of equilibrium to various specific cases. Because of the importance of this chapter, it is strongly emphasized that students should thoroughly master its contents before moving on. The procedure used here constitutes a comprehensive approach to the first step in the design and analysis of structures and their individual members.

The principles of equilibrium are spelled out by Newton's laws, which constitute the mathematical conditions for equilibrium and are therefore known simply as *equations of equilibrium*. Before these conditions can be applied to structural components, however, they have to be isolated. If an entire structure (such as a building, bridge, etc.) is under question, it must be isolated from its surroundings, namely its supports, or, if an individual component of a structure is under consideration, it must be isolated from other parts to which it was originally attached. A diagram showing this isolated body is called a *free-body diagram* and will be covered in detail in the following sections.

## 4.2 FREE-BODY DIAGRAMS

A free-body diagram of an object is a diagram that is geometrically identical to the object itself and shows all the forces (and moments) that originally acted on it. Nevertheless, it is an isolated object that is disconnected from any part to which it was originally attached; in other words, it floats freely in space. It is called a *free-body diagram* (FBD) for this very reason, that it is a diagrammatical representation of a body free from all original attachments.

## Steps in Construction of a Free-Body Diagram

Construction of the free-body diagram starts with freeing the body in question (a complete structure or an individual member) from any attachments. Next, the effect of the former attachments to the body must be properly indicated on the diagram. The step-by-step construction of free-body diagrams will be demonstrated by means of the following typical examples.

EXAMPLE 1: Draw the free-body diagram of the beam shown in Figure 4.1(a).

SOLUTION: Since we are concerned about the beam (independent of its supports), our object (body) is solely the beam *AB*. The step-by-step construction is as follows:

1. Redraw beam *AB* without altering its geometry or its orientation, as shown in Fig. 4.1(b). Notice that the beam is drawn with exactly the same dimensions as those originally given.
2. Add all the external (applied) forces to beam *AB* as they were originally given. Notice that in the original problem, the two applied forces of magnitudes 1000 and 2000 lb were both downward forces applied at *C* and *D*. In this step, therefore, indicate these two forces at the same points and in the same direction as in Fig. 4.1(a) and display their magnitudes next to them, as shown in Fig. 4.1(c).
3. Add all the reactions. Once the beam has been freed from the supports at *A* and *B*, the reactions at each support should be indicated on the diagram. These reactions can be thought of as the forces that the supports were subject to when the beam was resting on them. Since the beam was partially supported at *A*, say, there was an interacting force between the beam and the support (foundation) at this point. In general, we do not know the direction and magnitude of this force. For the time being, therefore, let us say it has some general direction as shown by $R_A$ (reaction at *A*) in Fig. 4.1(d). Similarly, $R_B$ signifies the reaction at *B*. At this point, do

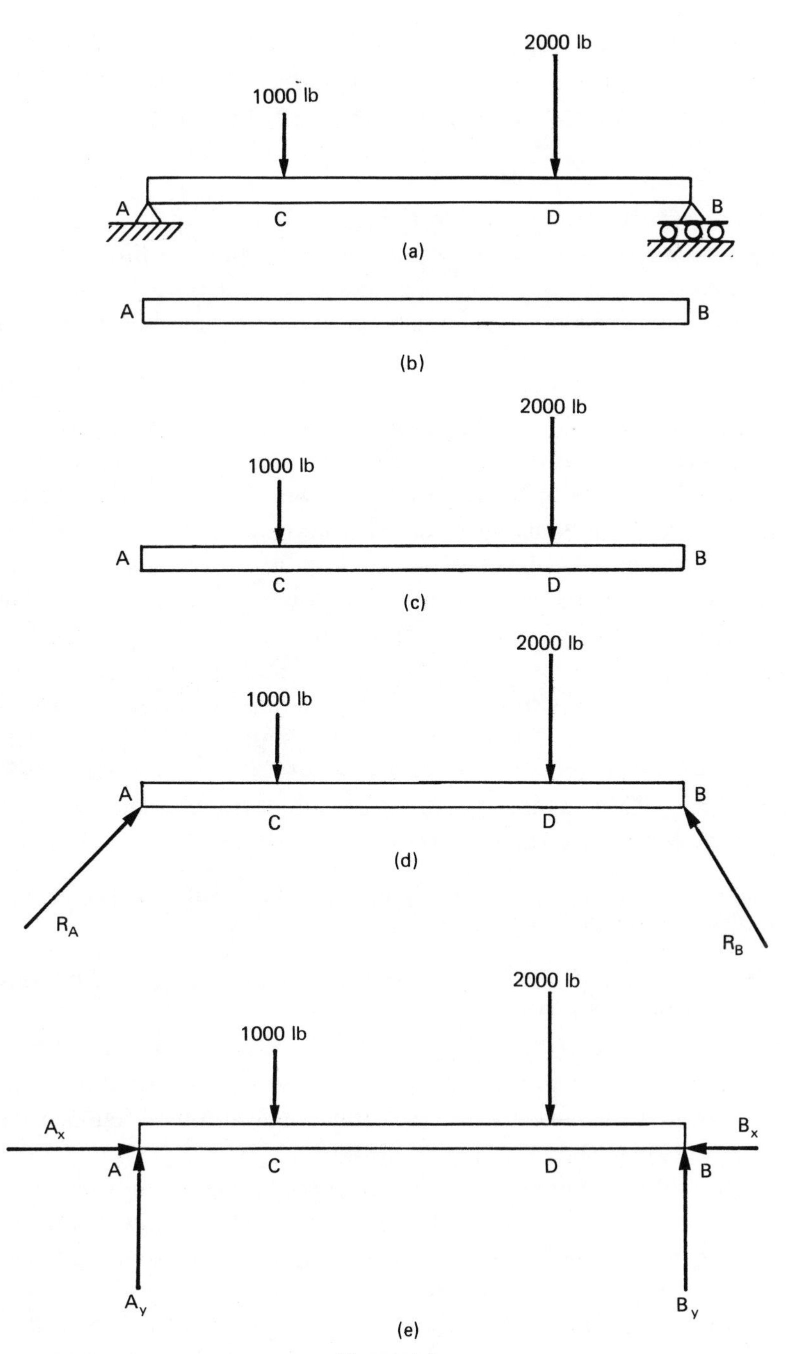
2000 lb
1000 lb
A
C
D
B
(a)
A
B
(b)
2000 lb
1000 lb
A
B
C
D
(c)
2000 lb
1000 lb
A
B
C
D
(d)
$R_A$
$R_B$
2000 lb
1000 lb
$A_x$
$B_x$
A
C
D
B
$A_y$
$B_y$
(e)

**Figure 4.1**

not worry about the correct directions and lines of action of these forces; they are only schematically shown in this figure. In this example, moreover, we have neglected the weight of the beam itself.

Figure 4.1(d) is the complete free-body diagram of beam *AB*. Later in the chapter we will use the equilibrium equations to find the correct directions, lines of action, and magnitudes of reactions $R_A$ and $R_B$. At this stage, however, we can furnish yet one more simplification to Fig. 4.1(d).

We saw in the previous chapter that any force can be decomposed into its components along any direction, in particular along the coordinate axes. Instead of representing the unknown force $R_A$ in an arbitrary direction as shown in Fig. 4.1(d), therefore, we can alternatively represent it in equivalent horizontal and vertical components that have known directions and concern ourselves about their unknown magnitudes later. Such a simplified free-body diagram is shown in Fig. 4.1(e). The way the forces are labelled here is more or less standard, since, for bookkeeping purposes, it is an easy way to keep track of the various forces, especially when many are involved. The method of presentation of reactions is as follows. The reaction force at support *A* ($R_A$) is shown by its horizontal and vertical components, $A_x$ and $A_y$, and at support *B* ($R_B$), by $B_x$ and $B_y$. Once again, directions are arbitrarily chosen. The correct directions will be determined later.

EXAMPLE 2: Draw the free-body diagram of the roof truss shown in Fig. 4.2(a).

SOLUTION: Steps similar to those in the previous example will be used in this problem, as follows:

1. Since we are interested in the entire truss, not its elements, redraw it so that its overall shape is outlined and all other details are left out. This is shown in Fig. 4.2(b).
2. Add all the external forces to the truss as they were originally indicated in Fig. 4.2(a). In this case, there are four downward forces and one inclined force. They are duplicated in Fig. 4.2(c) with their proper magnitudes shown at their side.
3. Add all reactions. The roof truss is supported by two walls (or supports) at *A* and *B*, and at each there exists a reaction force.

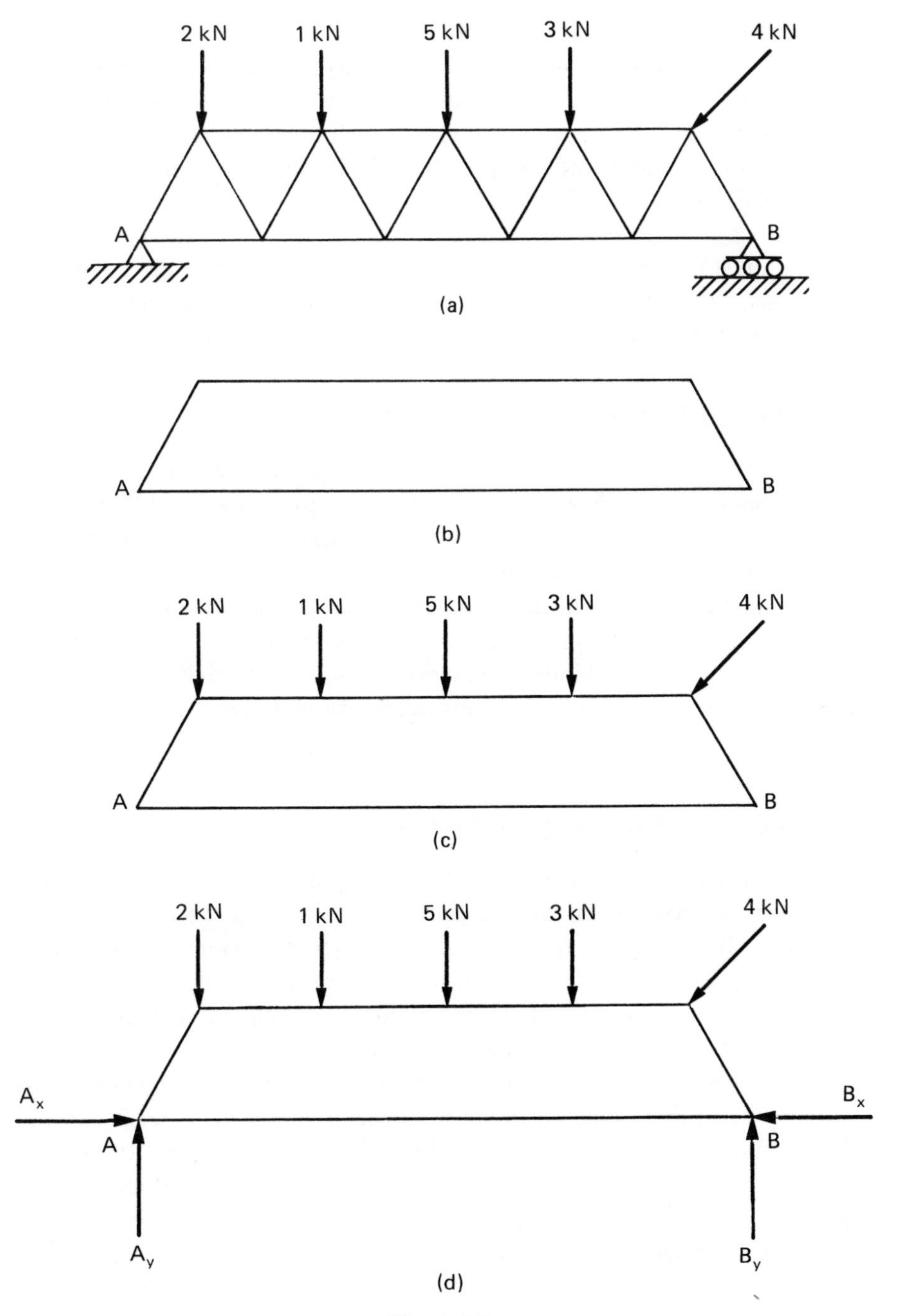
2 kN
1 kN
5 kN
3 kN
4 kN
A
B
(a)
A
B
(b)
2 kN
1 kN
5 kN
3 kN
4 kN
A
B
(c)
2 kN
1 kN
5 kN
3 kN
4 kN
$A_x$
$B_x$
A
B
$A_y$
$B_y$
(d)

**Figure 4.2**

Their directions can be arbitrarily chosen, or their $x$ and $y$ components can be shown as in Fig. 4.2(d). The correct directions of these forces will be studied later. Otherwise, all reaction forces are assumed to be correct and can be added in any direction you wish. Figure 4.2(d) is the complete free-body diagram of the roof truss of Fig. 4.2(a).

One must use the equations of equilibrium to determine the unknown reactions, but before we go into this, let us examine various support conditions that will further simplify the free-body diagram.

## Reactions of Various Types of Supports

In the two previous examples, we merely acknowledged the existence of a support at $A$ or $B$ without any reference to its kind. Although the solutions given were valid, we can further simplify the free-body diagram and consequently the calculations by taking into account the nature of the various support mechanisms involved. In the following sections, different types of support mechanisms will be studied in detail. Students should memorize the kinds of reactions involved in each case but not forget that common sense is always necessary in determining the proper reactions for each support configuration.

### Pin connection (hinge support)

If two pieces are connected to each other by means of a pin, as shown in Figure 4.3(a), this is called a *pin* (or *hinge*) *connection*. Such a pin has the following properties: (1) The member $AB$, which is connected to the support at point $A$, is free to rotate about $A$; i.e., it does not resist rotation (or moment). (2) If $AB$ is pulled, it will not be separated from support $A$ because of an interacting force between the support and $AB$ that is called a *reactive force* or, simply, *reaction*. This reaction has some general direction, but whatever that direction may be, it can be represented by its two components in the $x$ and $y$ directions. In fact, it is easier to talk about two force components along known directions ($x$ and $y$) rather than about a single force in a general direction. These resistive or reaction forces, $A_x$ and $A_y$, at pin or hinge support $A$ are shown in Fig. 4.3(b).

In this discussion, it has been assumed that there is no friction in the

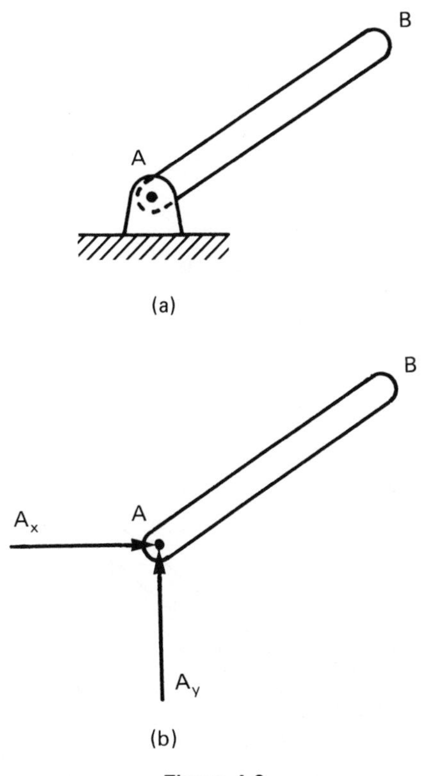


Figure 4.3

pin. Throughout this book, the same assumption will be followed, so that any reference to a pin or hinge support will mean simply a *frictionless pin*.

## Roller support

In practice, a structure such as a beam will not be pin-supported at both ends. One reason is that the length of the beam would likely be subject to change because of environmental temperature changes, and if the beam is fixed at both ends by pins, the thermal stresses that develop will tend to damage it. Consequently, one end will be hinged and the other put on a roller so that it will be free to move along the axis of the beam. This kind of support, called *roller support*, is schematically shown in Fig. 4.4(a). If the beam is pulled or pushed in a horizontal direction, the support will simply let the beam ride it out, and therefore no reactive

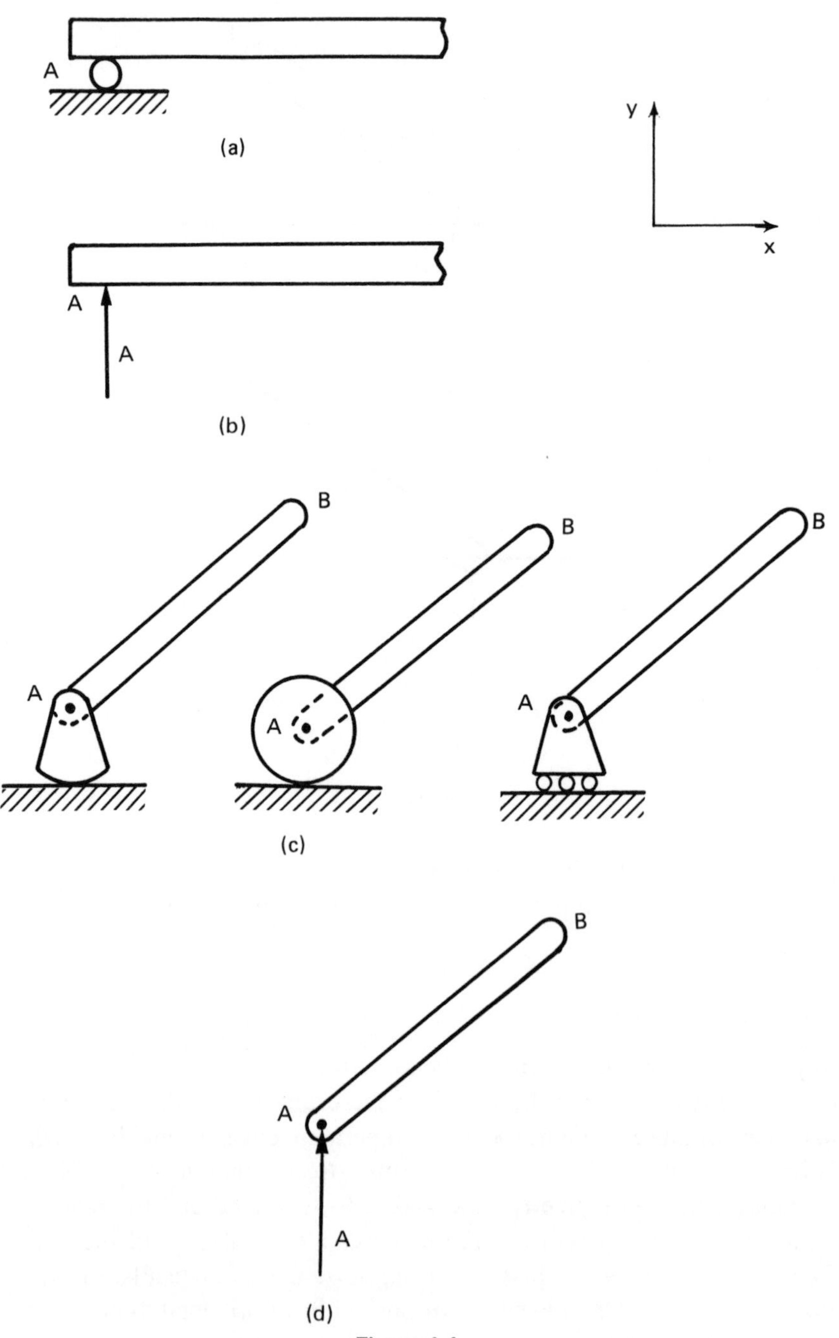


**Figure 4.4**

force will develop in the horizontal direction ($A_x = 0$). A free-body diagram of roller support is shown in Fig. 4.4(b). Since there is only one component of reaction, the subscript $y$ can be eliminated without any confusion.

Other supports of a similar nature are called *ball supports* or *rocker supports*, as shown in Fig. 4.4(c). Their free-body diagrams will also have only vertical reactions, as is shown in Fig. 4.4(d). The very fact that the balls, rollers, or rockers shown in Fig. 4.4 all rest on a smooth horizontal surface means that their reactions will be vertical only. Even if the resting surface is a smooth inclined one, as shown in Fig. 4.5(a), the reaction force will be normal to the inclined surface as shown in Fig. 4.5(b), and no reaction force (resisting force) will exist parallel to the surface. If the surface is rough, however, a component of reaction parallel to the surface would exist.

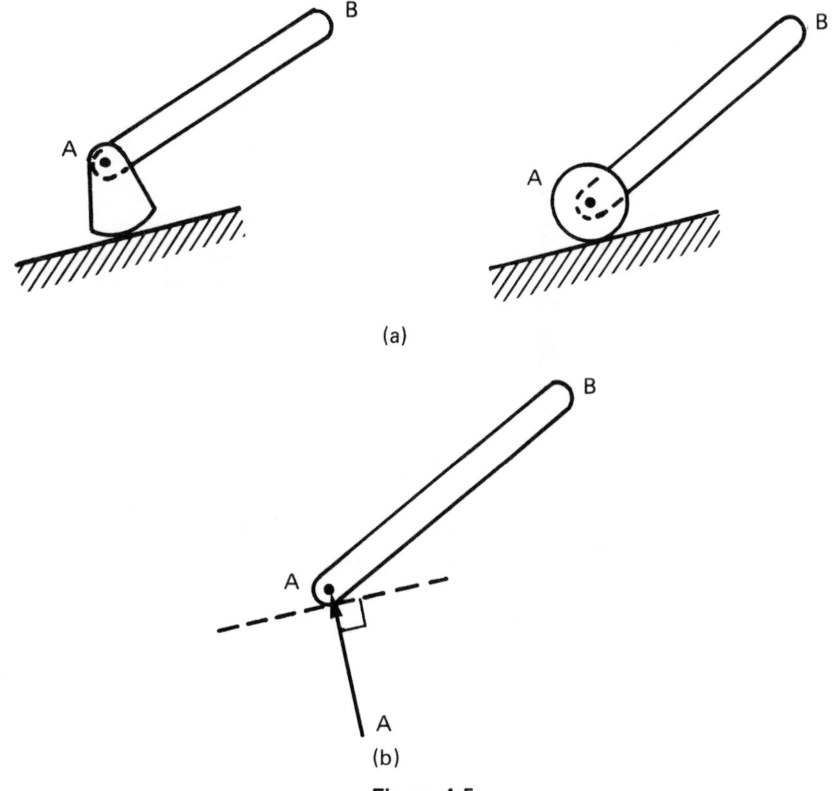


**Figure 4.5**

## Built-In, fixed, or cantilever support

The third type of support, called *built-in*, *fixed*, or *cantilever support*, is shown in Figs. 4.6(a) and (b). In such supports, the beam is rigidly fixed into a wall (or foundation). In addition to preventing the movement in horizontal and vertical directions that hinge supports allow, these supports also resist any rotation (contrary to the other types of supports). Therefore, there will be three reactions, two of which are forces—one horizontal and one vertical—and one of which is a moment, as shown in Fig. 4.6(c). Once again, the correct directions of the reactive forces and moment are assumed to be horizontally to the right, upward, and coun-

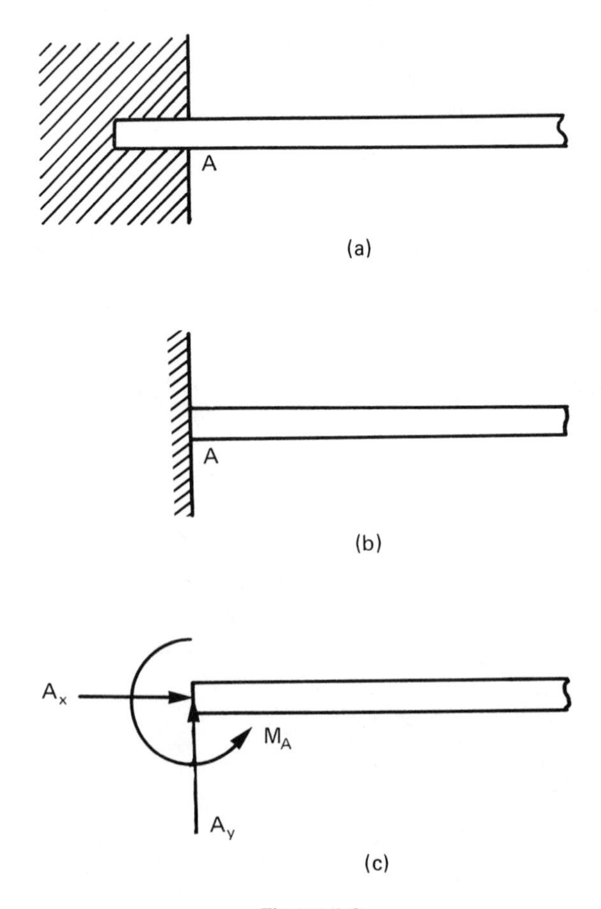


**Figure 4.6**

terclockwise, respectively. If there is a good reason to predict the correct directions, use such predictions; otherwise, assume arbitrary directions. If the assumed direction is indeed correct, the corresponding calculations will validate this assumption. The same approach holds true in assuming direction for the reactive moment as well.

The end conditions discussed here for two-dimensional supports can be easily extended to three-dimensional supports. In the case of the simple three-dimensional support shown in Fig. 4.7(a), there will be reaction forces in the $x$, $y$, and $z$ directions, respectively, as shown in Fig. 4.7(b). In the case of a three-dimensional built-in (or cantilever) support, as shown in Fig. 4.7(c), there will be three reaction forces—one in each of the $x$, $y$, and $z$ directions—as well as three moments, one moment about each of the coordinate axes as shown in Fig. 4.7(d).

## 4.3 EQUILIBRIUM CONDITIONS

If a force or a moment is applied to a free body, the body will move; as a result, it will not be in equilibrium. If many forces and moments are applied to such a body, the resultant of all the forces ($\Sigma F$) and all the moments ($\Sigma M$), will cause the body to move.* The only condition for a body subjected to loads and moments to remain in equilibrium, therefore, is that the sum of all the forces and the sum of all the moments be equal to zero. Intuitively, this condition also includes reactive (reaction) forces and moments. In other words, a body is in equilibrium if the forces and moments acting on it are completely balanced by counteracting forces and moments produced at its supports.

The condition that the sum of all the forces (applied and reactive) as well as the sum of all the moments (applied and reactive) be equal to zero is mathematically expressed as follows:

$$\Sigma F = 0 \tag{4.1}$$

$$\Sigma M = 0 \tag{4.2}$$

In the following discussion, this condition will be applied to various specific cases.

*The capital Greek letter sigma, $\Sigma$, is used mathematically to signify "the sum of."

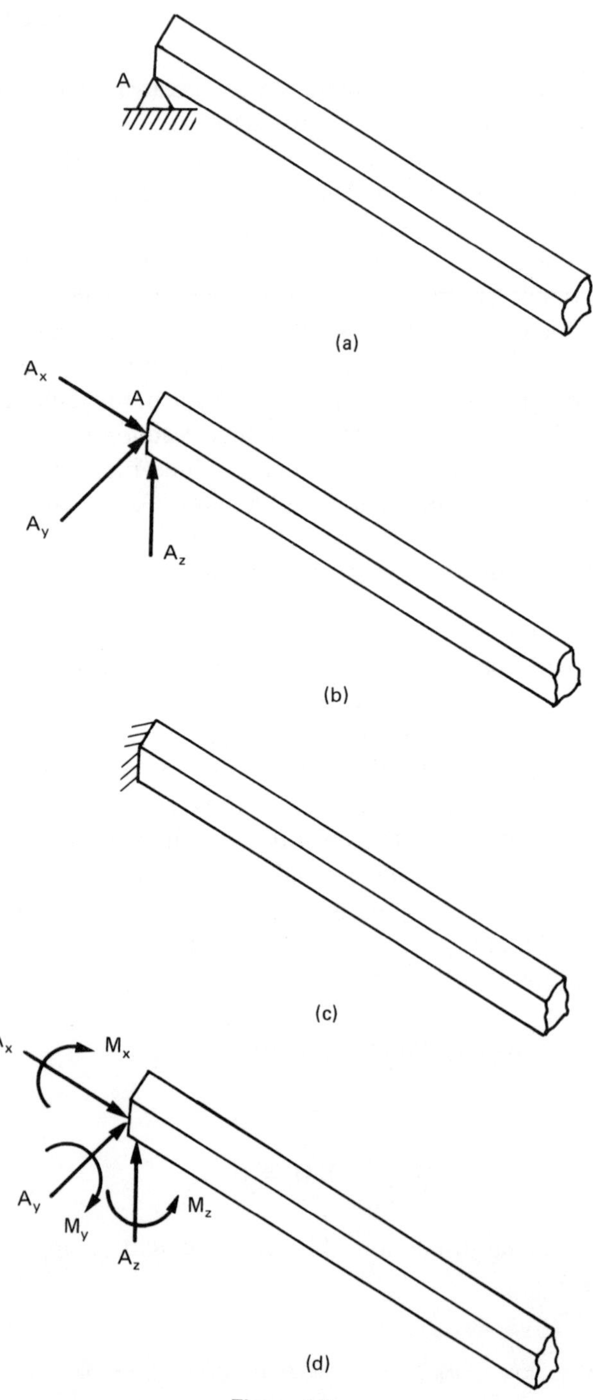


**Figure 4.7**

## General Case of Equilibrium in Two Dimensions

Consider the two-dimensional body shown in Fig. 4.8(a), with forces applied to it as indicated. The fact that all the forces are in the same plane as the body makes this a two-dimensional problem. For the sake of simplicity, only four applied forces ($\vec{F}_1$ through $\vec{F}_4$) and an $x$–$y$ coordinate system have been chosen. In Fig. 4.8(b), each of these forces

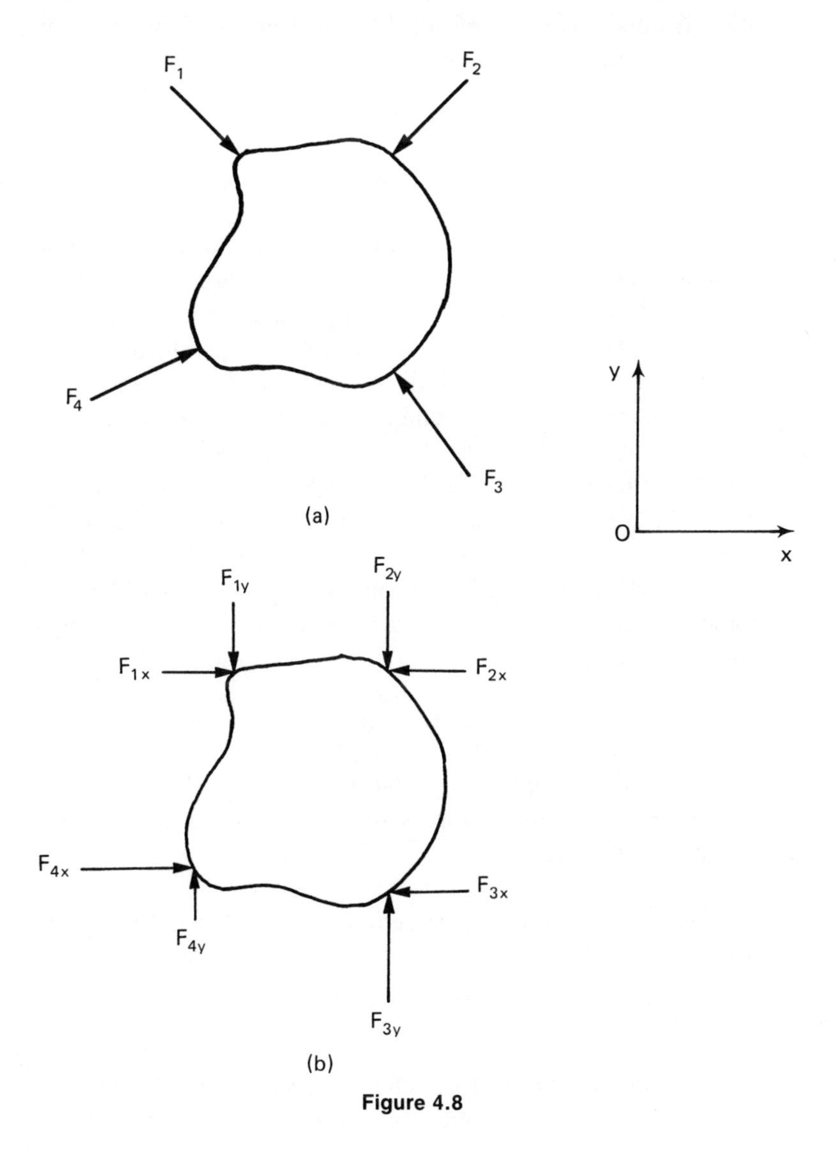


**Figure 4.8**

has been projected along the $x$ and $y$ axes, and we will use these components instead of the initial forces.

To maintain equilibrium, the following conditions must be satisfied:

1. The resultant of forces $\vec{F}_1 + \vec{F}_2 + \vec{F}_3 + \vec{F}_4$, or $\vec{R}$, must equal zero. If resultant $\vec{R}$ is equal to zero, then its components along the $x$ and $y$ axes must also be equal to zero. As far as the force condition is concerned, therefore, the components of the forces must satisfy the following equations:

$$R_x = \Sigma F_x = 0 \tag{4.3}$$

$$R_y = \Sigma F_y = 0 \tag{4.4}$$

   The first of these equations simply states that the $x$ component of the resultant ($R_x$) is equal to the algebraic sum of the $x$ components of all the forces ($\Sigma F_x$); similarly, the second equation states that the $y$ component of the resultant ($R_y$) is equal to the algebraic sum of the $y$ components of all the forces ($\Sigma F_y$). To satisfy equilibrium, moreover, both $R_x$ and $R_y$ must be equal to zero.

2. The moment of all forces about any given point (say point $O$) should also be equal to zero. In Fig. 4.8(b), there are eight forces, four in the $x$ and four in the $y$ direction. Each of these eight forces will have a moment about point $O$, and there will therefore be eight moments whose algebraic sum must be equal to zero (that is, the moments of the forces balance each other out). Mathematically this can be written as follows:

$$\Sigma M_O = 0 \tag{4.5}$$

   This equation simply states that, to satisfy equilibrium, the sum of the moments of all the forces about point $O$ (or any other point) must be equal to zero.

In the following section, the preceding concepts will be applied to more specific cases.

## Colinear Forces

In a *colinear force* system, all forces have the same line of action. Consider the two colinear forces applied to the body shown in Fig. 4.9. Since

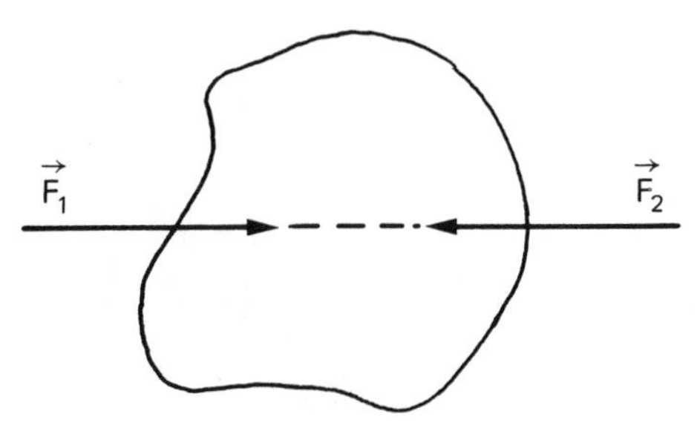


Figure 4.9

the forces are aligned, no rotation is involved. If both forces are directed to the right or to the left, however, the body will move to the right or left, respectively, and cease to be in equilibrium. Also consider that if the magnitude of force $\vec{F}_1$ were greater than that of $\vec{F}_2$, there would be a net force, $\vec{F}_1 - \vec{F}_2$, and the body would move in the direction of $\vec{F}_1$; similarly, if the magnitude of $\vec{F}_2$ were greater than that of $\vec{F}_1$, the body would move in the direction of $\vec{F}_2$. In both cases, however, there would be no equilibrium.

In order to satisfy equilibrium when only two forces are applied to a body, therefore, the following conditions must hold true:

1. The two forces must have the same line of action.
2. They must have opposite directions.
3. They must have equal magnitudes.

Notice that because of the first of these conditions, the condition, $\Sigma M = 0$, is automatically satisfied, and one has only to meet the force equilibrium condition ($\Sigma F_x = \Sigma F_y = 0$). In this simple case, the latter reduces to the formulation, $\Sigma F = F_1 - F_2 = 0$ (conditions 1 and 2), or simply to the equality, $F_1 = F_2$ (condition 3). Note that in this case only one force equilibrium equation ($\Sigma F_x = 0$) is needed, the second ($\Sigma F_y = 0$) being automatically satisfied.

## Concurrent Forces

In a *concurrent force system*, all forces pass through a common point. In the previous case involving the application of two forces to a body, it was necessary for them to be colinear, opposite in direction, and equal in magnitude for the body to be in equilibrium. If three forces are applied to a body, as shown in Fig. 4.10, they must pass through a common

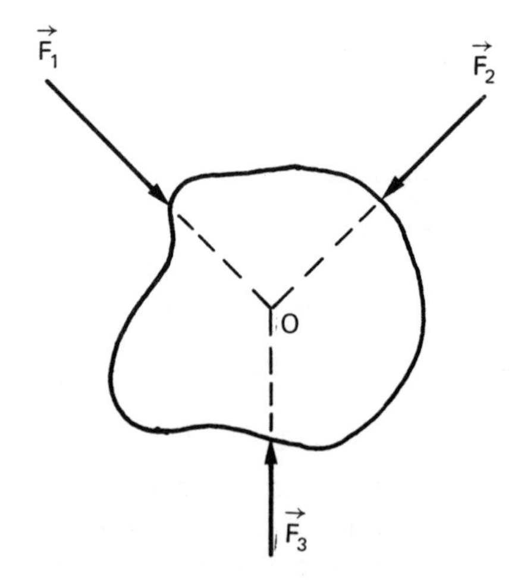


**Figure 4.10**

point ($O$), or else the condition, $\Sigma M_O = 0$, will not be satisfied and the body will rotate because of unbalanced moment. Moreover, the magnitudes of the forces must be such that the force equilibrium equations, $\Sigma F_x = 0$ and $\Sigma F_y = 0$, are satisfied.

It is fairly easy to see the reasoning for the first condition. Consider the two forces, $\vec{F}_1$ and $\vec{F}_2$, intersecting at point $O$ in Fig. 4.11. The sum of moments of these two forces about point $O$ is obviously equal to zero because they both pass through $O$. If $\vec{F}_3$ does not pass through $O$, on the other hand, it will have some nonzero moment about that point. Since this nonzero moment will cause the body to rotate, the body will not be in equilibrium.

Therefore, not only do three nonparallel forces applied to a body have to be concurrent for the body to be in an equilibrium state, but their magnitudes and directions must be such that the force equilibrium conditions are satisfied ($\Sigma F_x = \Sigma F_y = 0$). Notice that there is no need for the moment equilibrium equation in this case since it is automatically satisfied.

## Parallel Forces

When all the applied forces are parallel, as they are in Fig. 4.12, the sum of the forces in that particular direction must be equal to zero. The fact that they are parallel along the $y$ axis in this case causes the first

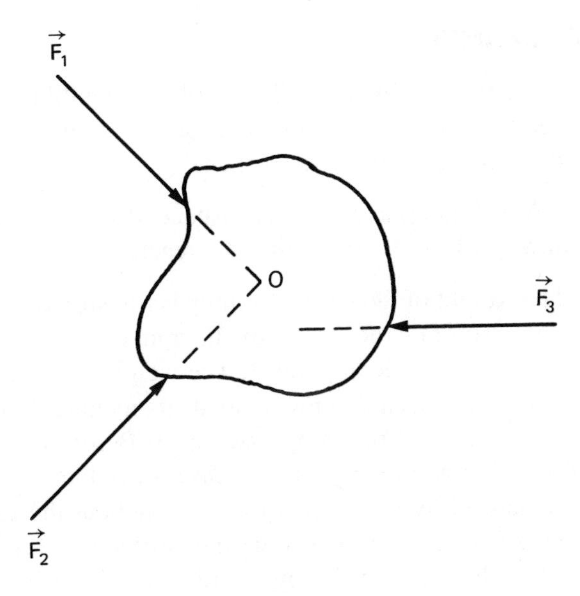


Figure 4.11

force equilibrium equation ($\Sigma F_x = 0$) to be automatically satisfied. Therefore, the equilibrium equations will be reduced to $\Sigma F_y = 0$ and $\Sigma M = 0$. Notice that the moment equilibrium equation must be satisfied in this case to prevent the body from rotating.

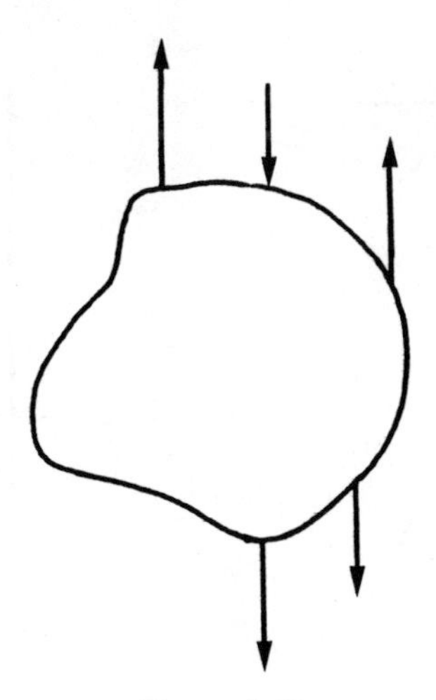

Figure 4.12

## 4.4 SOME EXAMPLES

In this section, some two-dimensional examples of equilibrium problems will be completely solved. Notice that the use of the free-body-diagram technique is the starting point in all of them.

EXAMPLE 3: A 5-lb lamp assembly is suspended from the ceiling with a cable, as shown in Fig. 4.13(a). Find the reactive force in the cable.

SOLUTION: The weight of the lamp assembly is transferred to the ceiling through the cable. To draw the free-body diagram of the cable, we make an imaginary cut on the cable at point $A$. The weight of the lamp assembly is applied at point $A$ as a downward load, its magnitude being equal to 5 lb (the total weight of the lamp assembly). Before introduction of the cut, all forces below point $A$ (in this case, only the 5-lb downward weight) were balanced by an internal force throughout the cable. Upon the cable's being cut, this internal force transformed into a resistive or reactive force to balance the 5-lb downward force. It is shown as $F_A$ in the free-body diagram of Fig. 4.13(b). In general, the direction of force $F_A$ is unknown, but for simple cases its direction can be correctly predetermined. In this problem, since all external forces are downward forces ($F_B = 5$ lb), $F_A$, to balance downward force $F_B$, has to be an upward force.

Notice that in the free-body diagram of Fig. 4.13(b) we have also predetermined that the lines of action of forces $F_A$ and $F_B$ coincide sim-

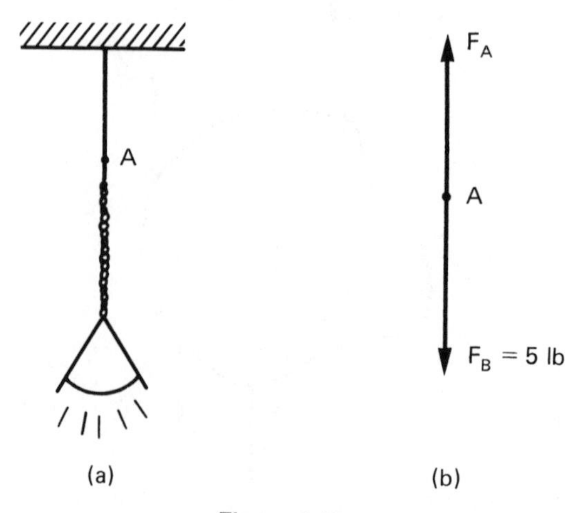


Figure 4.13

ply because we have a two-force system and, to satisfy equilibrium, they have to be colinear; furthermore, they have to be opposite in direction and equal in magnitude. In fact, since the lamp cable is a flexible cable, forces $F_A$ and $F_B$ must be tensile forces because a flexible cable can only be pulled.

In later examples we will see that if the correct direction of reaction forces cannot be predetermined, we can *assume* some direction as the correct direction. If, after solving the problem, this assumed direction turns out to be the correct one, the magnitude of the reaction force will have a positive sign; a negative sign, on the other hand, will warn that the assumed direction has to be changed.

Since the complete free-body diagram of this problem shown in Fig. 4.13(b) is known to be in equilibrium, the equations of equilibrium can be applied, and since there is only one unknown force, $F_A$, just one equation will suffice. Moreover, since the sum of forces in the $y$ direction must be equal to zero ($+\uparrow\Sigma F_y = 0$, or $F_A + F_B = 0$), $F_A + (-5 \text{ lb}) = 0$, or $F_A = 5$ lb. Finally, the fact that this answer for unknown force $F_A$ is positive indicates that the assumption of upward direction for $F_A$ was correct.

EXAMPLE 4: Consider the same lamp suspended from the ceiling by two cables, as shown in Fig. 4.14(a).

SOLUTION: The lamp is attached to cables $AB$ and $AC$ at point $A$. To draw the free-body diagram, one has to introduce cuts in cables $AB$ and $AC$. As before, once the cable $AB$ has been cut, a reaction force, $F_{AB}$, will develop. This reaction force should be aligned with cable $AB$, and its direction will be assumed to be away from $A$. Similarly, cable $AC$ will have a reaction, $F_{AC}$. The complete free-body diagram of this problem is shown in Fig. 4.14(b). Since the free-body diagram is in equilibrium, the equations of equilibrium can now be applied to it. Also, since there are only two unknown forces, $F_{AB}$ and $F_{AC}$, two equilibrium equations, $\Sigma F_x = 0$ and $\Sigma F_y = 0$, will be enough to determine them. Remembering that $\Sigma F_x = 0$ simply means that the algebraic sum of all the forces in the $x$ direction equals zero, we note that the 5-lb force is a downward force and does not have an $x$-direction component, but that the other two forces do. The $x$ component of force $F_{AC}$ is equal to $F_{AC}$ cos 45° with a positive sign since its component along the $x$ axis lies in the direction of positive $x$. Similarly, the $x$ component of force $F_{AB}$ is equal to $F_{AB}$ cos 60° with a negative sign since its component along the

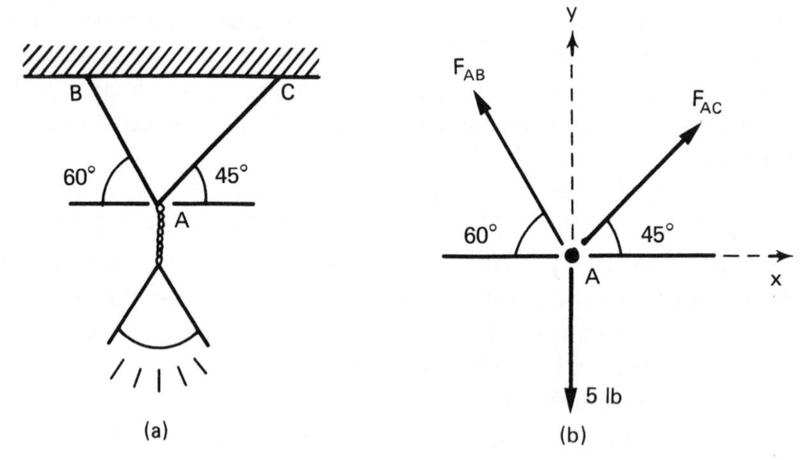


**Figure 4.14**

$x$ axis lies in the opposite direction. Substituting these forces into the equilibrium equation, $\Sigma F_x = 0$, will result in the following:

$$F_{AC} \cos 45° - F_{AB} \cos 60° = 0 \tag{4a}$$

Now we have to consider the $y$ components. Since the downward force is along the negative $y$ axis, it will appear as $-5$ in the equation, $\Sigma F_y = 0$. The $y$ component of force $F_{AC}$ is $F_{AC} \cos 45°$ and the $y$ component of force $F_{AB}$ is $F_{AB} \cos 30°$. Since both $y$-component forces lie along the positive $y$ direction (upward), they will both appear as positive values in the equation, which will thus take the form,

$$F_{AB} \cos 30° + F_{AC} \cos 45° - 5 = 0 \tag{4b}$$

To evaluate $F_{AB}$ and $F_{AC}$, Eqs. 4a and 4b must be solved simultaneously. To do so, we can find $F_{AB}$ from Eq. 4a in terms of $F_{AC}$, or

$$F_{AB} = F_{AC} \frac{\cos 45°}{\cos 60°}$$

and substitute it into Eq. 4b, which results in the following:

$$F_{AC} \frac{\cos 45°}{\cos 60°} \cos 30° + F_{AC} \cos 45° - 5 = 0$$

The above equation has only one unknown, $F_{AC}$, which solution proves to be 2.59 lb. Substituting this value into Eq. 4a determines $F_{AB}$ to be 3.66 lb. Since both $F_{AB}$ and $F_{AC}$ turn out positive, it indicates that the preassumed directions for $F_{AB}$ and $F_{AC}$ were indeed correct.

EXAMPLE 5: A beam *AB* resting at points *A* and *B* supports 100- and 200-lb loads as shown in Fig. 4.15(a). Find the reactions at *A* and *B*.

SOLUTION: The free-body diagram of beam *AB* is shown in Fig. 4.15(b). Since support *A* is a simple (pin) support, it has two components, $A_x$ and $A_y$, as reactions; since support *B* is a roller support, it has only one component, *B*, as a reaction. Therefore, there are three unknowns—$A_x$, $A_y$, and *B*—and we need three equations to solve them. Since the sum of forces in the *x* direction, $\Sigma F_x$, must equal 0, $A_x = 0$. This was obvious from the free-body diagram, since $A_x$ is the only force in the *x* direction. Now, therefore, there are only two unknown forces, $A_y$ and *B*.

Using the sum of forces in the *y* direction will result in an equation with two unknowns. Another equation involving one or both of these

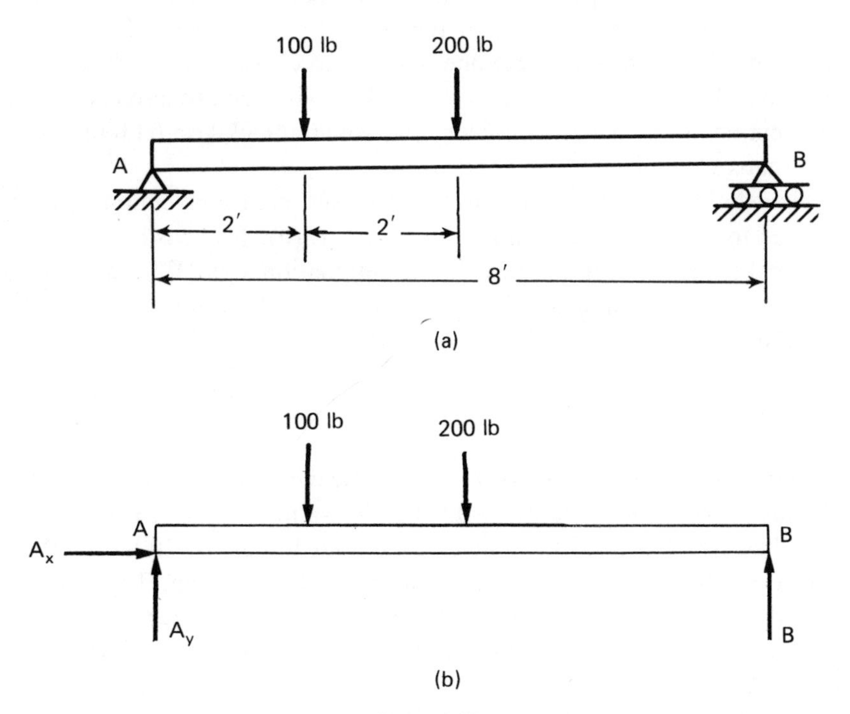


**Figure 4.15**

would thus be required so that the two equations may be solved simultaneously. Although there is nothing wrong with this method, there is a better way to solve for two unknowns in this kind of problem. That is through the use of the moment equilibrium equation. Since the free-body diagram of Fig. 4.15(b) is in equilibrium, the sum of the moments of all forces about any point must also be equal to zero. Any point can be chosen to serve as this moment center, but if we take a point that one of the unknown forces passes through (such as point $A$ or $B$), the equation will be simpler and contain only one unknown; otherwise, we will again end up with an equation with two unknowns.

Let us choose point $A$ as the moment center and use the moment equilibrium equation,

$$\Sigma M_A = 0$$

This equation says that the moments of all forces present in the free-body diagram about point $A$ are to be taken one by one and their algebraic sum then set equal to zero. Note that in considering one force at a time, we temporarily ignore all the other forces present. To find the moment of a force, however, one first has to choose some arbitrary direction as being positive. Let's choose clockwise (cw) rotation about the moment center as positive moment and counterclockwise (ccw) rotation as negative moment.

Now let's apply the moment equation and sign convention just described to the free-body diagram of Fig. 4.15(b). It is customary to indicate the sign convention next to the summation sign, $\Sigma$, as a rounded or circular arrow, and usually the direction of the arrow indicates the assumed positive direction, as shown here:

$$+\circlearrowright \Sigma M_A = 0$$

In the free-body diagram of Fig. 4.15(b), there are four forces ($A_x$ has already been found to be zero; if it had had a nonzero value, we would have had five forces). Each one of these four forces ($B$, 200 lb, 100 lb, and $A_y$) will produce a moment about point $A$ of its own. The moment of force $B$ about point $A$ is equal to the magnitude of force $B$ (which is unknown) times its normal distance from point $A$, which, we know from Fig. 4.15(a), is 8 ft, and since this force causes a ccw rotation about $A$, its moment about $A$ will be $(-)B \times 8$ (negative moment). Similarly, the moment of the 200-lb force about point $A$ is equal to the magnitude of this force (200 lb) times its normal distance from point $A$, which is 4 ft, and since this force causes a cw rotation about $A$, its moment about $A$ will be $(+)200 \times 4$. Finally, the moment of the 100-lb force about $A$ is equal to $(+)100 \times 2$. The moment of $A_y$ about $A$ will be zero since force $A_y$ passes through moment center $A$ and thus does not cause any rotation about $A$. Adding up all these moments algebraically, we have

$$+\circlearrowright \Sigma M_A = 0$$

$$(-B \times 8) + (200 \times 4) + (100 \times 2) + (0) = 0 \qquad \text{(4c)}$$

which is a single equation with but one unknown. Solving for $B$ gives us the value, $B = 125$ lb, and the fact that it is positive indicates that the assumed direction for $B$ was the correct one.

Notice that the assumption of clockwise moment as positive moment does not alter the sign, magnitude, nor direction of force $B$. To see the truth of this, let's recast the moment equation, this time assuming that counterclockwise moment is positive. We will then have the following:

$$+\circlearrowleft \Sigma M_A = 0$$

$$(B \times 8) - (200 \times 4) - (100 \times 2) + (0) = 0 \qquad \text{(4d)}$$

Solving for $B$, we get $B = 125$ lb, which is the same value as before. In fact, multiplying Eq. 4c by $(-1)$ will result in Eq. 4d, and that's why the value of $B$ will not be altered by using different sign conventions. Notice that, although we have a choice of sign conventions, for a given equation we must use only one of them. Although this would seem to be a very simple, straightforward matter, it has been observed that many students, especially beginners, are prone to make mistakes in following sign conventions. To avoid this confusion, let us make two rules that will be followed throughout this book:

1. Assume clockwise moment to be positive moment.
2. Indicate this sign convention whenever the moment equation is used by placing a circular arrow next to the summation sign ($\Sigma$), as follows: $+\circlearrowright \Sigma M$. If this convention is observed, mistakes will be avoided (at least those resulting from sign convention).

Now, back to solving our problem. Since $B$ has been found, the only unknown to be determined is $A_y$. This solution can be achieved in two ways. For the sake of practice, let's do it both ways.

First, if we take moment of all forces about $B$ following the same procedure just performed, we get

$$+\circlearrowleft \Sigma M_B = 0$$

$$(A_y \times 8) - (100 \times 6) - (200 \times 4) = 0$$

Solving for $A_y$,

$$A_y = 175 \text{ lb}$$

Or, after solving for $B$, we can use the force equilibrium equation in the $y$ direction:

$$+\uparrow \Sigma F_y = 0$$

Choosing upward forces in this equation as positive , we get

$$A_y - 100 - 200 + B = 0$$

Since the value of $B$ has already been found to be 125 lb, we have

$$A_y - 100 - 200 + 125 = 0$$

Solving for $A_y$,

$$A_y = 175 \text{ lb}$$

Usually, only one of the methods is used to evaluate $A_y$, the other being reserved to check if all the values obtained are correct. More about this will be discussed in the next chapter.

EXAMPLE 6: A crane is used to lift heavy weights and move them from one end of a warehouse to the other. When it is lifting a 5-ton load at the latter's midpoint, as shown in Fig. 4.16(a), calculate the reactions of the crane. The weight of the crane itself is 3 tons.

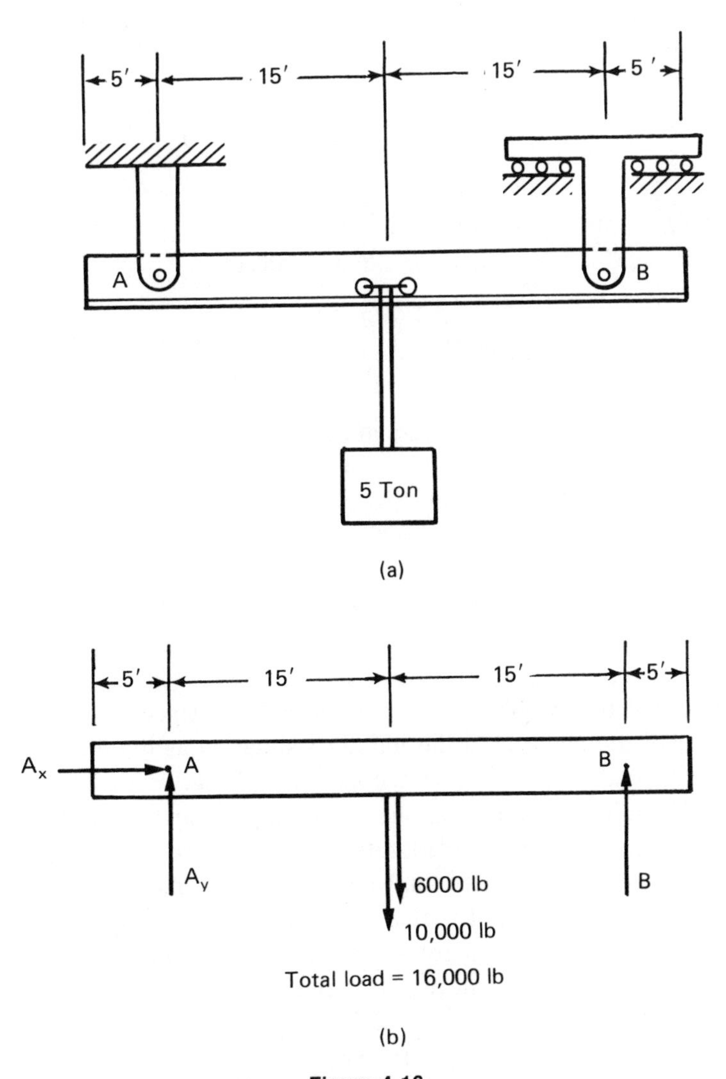


**Figure 4.16**

SOLUTION: In the free-body diagram of the crane shown in Fig. 4.16(b), support $A$ is a simple (pin) support with two reactions and support $B$, because of the roller, has only one reaction. The weight of the crane, 3 tons, is assumed to be concentrated at its midpoint. (The two weights at the midpoint are slightly separated for identification purposes only; note that one ton is equal to 2000 lb.) Using the free-body diagram, the following steps are taken:

1. $$\overset{+}{\rightarrow}\Sigma F_x = 0 \text{ or } A_x = 0.$$

2. $$+\circlearrowright \Sigma M_A = 0$$
$$(-B \times 30) + (16{,}000 \times 15) = 0$$
$$B = 8{,}000 \text{ lb}$$

3. $$+\circlearrowright \Sigma M_B = 0$$
$$(A_y \times 30) - (16{,}000 \times 15) = 0$$
$$A_y = 8{,}000 \text{ lb}$$

4. Check by using the third equation, $\Sigma F_y = 0$:
$$+\uparrow\Sigma F_y = 0$$
$$A_y - 16{,}000 + B = 0$$
$$8{,}000 - 16{,}000 + 8{,}000 = 0$$

   Since the left side of this equation is also zero, the values of $A_y$ and $B$ were correctly calculated.

Notice that this problem is a symmetrical one since $A_x = 0$ and the right half of the crane is in the identical situation as the left half, both as far as its loads and supports are concerned. Therefore, the left support is in the identical condition as the right support, and since there is no reason for one to carry more load than the other, both are under equal forces (equal reaction). The total applied load (force) on the crane is equal to 16,000 lb, half of which must be taken by support $A$ and the other half by support $B$; therefore, without taking the trouble to solve the above equations, we can readily write

$$A_y = B = \frac{16{,}000}{2} = 8{,}000 \text{ lb}$$

EXAMPLE 7: Consider the same crane but with the 5-ton load at point $C$, as shown in Fig. 4.17(a). What are the reactions of supports $A$ and $B$ now?

SOLUTION: The free-body diagram of the crane is shown in Fig. 4.17(b). Similarly, the following steps are taken to solve this problem:

1. $$\overset{+}{\rightarrow}\Sigma F_x = 0 \text{ or } A_x = 0$$

2. $$+\curvearrowright \Sigma M_A = 0$$
$$(-B \times 30) + (10000 \times 20) + (6000 \times 15) = 0$$
$$B = 9667 \text{ lb}$$

3. $$+\uparrow \Sigma F_y = 0$$
$$A_y - 6000 - 10000 + 9667 = 0$$
$$A_y = 6333 \text{ lb}$$

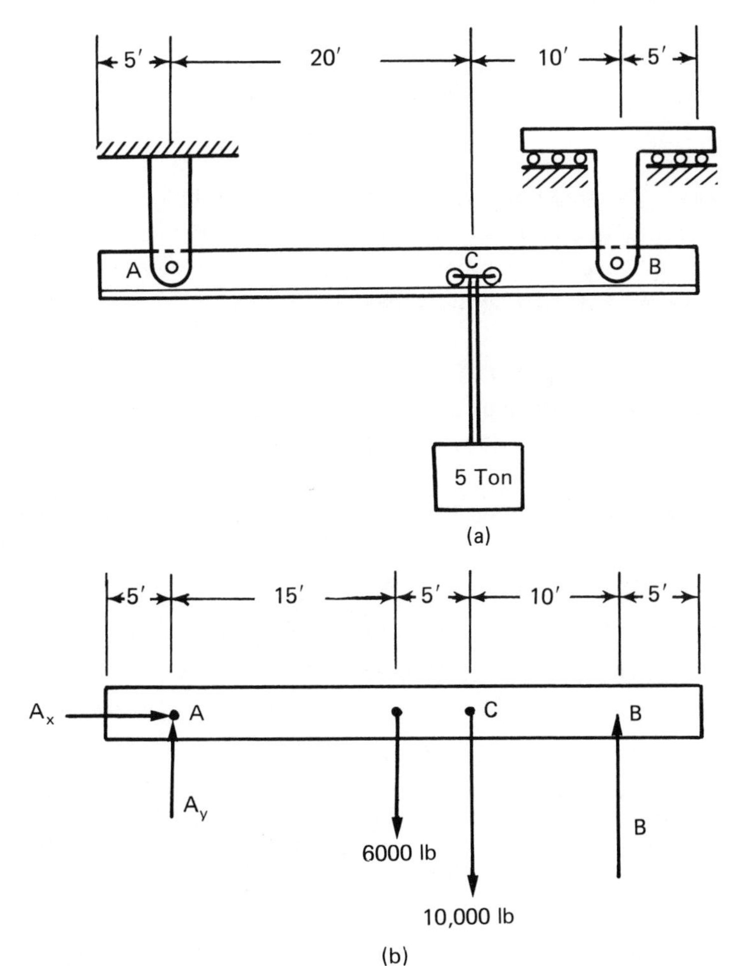


**Figure 4.17**

4. Check using another moment equation:

$$+\circlearrowleft \Sigma M_B = 0$$

$$(6333 \times 30) - (6000 \times 15) - (10000 \times 10) = 0$$

Notice that the left side of this equation is $-10$ lb, not zero. This discrepancy is due to approximating the values for $A_y$ and $B$. For all practical purposes, a 10-lb difference in an applied load of several thousand pounds is negligible.

EXAMPLE 8: Determine the reactions of the portable jib crane shown in Fig. 4.18(a). The weight of the crane is 1,000 kg and is assumed to act at point $C$.

SOLUTION: The free-body diagram of the crane is shown in Fig. 4.18(b). Unlike our practice in previous examples, we will not use $\Sigma F_x = 0$ first, since if we do, we will get a single equation with two unknowns. In this problem, $\Sigma F_y = 0$ will be used first, as follows:

1.
$$+\uparrow \Sigma F_y = 0$$
$$B_y - 9810 - 29{,}430 = 0$$
$$B_y = 39{,}240 \text{ N}$$

Notice that the weights in kilograms (kg) are multiplied by 9.81 to be converted to newtons (N), which is the unit of force in the SI system.

2.
$$+\circlearrowleft \Sigma M_B = 0$$
$$(-A \times 5) + (9810 \times 2) + (29{,}430 \times 10) = 0$$
$$A = 62{,}784 \text{ N}$$

3.
$$\overset{+}{\rightarrow} \Sigma F_x = 0$$
$$A - B_x = 0$$
$$62{,}784 - B_x = 0$$
$$B_x = 62{,}784 \text{ N}$$

Another moment equilibrium equation can be used for checking purposes, which is left to the student to solve.

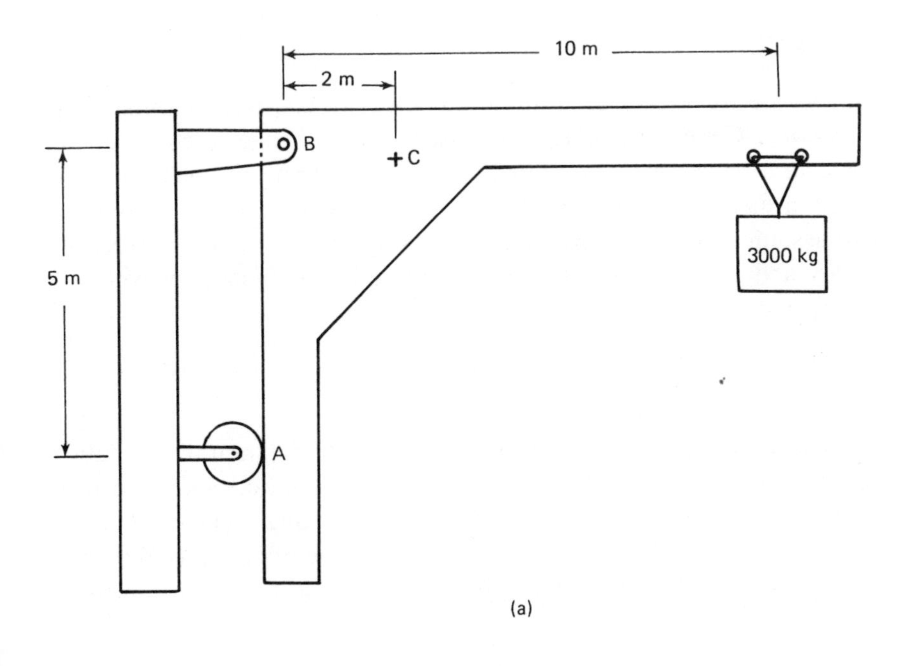


(a)

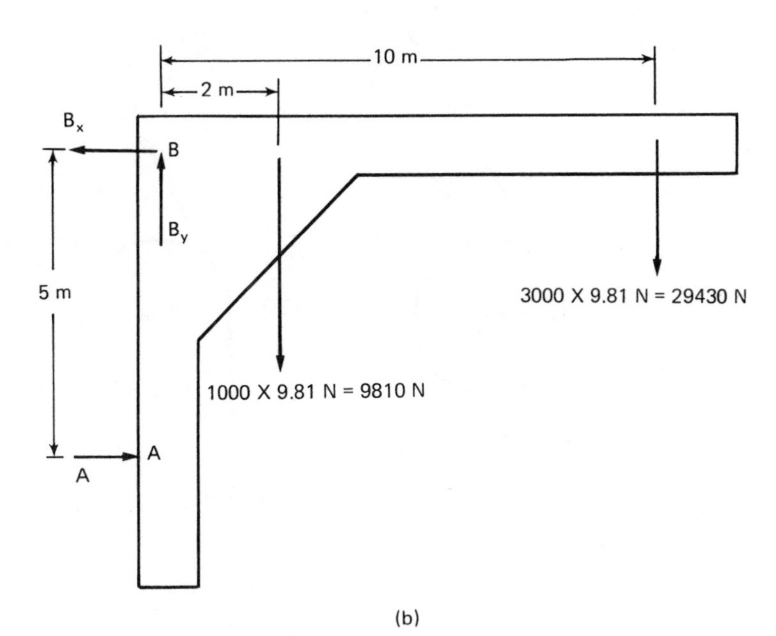


(b)

**Figure 4.18**

## 4.5 PULLEYS

Pulleys are frequently used in various types of construction-related applications. Generally, pulleys are used (1) to change direction of the force in a cable, belt, or chain; (2) in power transmissions, or (3) to obtain a mechanical advantage. In solving problems involving pulleys, two assumptions are usually made: (1) There is no friction between the pulley and its shaft (or pin), and (2) there is no slippage between the pulley and its cable.

EXAMPLE 9: Determine the pin reactions for the pulley shown in Fig. 4.19(a).

SOLUTION: Figure 4.19(b) shows the free-body diagram of the pulley. Since the pulley was pinned at point $A$, pin reactions consist of horizontal and vertical reactions $A_x$ and $A_y$. By introducing a cut in the upper part of the cable, a tension force $T$ along the line of action of the upper cable (60°) will be developed. For equilibrium to exist, tension forces

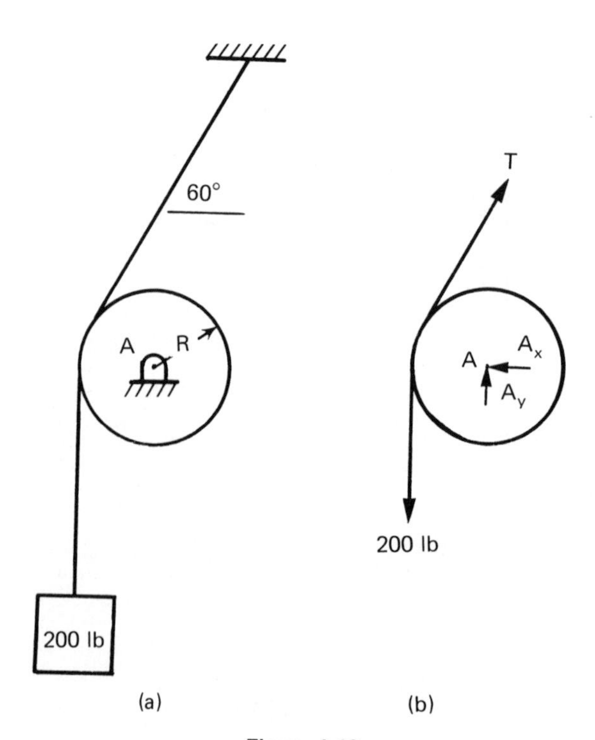


**Figure 4.19**

at both ends of the cable that goes over the pulley must be equal. That this is so can easily be seen by taking the moment of these two tension forces about the center of the pulley:

$$+\circlearrowright \Sigma M_A = 0$$

$$(T \times R) - (200 \times R) = 0$$

$$T = 200 \text{ lb}$$

It also means that all the pulley does is to change the direction of the force.

To calculate $A_x$ and $A_y$, the force equilibrium equations can be used as follows:

$$\Sigma F_x = 0$$

$$T \cos 60° - A_x = 0$$

$$200 \cos 60° - A_x = 0$$

$$A_x = 100 \text{ lb}$$

and

$$+\uparrow\Sigma F_y = 0$$

$$A_y - 200 + T \sin 60° = 0$$

$$A_y = 27 \text{ lb}$$

EXAMPLE 10: What is the required force, $F$, to lift a 500-lb weight using the three-pulley system shown in Fig. 4.20(a)?

SOLUTION: Tension forces on the cable at both ends over each pulley are equal; therefore, the right-side of the cable over pulley $A$ has tension equal to $F$. Similarly, the extension of this cable over pulley $C$ has tension equal to $F$ on both sides, and finally the cable on both sides of pulley $B$ is also under tension equal to $F$. Using the equilibrium of forces in the $y$ direction for pulley $C$ likewise results in the value of $F$. (The small angle that the middle part of the cable makes with the vertical $y$ axis is neglected.) Since

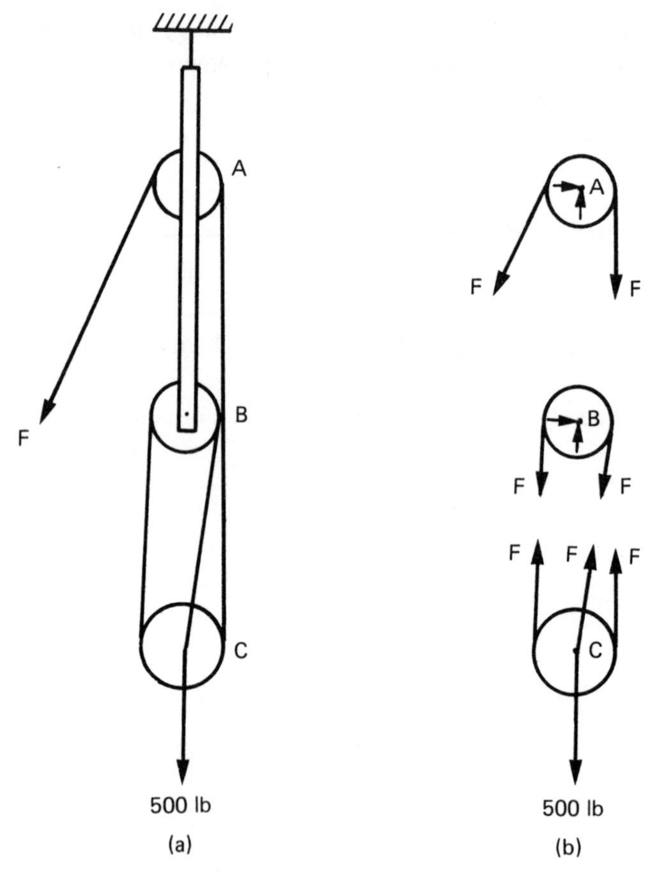


**Figure 4.20**

$$+\uparrow\Sigma\, F_y = F + F + F - 500 = 0$$

$$F = 167 \text{ lb}$$

this pulley can be used to considerable advantage: By applying a tension force of 167 lb, one can lift a 500-lb weight.

## 4.6 EQUILIBRIUM OF FORCES IN SPACE (THREE-DIMENSIONAL EQUILIBRIUM)

In Sec. 4.3, conditions of equilibrium in two dimensions were discussed. In general, however, structures exist in three-dimensional space, and applied forces may not be limited to *plane* (two-dimensional) *forces*.

Although the solution of equilibrium problems in three dimensions is a logical extension of the solution of the two-dimensional equilibrium problems discussed earlier, it has been generally observed that many students have difficulties in solving them. These problems usually require exercises in visualization rather than mechanics or mathematics. There is nothing really new or special about them except that one more dimension is added to the coordinate system, which in turn means that more equations must be added to the equilibrium equations discussed earlier. In this section, we will extend equilibrium equations 4.3, 4.4, and 4.5 to accommodate this added dimension.

When a body is under the action of a group of forces, the sum of these forces, which is called the *resultant*, is usually a three-dimensional force (or vector). This resultant force $R$ in three-dimensional space has a maximum of three rectangular components—$R_x$, $R_y$, and $R_z$. To satisfy equilibrium, as far as force equilibrium conditions are concerned, either this net resultant force must vanish (be equal to zero), or, alternatively, its components along the $x$, $y$, and $z$ axes must be equal to zero. The resultant force also has a general tendency to rotate the body in some general direction in three-dimensional space. This rotation can be decomposed into three components, one about each of the three coordinate axes. If we want the structure to remain in equilibrium, however, we cannot permit any rotation about any of the axes.

Therefore, to establish three-dimensional equilibrium, three force and three moment equations must be satisfied. The three force equations are

$$\begin{aligned} R_x &= \Sigma F_x = 0 \\ R_y &= \Sigma F_y = 0 \\ R_z &= \Sigma F_z = 0 \end{aligned} \tag{4.6}$$

and the three moment equations are

$$\begin{aligned} \Sigma M_x &= 0 \\ \Sigma M_y &= 0 \\ \Sigma M_z &= 0 \end{aligned} \tag{4.7}$$

In two dimensions, our use of two coordinate axes, $x$ and $y$, resulted in two force equilibrium equations, $\Sigma F_x = 0$ and $\Sigma F_y = 0$. In three dimensions, we have three coordinate axes, $x$, $y$, and $z$, and, quite logically, one more force equilibrium equation, $\Sigma F_z = 0$, has to be added to these. The moment equilibrium equation, however, does not follow this pattern. Remember that only one moment equilibrium equation was required in two dimensions, but three dimensions require two additional equations instead of just one. The reason for this is that when we talk about the moment of a force about point $O$ in a two-dimensional coordinate system ($Ox$, $Oy$), we really mean the rotational tendency about an axis that is perpendicular to the plane of $Ox$ and $Oy$, and this axis is really the $z$ axis of the coordinate system (or parallel to it). Therefore, prohibitive rotation about any of the three coordinate axes requires us to satisfy three moment equilibrium equations, one about each of the three coordinate axes.

As a result, we usually have to satisfy six equilibrium equations for the most complicated problems. Fortunately, most structural problems are such that some of these six equations can be automatically satisfied by inspection alone.

EXAMPLE 11: A traffic light post is installed at the corner of a cross section, as shown in Fig. 4.21(a). The two horizontal beams $BC$ and $BD$ are perpendicular to each other. Their weights are 50 and 75 lb, respectively, and are assumed to act at a point one-third of the length of the beams as measured from point $B$. The weight of each light box is 25 lb. The independent weight of post $AB$ is 250 lb. Calculate the reactions at $A$.

SOLUTION: The free-body diagram of the light system is shown in Fig. 4.21(b). Since support $A$ is a fixed support, there will be three components of reaction along the $x$, $y$, and $z$ axes and three moment reactions, one about each of these axes. The presence of three unknown forces and three unknown moments requires six equations for a solution.

First of all, since there is no applied force in the $x$ and $y$ directions, use of the equations, $\Sigma F_x = 0$ and $\Sigma F_y = 0$, will automatically give the component values, $A_x = 0$ and $A_y = 0$.

Then, using $+\uparrow\Sigma F_z = 0$,

$$A_z - 25 - 25 - 75 - 50 - 25 - 25 - 250 = 0$$

$$A_z = 475 \text{ lb}$$

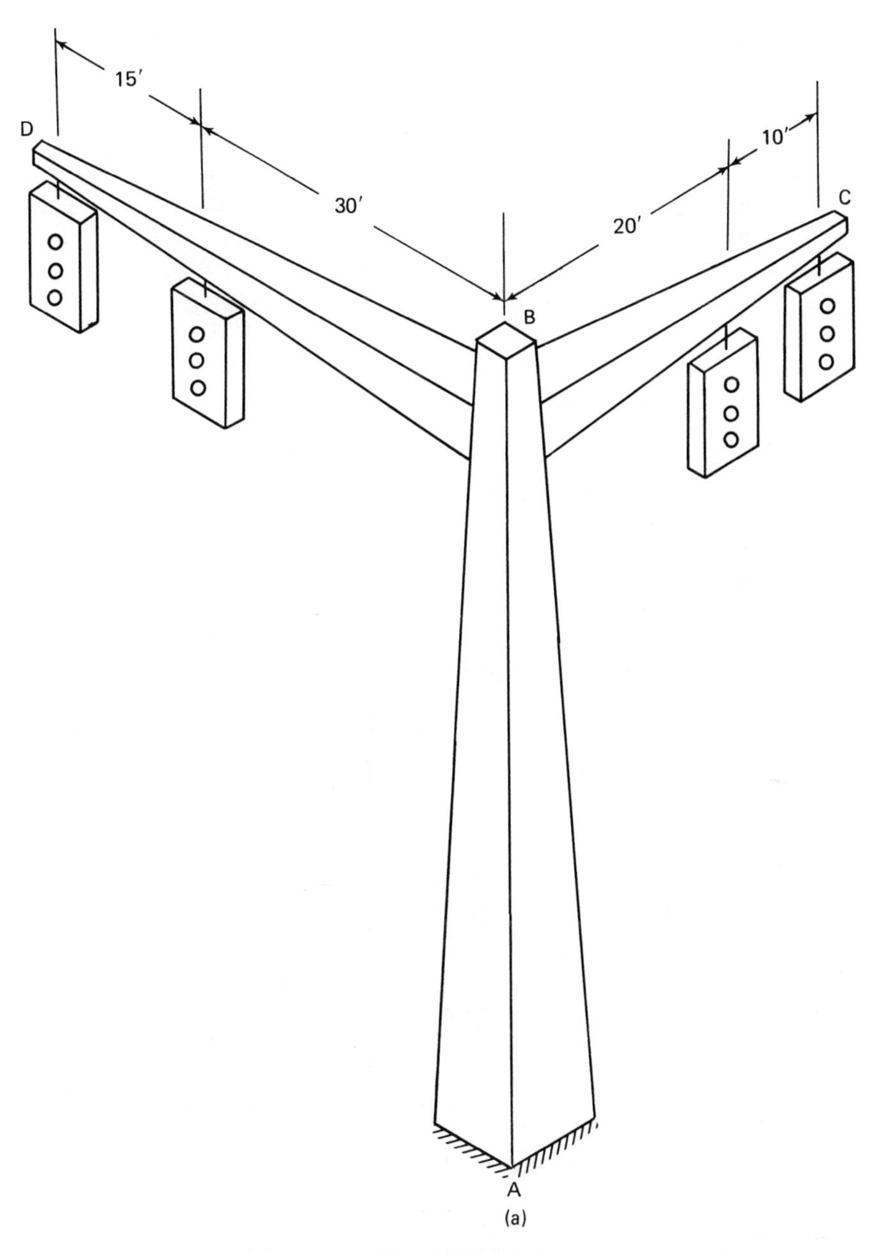


Figure 4.21

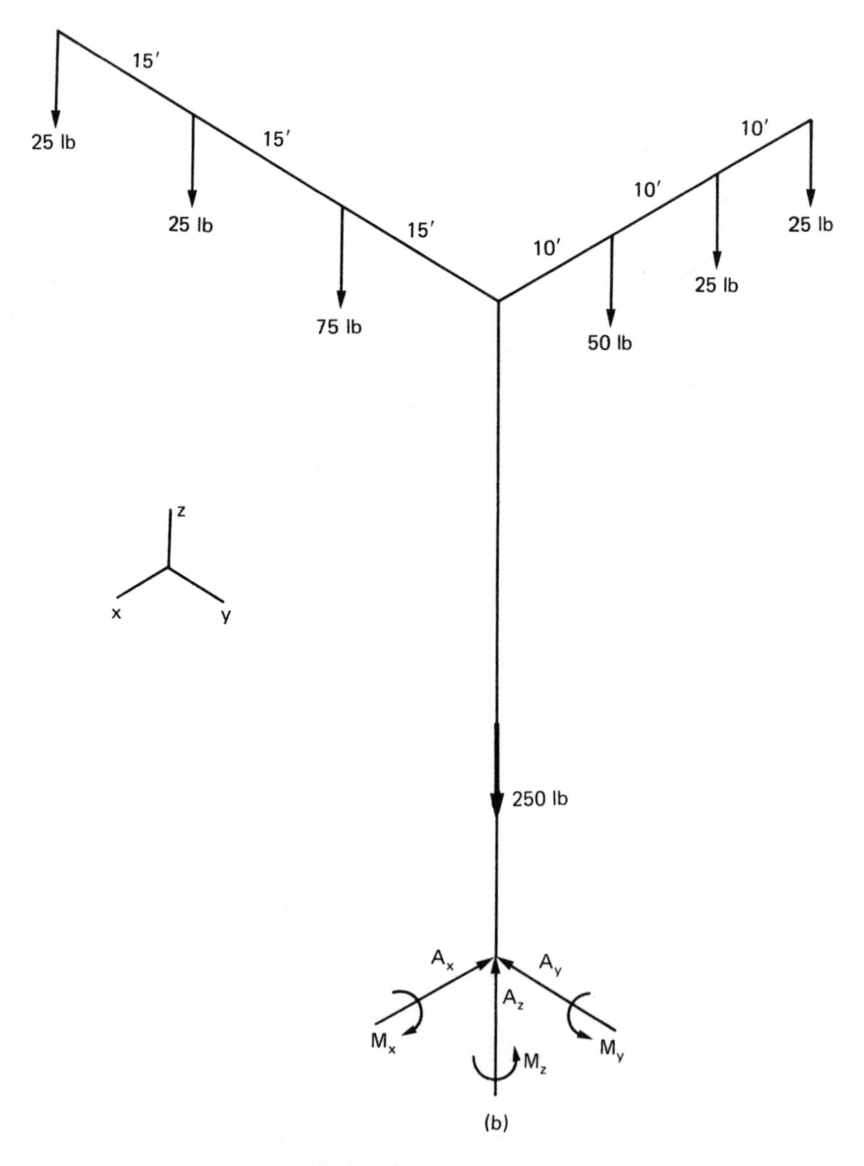


**Figure 4.21.** *(Continued)*

To calculate the three moment reactions, one has to use the moments of forces about each coordinate axis. Since all the applied forces are parallel to the $z$ axis, $M_z = 0$, and to get the moment of each force requires that they be multiplied by their perpendicular distance from the $z$–$x$ or $z$–$y$ planes. Therefore,

$$+ \quad \Sigma M_x = 0$$

$$M_x - (75 \times 15) - (25 \times 30) - (25 \times 45) = 0$$

$$M_x = 3000 \text{ lb ft}$$

Likewise,

$$+ \quad \Sigma M_y = 0$$

$$-M_y + (50 \times 10) + (25 \times 20) + (25 \times 30) = 0$$

$$M_y = 1750 \text{ lb ft}$$

## PROBLEMS

**4.1.–4.8.** Draw a complete free-body diagram for each of the problems presented by Figs. P4.1, P4.2, P4.3, P4.4, P4.5, P4.6, P4.7, and P4.8. Disregard the weight of the members themselves, and label all the known and unknown forces properly.

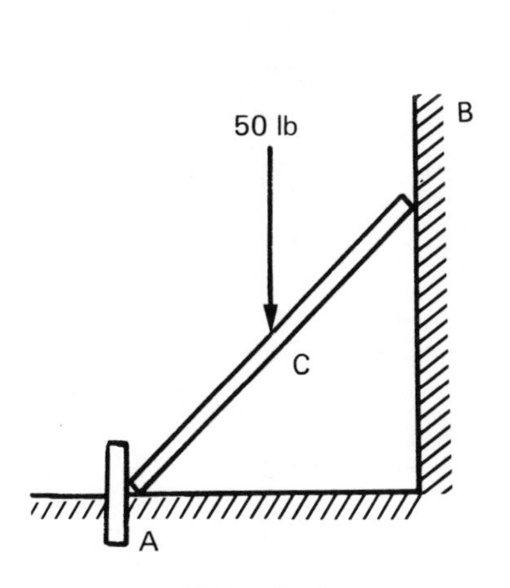


Figure P4.1

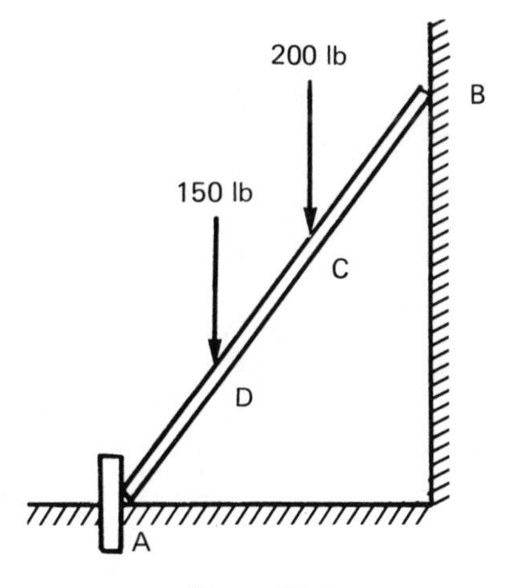


Figure P4.2

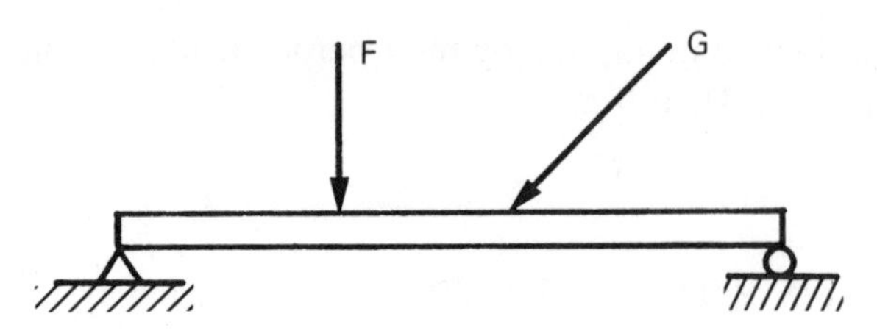


Figure P4.3

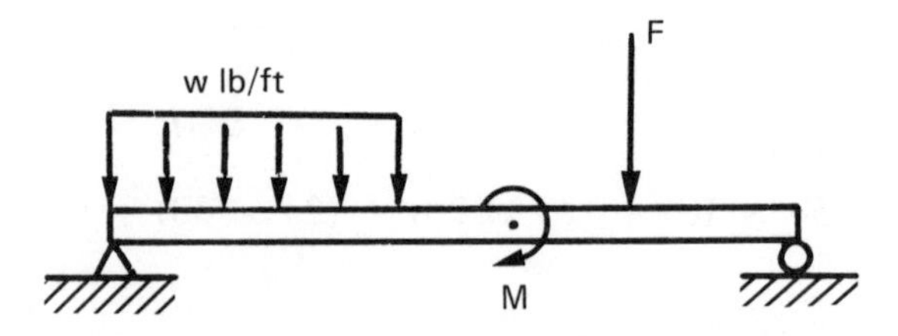


Figure P4.4

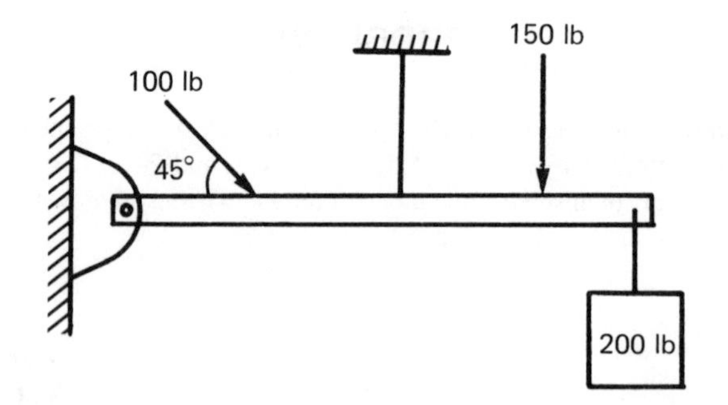


Figure P4.5

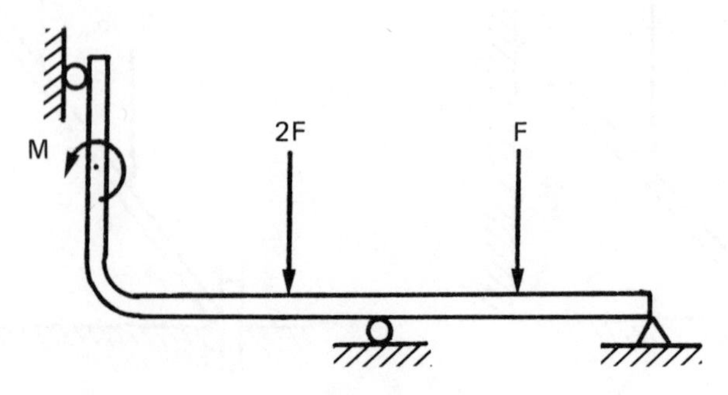


Figure P4.6

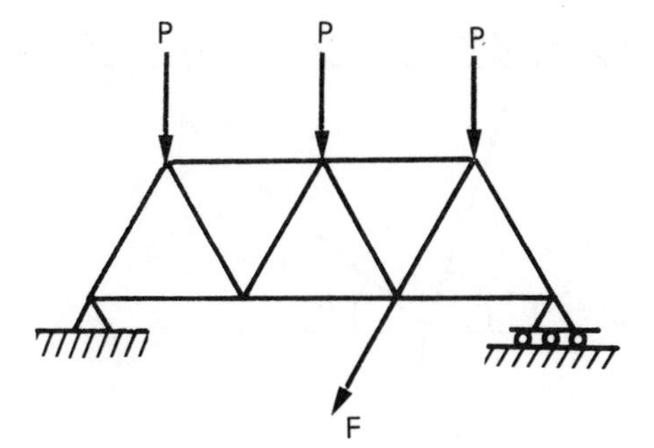


Figure P4.7

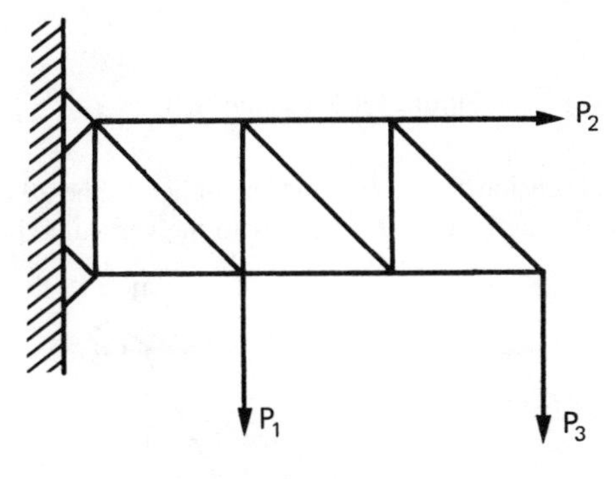


Figure P4.8

**4.9.** Calculate the tension force in each member of Figs. P4.9(a) and (b).

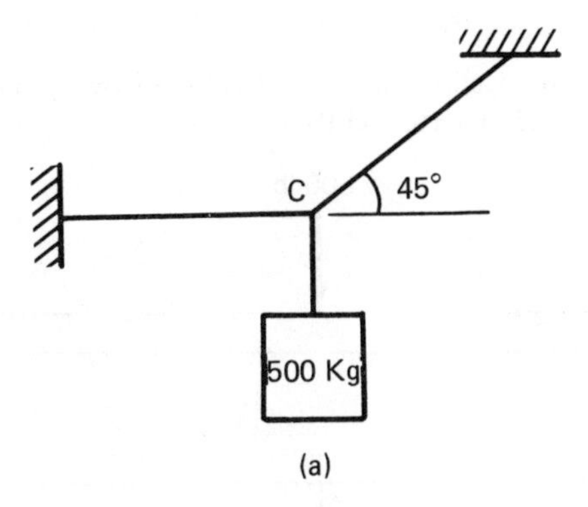


(a)

Figure P4.9

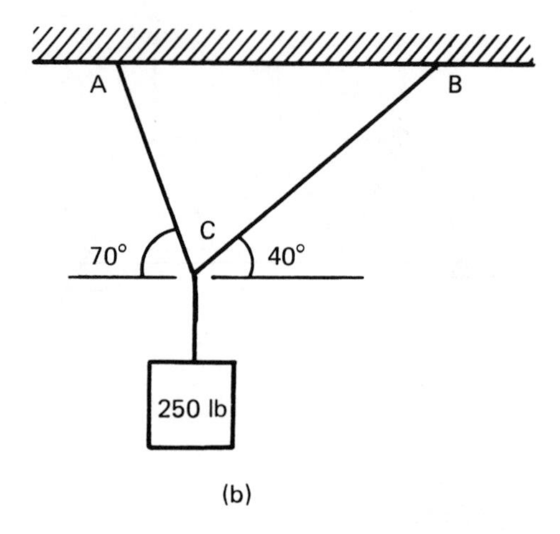


(b)

**Figure P4.9** *(Continued)*

**4.10.** What horizontal tension force, $F$, must be applied to the 200-lb weight shown in Fig. P4.10 for the angle that $AB$ makes with the vertical axis to be equal to 30°?

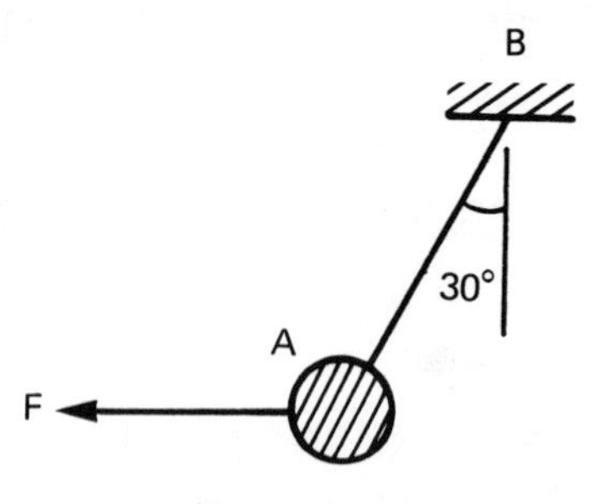


**Figure P4.10**

**4.11.–4.17.** Find the pin reactions at $A$ and $B$ in the following figures: P4.11, P4.12, P4.13, P4.14, P4.15, P4.16, and P4.17.

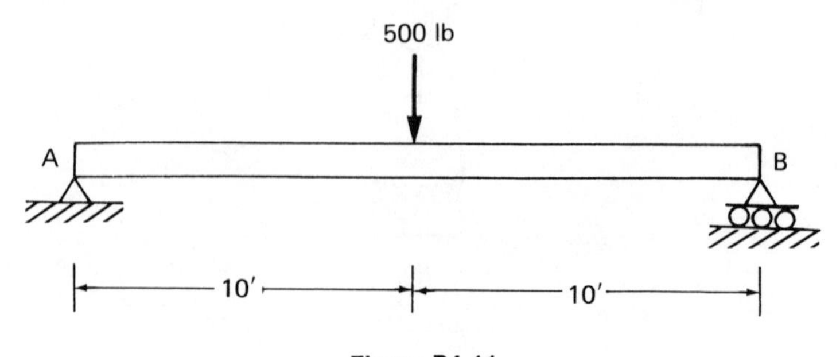


**Figure P4.11**

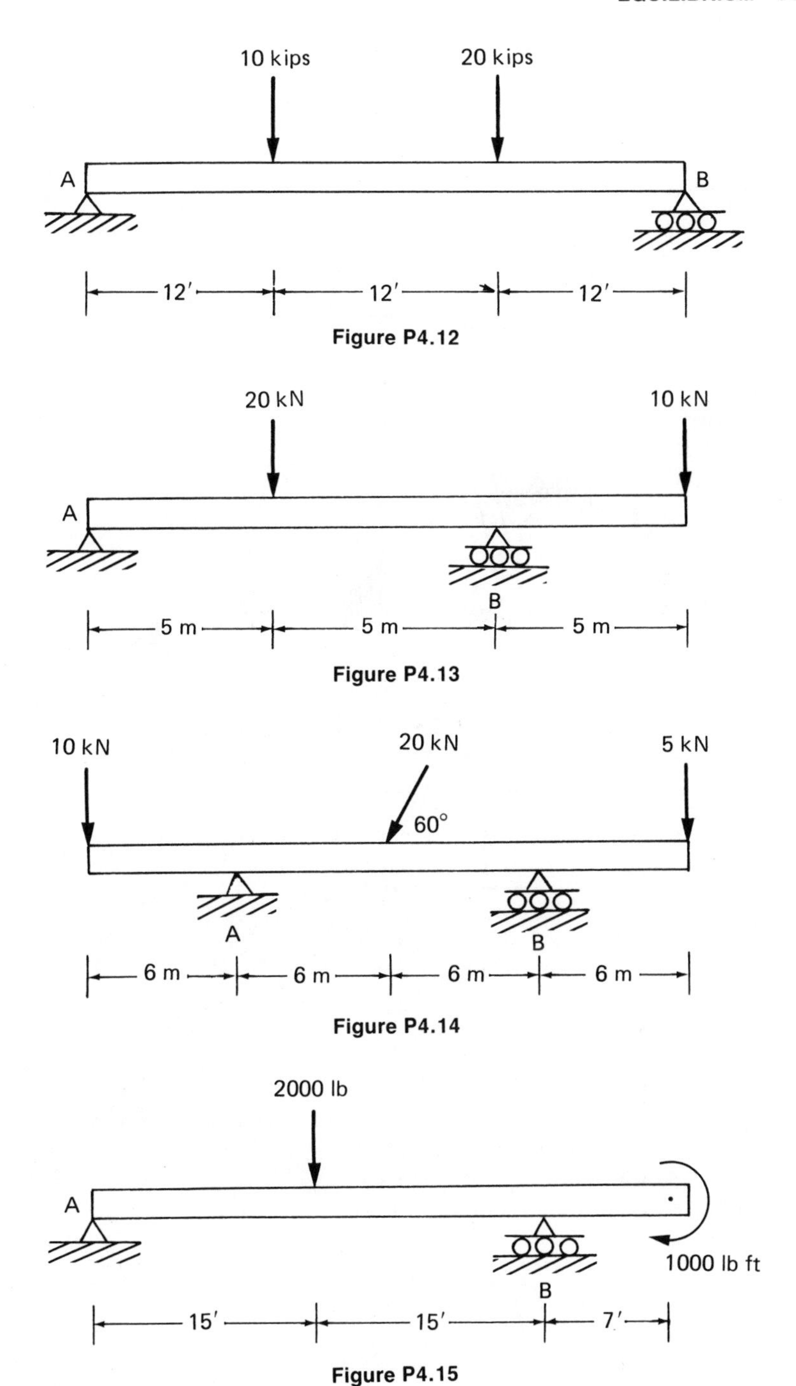


Figure P4.12



Figure P4.13



Figure P4.14



Figure P4.15

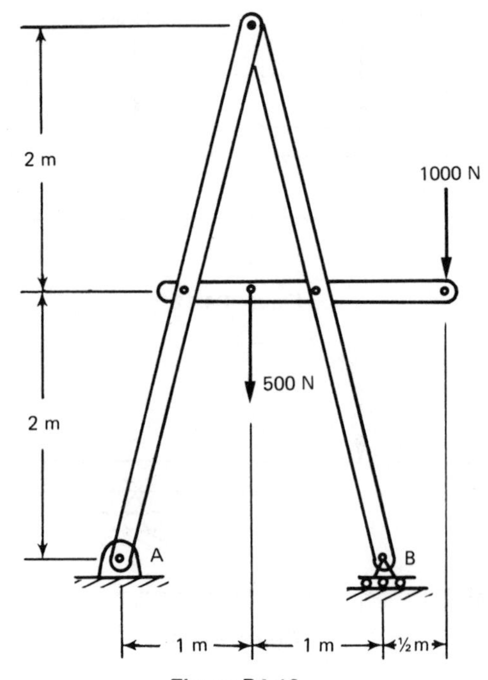


**Figure P4.16**

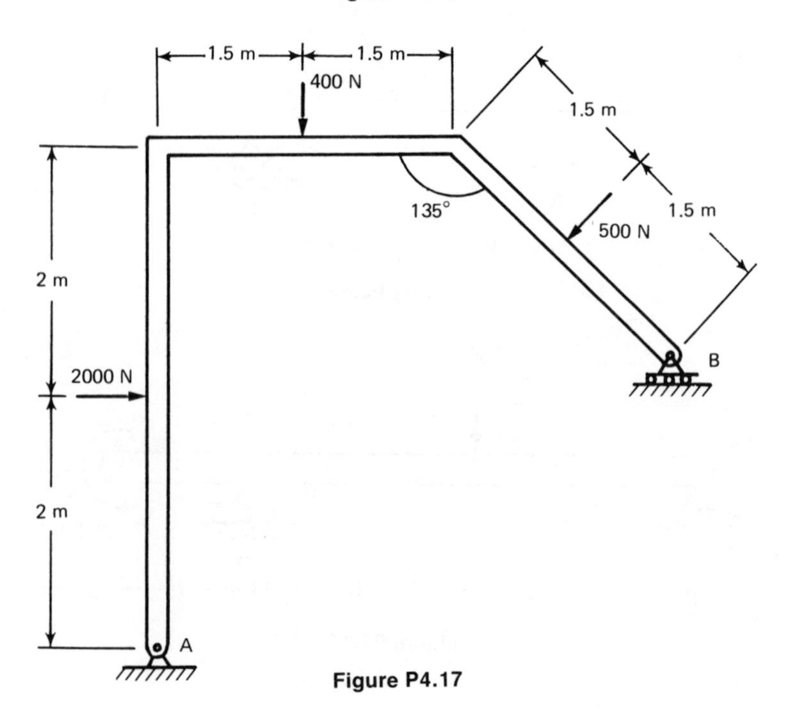


**Figure P4.17**

**4.18.–4.19.** Determine the force $F$ needed to lift the weights shown in the pulley systems of Figs P4.18 and P4.19.

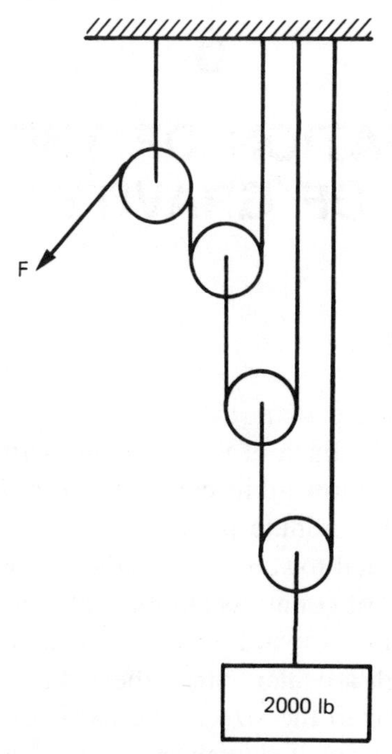


**Figure P4.18**

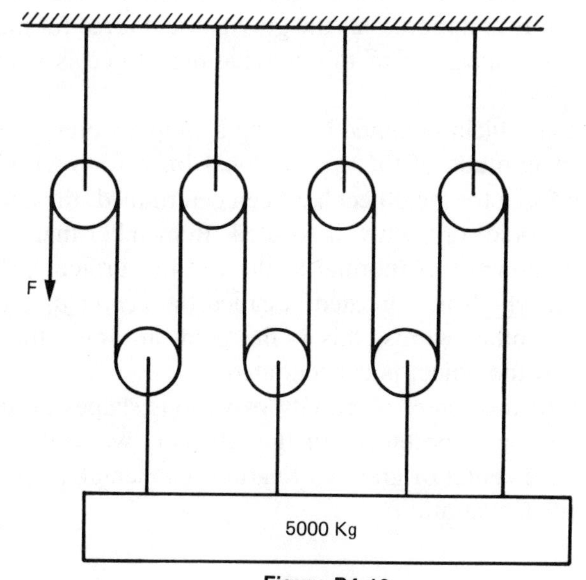


**Figure P4.19**

# 5

# DETERMINATION OF THE CENTER OF GRAVITY

## 5.1 INTRODUCTION

Every object is composed of a great number of minute particles that collectively constitute it. Each of these minute particles is continuously being pulled by the earth toward its center by a force called *gravitational force*. This force on the minute particle can be thought of as a small downward vector directed toward the earth's center. Now consider that the object is composed of countless numbers of these small particles, for each of which a small downward vector exists and all of which are directed toward the earth's center. Since the radius of the earth is enormously large compared to the size of the particles or the object itself, practically all of these small downward vectors can be thought of as being parallel. The sum of all these parallel vectors is also parallel to the individual vectors and represents the gravitational force for the whole object. The physical meaning of this gravitational force is simply the *weight* of the object.

The weight of an object obtained by using a simple weighing scale is nothing but the magnitude of this gravitational force (or vector). Once this gravitational force for the object has been determined, the next question is where this force (or vector) is located. Remember that we know its direction; it is *downward* (normal to the earth's surface). The very important point at which it is located is called the *center of gravity* or *center of mass*. In other words, this is the point at which the overall effect of gravity on the object is concentrated.

Determination of the center of gravity of various shapes pertinent to structures is of prime importance. In this chapter, we will discuss a method to locate the center of gravity of various elemental areas without the use of calculus (integration).

## 5.2 EXPERIMENTAL DETERMINATION OF THE CENTER OF GRAVITY

Consider the essentially two-dimensional object (an object with very small uniform thickness) shown in Fig. 5.1(a). To determine the location of the center of gravity, the following experiment should be performed. First, by using a cable or string, suspend the object from, say, point *A*, as shown in Fig 5.1(b). After equilibrium has been reached, the weight force (or vector) of the object that is a downward force must be aligned with line *AB*. This line *AB* (which lies along the extension of the string) is perpendicular to the surface of the earth and can be marked on the surface of the object either by using a square or by simply extending the string, as shown by the dotted line in Fig. 5.1(b). The location of the center of gravity will then lie somewhere on line *AB*. To locate its position on *AB*, repeat the same experiment by suspending the object again, but from a point different from *A* or *B*, say, point *C*, as shown in Fig. 5.1(c). Once again, after equilibrium has been reached, mark the extension of the string on the surface of the object (line *CD*). Obviously, the center of gravity must also be located somewhere on line *CD*. Since the center of gravity must be located on both lines *AB* and *CD*, it must lie at their intersection and is so indicated on the figure by the letters *CG* (for "Center of Gravity").

Physically, the moment of all the weight forces (or vectors) of the particles on the right side of line *AB* or *CD* must balance the moment of

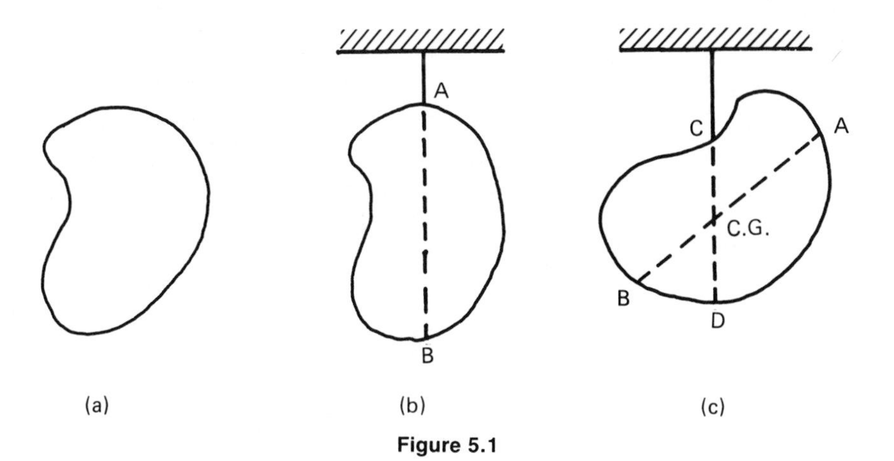


**Figure 5.1**

all the weight forces (or vectors) of the particles on the left side of line *AB* or *CD*; in other words, the sum of the moments of all weight forces about points on any axis passing through the center of gravity is equal to zero. This fact will provide an analytical method of obtaining the location of the center of gravity. In the following sections, it will be demonstrated first on groups of one- and two-dimensional discrete particles and then on continuous objects and areas pertinent to structures.

## 5.3 DETERMINATION OF THE CENTER OF GRAVITY OF A GROUP OF DISCRETE PARTICLES

### One-Dimensional Cases

Consider the group of weights—$w_1$, $w_2$, $w_3$, and $w_4$—shown in Fig. 5.2(a). In this case, discrete particles are arranged in a straight line (one-dimensional case) and connected to a rigid straight wire whose weight

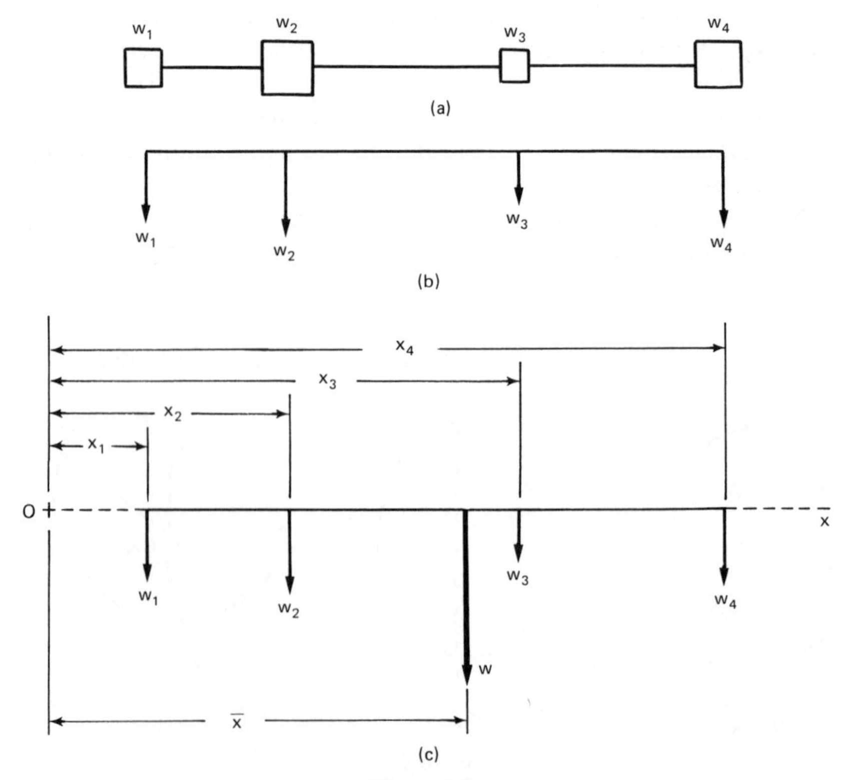


**Figure 5.2**

is negligible. We want to determine the center of gravity of this group of weights; in other words, we want to replace this system of four weights with one weight and locate it at a point where its overall effect will be the same as it was for all the four weights collectively in the way in which they were originally arranged.

To solve this problem, one can assign downward vectors to each weight of a magnitude equal to the weight itself, as shown in Fig. 5.2(b). Obviously, the equivalent weight, $W$, must be equal to the sum of all the individual weights involved (because all are parallel downward vectors), or

$$W = w_1 + w_2 + w_3 + w_4$$

and its moment should be equal to the moments of all the individual weight vectors about any arbitrary point. Let's take this arbitrary point (or moment center) to be $O$, as shown in Fig. 5.2(c), which is located on the same axis as the individual weights, and let's denote the distances from point $O$ to individual weights as $x_1$, $x_2$, $x_3$, and $x_4$ and the unknown distance of the total weight $W$ from $O$ as $\bar{x}$. (It is customary to denote the coordinates of the center of gravity with a bar at the top of each coordinate.)

Now apply the principle of moments by making the moment of $W$ about point $O$ equal to the sum of moments of the individual weights about the same point, as follows:

$$W\bar{x} = w_1x_1 + w_2x_2 + w_3x_3 + w_4x_4$$

After substituting $w_1 + w_2 + w_3 + w_4$ for $W$ and solving for $\bar{x}$, we get

$$\bar{x} = \frac{w_1x_1 + w_2x_2 + w_3x_3 + w_4x_4}{w_1 + w_2 + w_3 + w_4}$$

## Two-Dimensional Cases

Consider a group of particles arranged on a plane rather than a straight line as shown in Fig. 5.3(a). For the sake of simplicity, let's assume that the particles $w_1$, $w_2$, and $w_3$ are located on the $x$ axis and $w_1$, $w_4$, and $w_5$ on the $y$ axis and that these two axes are perpendicular to one another.

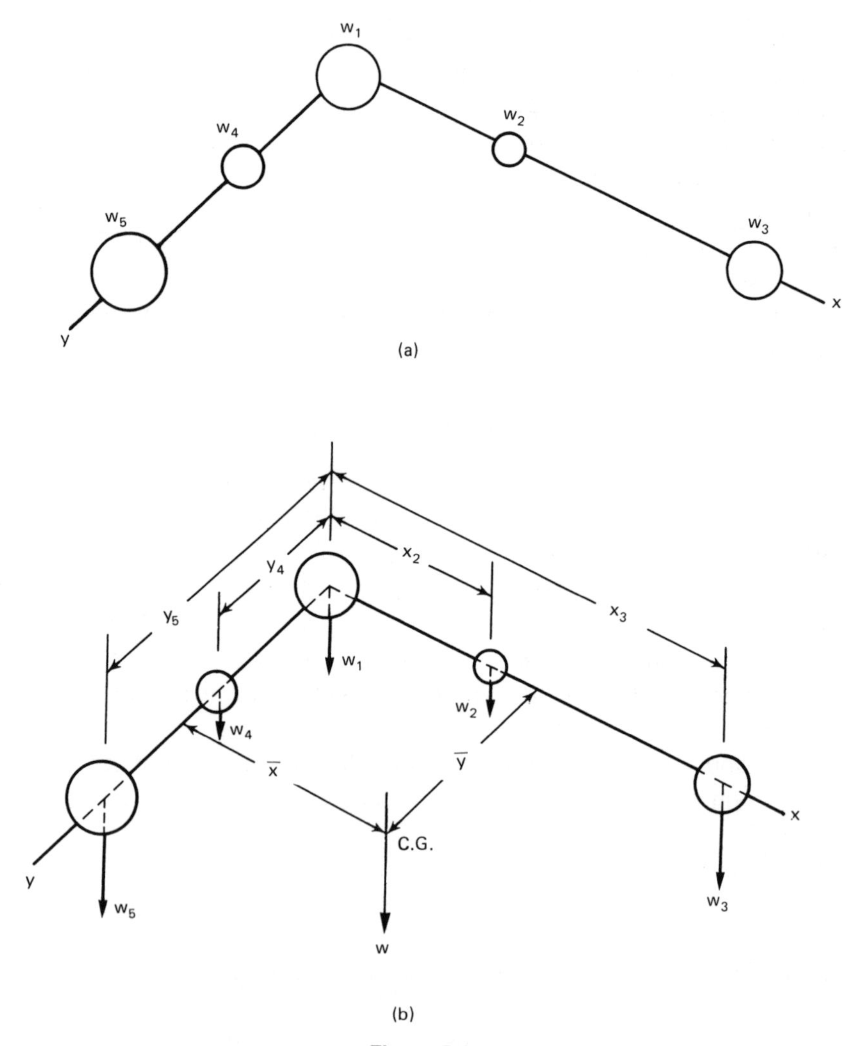


**Figure 5.3**

The solution to this problem is similar to that of the previous problem except that the principle of moment has to be applied twice because two coordinates are now needed to locate the center of gravity. Let's take the location of $w_1$ as the moment center. Once again, the resultant of all the individual weights, $W$, must be equal to the sum of all the weights:

$$W = w_1 + w_2 + w_3 + w_4 + w_5$$

Let's assume that the coordinates of the center of gravity are $\bar{x}$ and $\bar{y}$ as shown in Fig. 5.3(b). Applying the moment equation about the $y$ axis will result in the following:

$$W\bar{x} = w_2x_2 + w_3x_3$$

Notice that since the moment arms (distances) for weights $w_1$, $w_4$, and $w_5$ are zero, they do not enter into this equation.

Similarly, taking moment about the $x$ axis will result in the following:

$$W\bar{y} = w_4y_4 + w_5y_5$$

In this case, since the moment arms for weights $w_1$, $w_2$, and $w_3$ are zero, they did not enter into this equation.

Solving the two preceding equations for $\bar{x}$ and $\bar{y}$ and substituting $w_1 + w_2 + w_3 + w_4 + w_5$ for $W$ will result in the following coordinates for the center of gravity:

$$\bar{x} = \frac{w_2x_2 + w_3x_3}{w_1 + w_2 + w_3 + w_4 + w_5}$$

$$\bar{y} = \frac{w_4y_4 + w_5y_5}{w_1 + w_2 + w_3 + w_4 + w_5}$$

EXAMPLE 1: Locate the center of gravity of the three-weight system shown in Fig. 5.4.

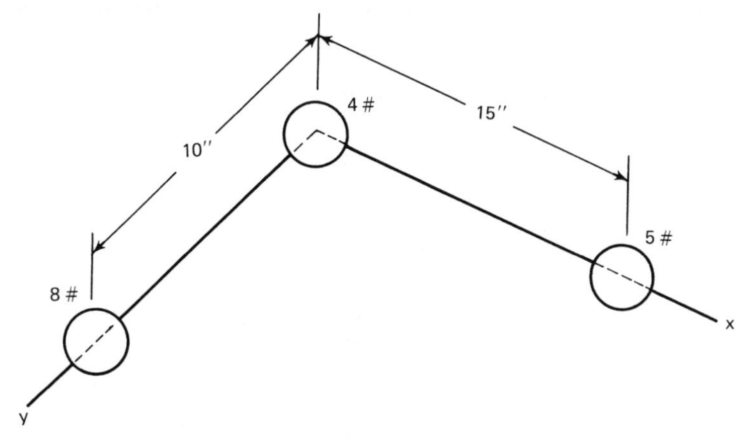


**Figure 5.4**

SOLUTION: The total weight, $W$, is 4 + 5 + 8, or 17 lb. Since this is a two-dimensional problem, we need to determine $\bar{x}$ and $\bar{y}$ and therefore the moment equation has to be applied twice—once about the $y$ axis and the second time about the $x$ axis, as follows:

$$17\bar{x} = 5 \times 15$$

$$\bar{x} = \frac{5 \times 15}{17} = 4.41''$$

and

$$17\bar{y} = 8 \times 10$$

$$\bar{y} = \frac{8 \times 10}{17} = 4.71''$$

## Three-Dimensional Cases

If the group of particles occupies a three-dimensional space rather than a line or plane, a similar approach will be taken, but the moment equation must be used three times so as to find $\bar{z}$ as well as $\bar{x}$ and $\bar{y}$.

EXAMPLE 2: Locate the center of gravity of the five-weight system shown in Fig. 5.5.

SOLUTION: The total weight $W$, is 2 + 3 + 4 + 5 + 6, or 20 lb. Since this is a three-dimensional problem, we need to determine $\bar{x}$, $\bar{y}$, and $\bar{z}$. Applying the moment equation three times, once about each coordinate axis, will result in the following:

$$20\bar{x} = (4 \times 5) + (6 \times 8) + (5 \times 2)$$

$$\bar{x} = \frac{(4 \times 5) + (6 \times 8) + (5 \times 2)}{20} = 3.9''$$

$$20\bar{y} = (4 \times 4) + (5 \times 3) + (6 \times 2)$$

$$\bar{y} = \frac{(4 \times 4) + (5 \times 3) + (6 \times 2)}{20} = 2.15''$$

$$20\bar{z} = (3 \times 4) + (4 \times 3) + (5 \times 3) + (6 \times 6)$$

$$\bar{z} = \frac{(3 \times 4) + (4 \times 3) + (5 \times 3) + (6 \times 6)}{20} = 3.75''$$

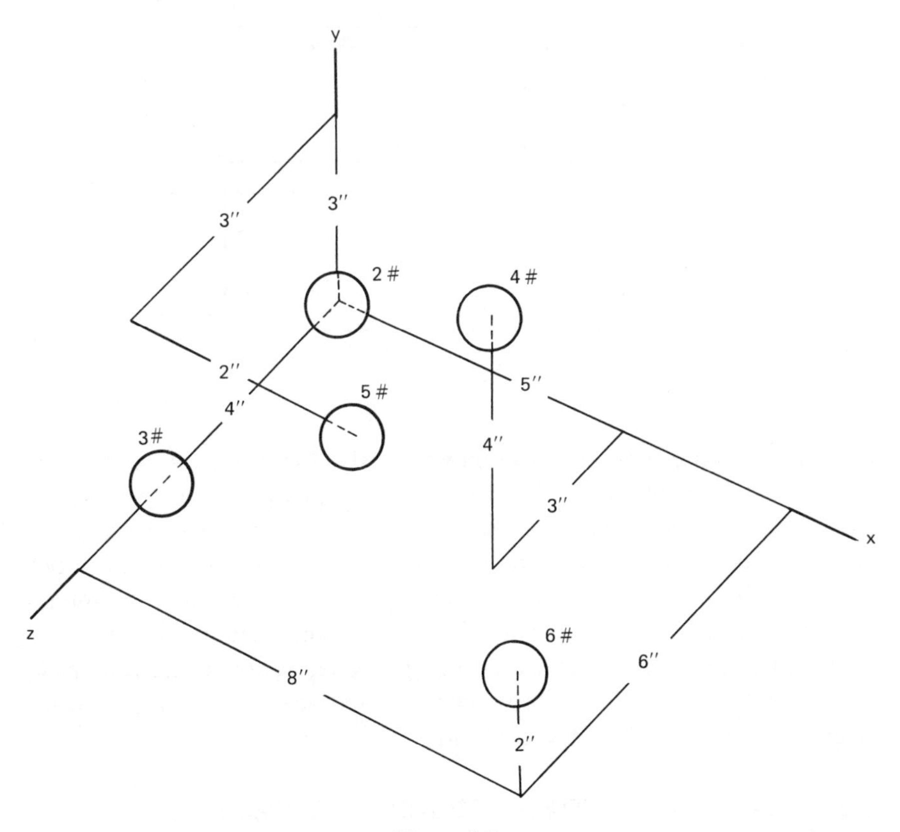


Figure 5.5

## 5.4 CENTROID

The words *centroid* and *center of gravity* are often used interchangeably. Usually, *center of gravity* is used in reference to objects or particles that have physical mass (or weight). The idea of the center of gravity can just as well be applied to figures that do not have physical weight, however, such as areas and volumes in their purely geometric sense. In these cases, it is more appropriate to talk about the *centroid* rather than the *center of gravity*, simply because no weight or gravity is involved. In other words, we can think of a geometric block and use the equations discussed earlier to get its $\bar{x}$, $\bar{y}$, and $\bar{z}$ coordinates. These values of $\bar{x}$, $\bar{y}$, and $\bar{z}$ then represent the *centroid* of the block. It is only when we think of the block as having weight that $\bar{x}$, $\bar{y}$, and $\bar{z}$ represent the *center of gravity*.

For homogeneous objects, the center of gravity and centroid coincide;

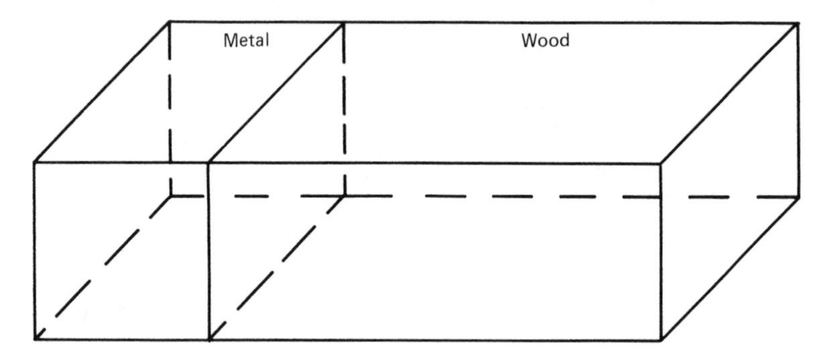


**Figure 5.6**

for objects whose mass is not homogeneously distributed, the center of gravity and the centroid are located at two distinct points. Consider the example in Fig. 5.6 where two solid blocks, one wood and one metal, are cemented to each other. The centroid of the composite block can be located as the geometric center of the entire block, located at the intersection of three planes half way through its entire length, width, and depth. (The centroid is also called the *geometric center*.) The center of gravity of the entire block is obviously located nearer the heavy portion (metal part) to the left of the centroid.

## 5.5 CENTROID OF TWO-DIMENSIONAL FIGURES (AREAS)

We will see later that in the calculation of stresses, strains, and deflections of structural shapes, one has to know the exact location of the centroid of the cross section of each member involved. The members used in structures are usually chosen from a handful of available cross-sectional areas in the shape of squares, rectangles, *I*-beams, *T*-beams, etc. We will start by considering an idealized cross-sectional shape for demonstration purposes, and later on we will examine the commercially available structural shapes.

The centroid of a symmetric shape is located on its *axis of symmetry* [see Fig. 5.7(a)]. If the area happens to have two axes of symmetry, then the location of the centroid is the intersecting point of the two axes of symmetry [see Figs. 5.7(b) through (e)]. The centroid of a rectangular area is therefore located at the geometric center of the rectangle, which can be located by finding the intersection of its two diagonals, as shown.

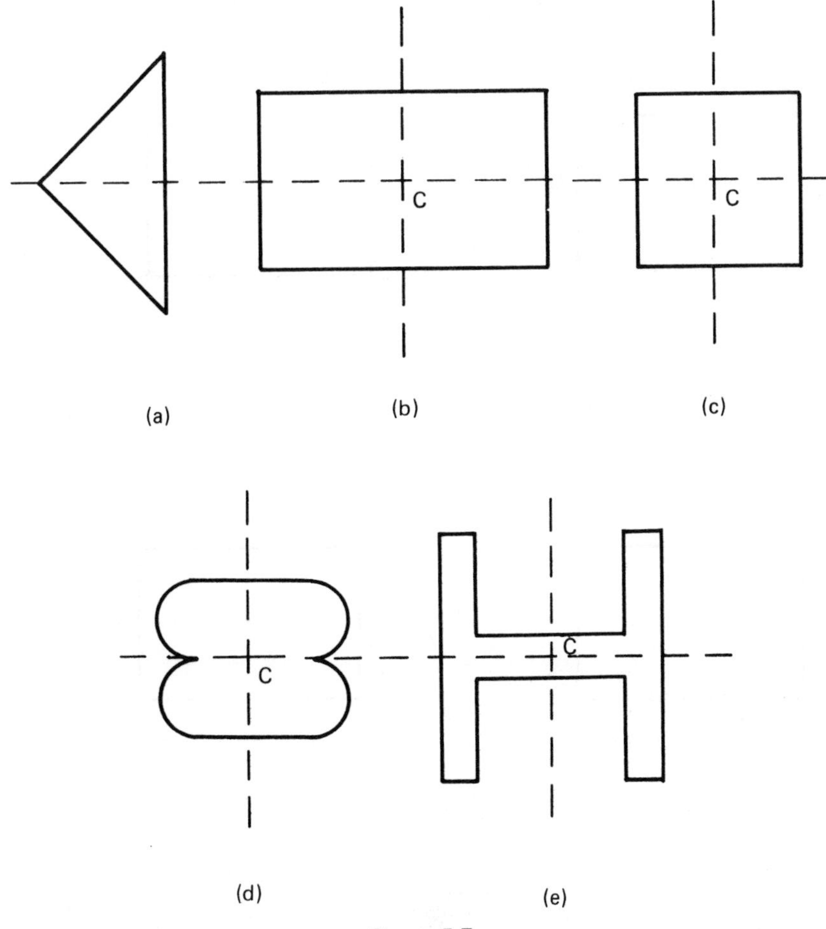


Figure 5.7

in Fig. 5.8(a). A more complicated area can usually be subdivided into areas whose centroids are known. Therefore, one should remember the location of the centroids of simple geometric figures frequently used in structures, such as those given in Table 5.1. It will be demonstrated in the next section that this table can be used to locate the centroid of most other irregular shapes.

The location of the centroid for each geometric shape is identified by its $\bar{x}$ and $\bar{y}$ coordinates in the rectangular coordinate system chosen. It is important to notice that if another coordinate system is used, $\bar{x}$ and $\bar{y}$ will have different values. As an example, the centroid of the rectangle shown

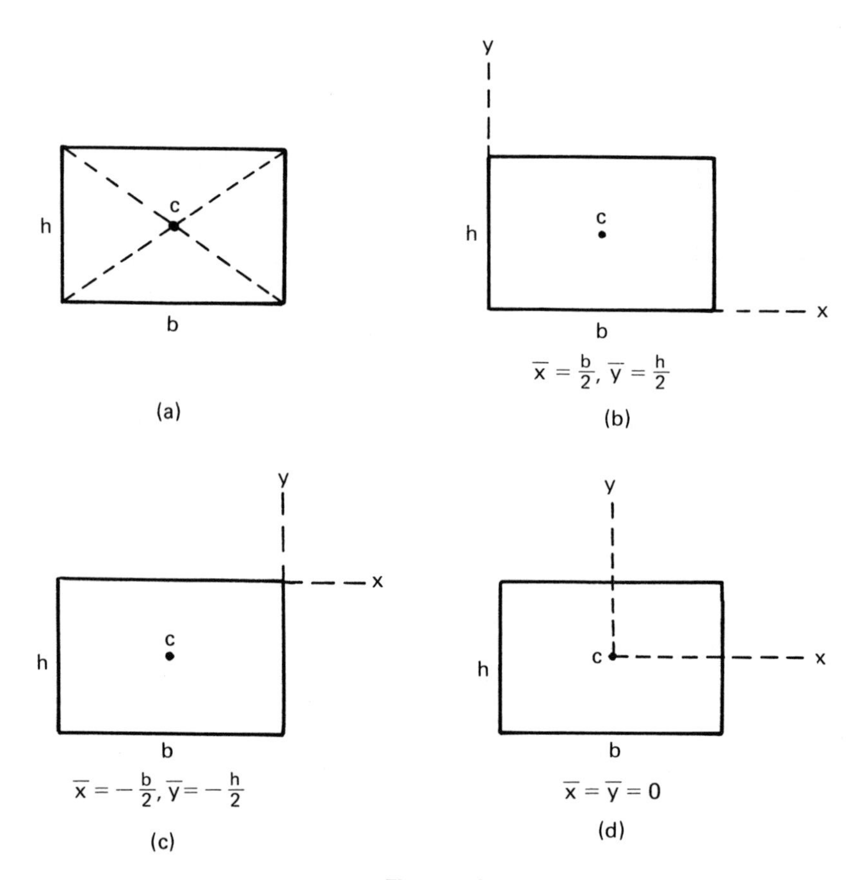


**Figure 5.8**

in Fig. 5.8(a) is at its geometric center. If the coordinate system chosen is like that shown in Fig. 5.8(b), then $\bar{x} = b/2$ and $\bar{y} = h/2$. If the coordinate system is that shown in Fig. 5.8(c), then $\bar{x} = -b/2$ and $\bar{y} = -h/2$. If the coordinate axes intersect at the centroid, as shown in Fig. 5.8(d), then $\bar{x}$ and $\bar{y}$ are both equal to zero. All such axes passing through the centroid are of particular interest and are referred to as *centroidal axes*; they will be frequently used—for example, in the study of the bending of a beam.

EXAMPLE 3: Determine the centroid of the area shown in Fig. 5.9(a).

SOLUTION: This figure can be subdivided into two rectangles, as shown in Fig. 5.9(b), with areas $A_1$ and $A_2$. Both $A_1$ and $A_2$ are rectangles and

## TABLE 5.1. Centroids of Common Areas.

| | Area | $\bar{x}$ | $\bar{y}$ |
|---|---|---|---|
| 1. Rectangle | 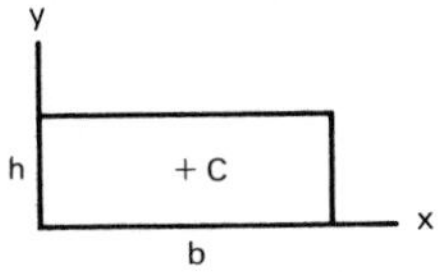 $bh$ | $\frac{b}{2}$ | $\frac{h}{2}$ |
| 2. Square | 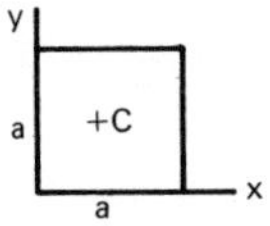 $a^2$ | $\frac{a}{2}$ | $\frac{a}{2}$ |
| 3. Circle | 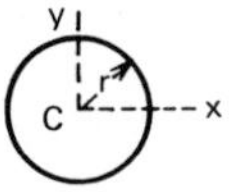 $\pi r^2$ | 0 | 0 |
| 4. Semicircle | 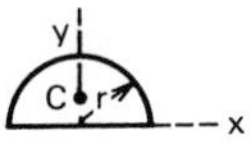 $\frac{1}{2}\pi r^2$ | 0 | $\frac{4r}{3\pi}$ |
| 5. Quadrant | 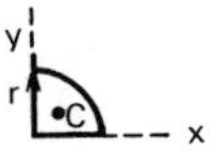 $\frac{1}{4}\pi r^2$ | $\frac{4r}{3\pi}$ | $\frac{4r}{3\pi}$ |
| 6. Right triangle | 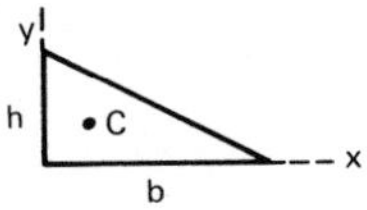 $\frac{1}{2}bh$ | $\frac{b}{3}$ | $\frac{h}{3}$ |
| 7. General triangle | 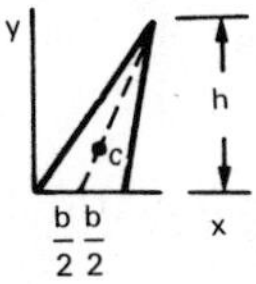 $\frac{1}{2}bh$ | — | $\frac{h}{3}$ |
| 8. Parabolic spandrel | 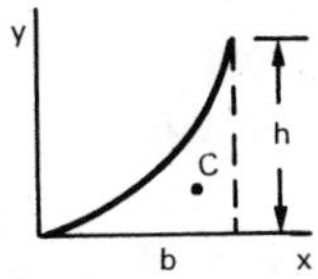 $\frac{bh}{3}$ | $\frac{3b}{4}$ | $\frac{3h}{10}$ |
| 9. Cubic spandrel | 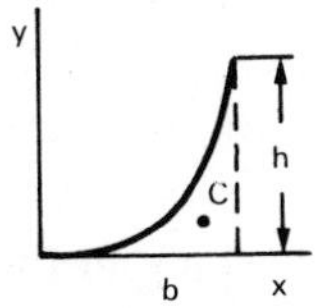 $\frac{bh}{4}$ | $\frac{4b}{5}$ | $\frac{2h}{7}$ |

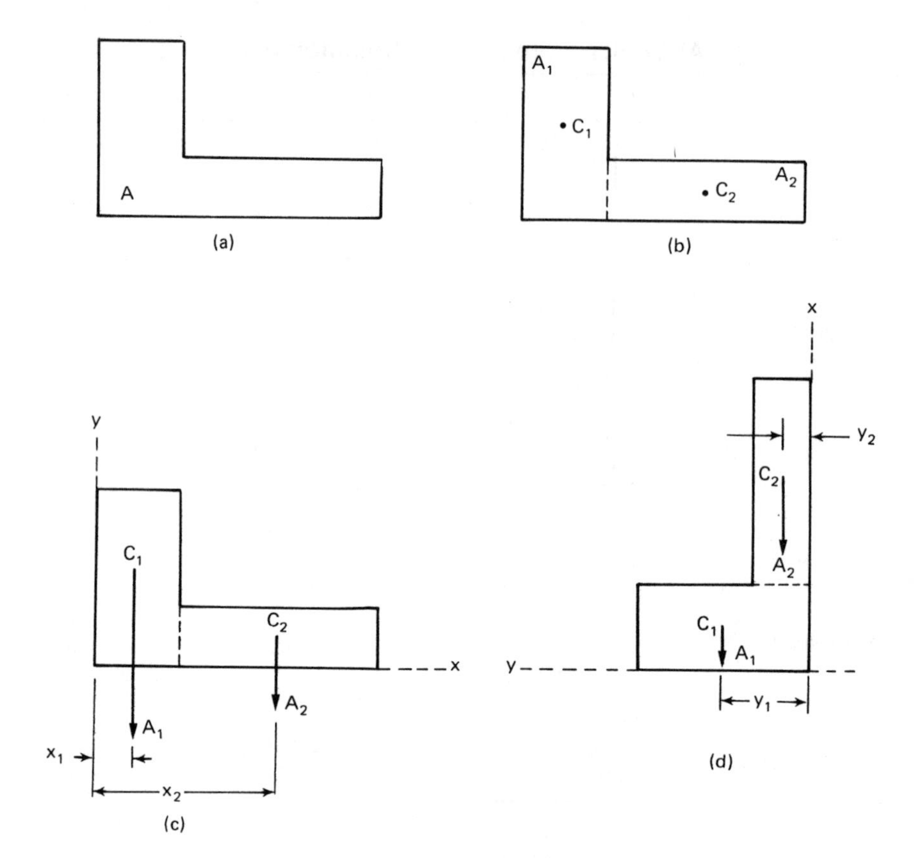


Figure 5.9

their individual centroids are in their geometric centers, shown in Fig. 5.9(b) as $C_1$ and $C_2$. Now, let's artificially assign vectors $A_1$ and $A_2$ to these areas and think of them as weight vectors (forces), their magnitudes being set equal to the values of areas $A_1$ and $A_2$. Since we think of them as weight vectors, they will be shown as downward vectors at $C_1$ and $C_2$, as in Fig. 5.9(c).

To determine the centroid of area $A$ is to determine the point of action of the resultant of vectors $A_1$ and $A_2$. Obviously, the resultant of vectors $A_1$ and $A_2$ is a downward vector with magnitude, $A_1 + A_2$, which is equal to area $A$ in Fig. 5.9(a). Let's choose a coordinate system like that shown in Fig. 5.9(c). Using the moment principle, the sum of the moments of the individual areas (or forces) $A_1$ and $A_2$ about the $y$ axis must be equal to the moment of the total area, $A = A_1 + A_2$, about the same axis, or

$$(A_1 + A_2)\,\bar{x} = A_1x_1 + A_2x_2 \tag{5.1}$$

where $\bar{x}$ represents the distance of the centroid of area $A$ from the $y$ axis.

Since we attached meaning of weight to each area and since weight is a downward vector (force), to calculate $\bar{y}$, let's artificially rotate the original area in Fig. 5.9(c) counterclockwise by 90°, as shown in Figure 5.9(d). In doing this, weight vectors $A_1$ and $A_2$ become parallel to the new $y$ axis [the $x$ axis in Fig. 5.9(d)] and, if we repeat the same process as before, this time, by applying the moment principle about the $x$ axis, the result will be

$$(A_1 + A_2)\,\bar{y} = A_1y_1 + A_2y_2 \tag{5.2}$$

Solving Eqs. 5.1 and 5.2 for $\bar{x}$ and $\bar{y}$ will result in the values of the coordinates of the centroid of the irregular area shown in Fig. 5.9(a), as follows:

$$\bar{x} = \frac{A_1x_1 + A_2x_2}{A_1 + A_2} \qquad \bar{y} = \frac{A_1y_1 + A_2y_2}{A_1 + A_2} \tag{5.3}$$

For determination of the centroid of three-dimensional figures, one will have to use the moment principle three times, of course, to get $\bar{z}$ as well as $\bar{x}$ and $\bar{y}$.

Notice that in going from Fig. 5.9(c) to 5.9(d), the coordinate system assigned to the initial figure was also rotated 90°. After solving several problems of this kind, students will be able to use the moment principle automatically and write Eq. 5.3 straight off. In practice, Figs. 5.9(c) and 5.9(d) are not usually drawn, merely constructed in the thinking process.

In this example, area $A$ was subdivided into only two simple rectangles, but if the initial area is more complex, it will have to be divided into many simple elements, and the same process just discussed has to be repeated. Equations 5.3 can then be generalized as follows:

$$\bar{x} = \frac{A_1x_1 + A_2x_2 + A_3x_3 + \cdots}{A_1 + A_2 + A_3 + \cdots} \qquad \bar{y} = \frac{A_1y_1 + A_2y_2 + A_3y_3 + \cdots}{A_1 + A_2 + A_3 + \cdots} \tag{5.4}$$

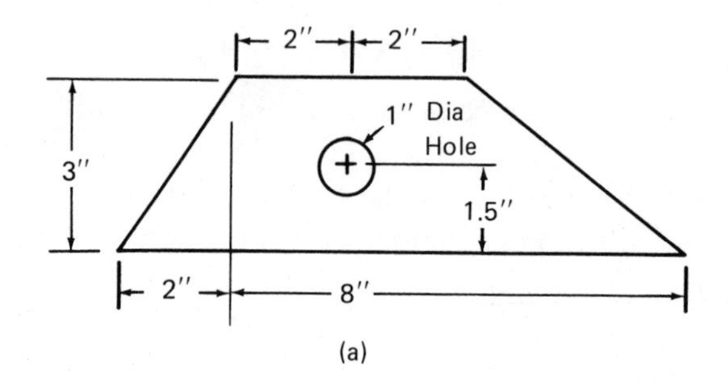


(a)

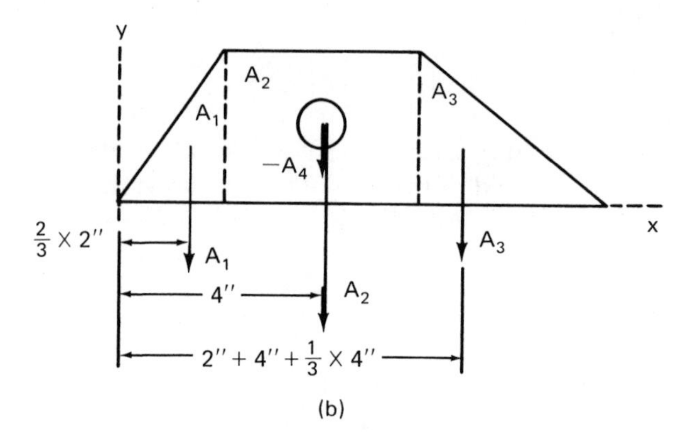


(b)

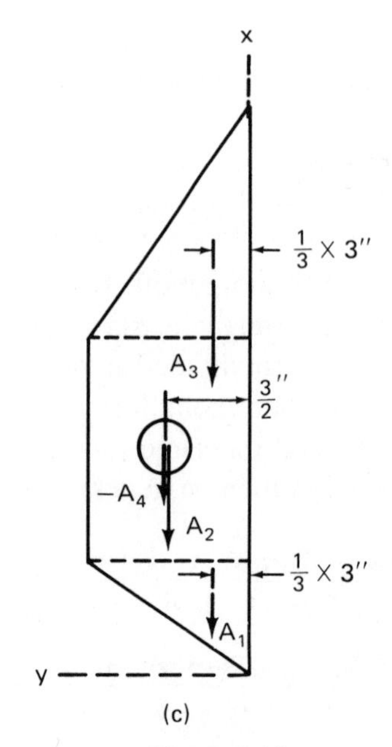


(c)

**Figure 5.10**

EXAMPLE 4: Determine $\bar{x}$ and $\bar{y}$ for the composite area shown in Fig. 5.10(a).

SOLUTION: The given area can be subdivided into three areas: a 2″ × 3″ right triangle on the left side, a 4″ × 3″ rectangle, and a 4″ × 3″ right triangle on the right side. To do so, however, will overestimate the initial area because of the existence of the 1″ diameter hole. Therefore, a fourth area (that of the circular hole) has to be added, but with a negative sign. All four areas have simple shapes, the centroid of each is known, and their distances from any coordinate axes chosen can be easily determined, as shown in Figs. 5.10(b) and (c). Sizes of each area as well as the coordinates of the centroid of each are calculated below:

$$A_1 = \tfrac{1}{2} \times 2'' \times 3'' = 3 \text{ in.}^2 \qquad A_3 = \tfrac{1}{2} \times 4'' \times 3'' = 6 \text{ in.}^2$$
$$A_2 = 4'' \times 3'' = 12 \text{ in.}^2 \qquad A_4 = \pi(0.5'')^2 = 0.785 \text{ in.}^2$$
$$\bar{x}_1 = \tfrac{2}{3} \times 2'' = 1.33'' \qquad \bar{x}_3 = 2'' + 4'' + \tfrac{1}{3} \times 4'' = 7.33''$$
$$\bar{x}_2 = 2'' + 2'' = 4'' \qquad \bar{x}_4 = 2'' + 2'' = 4''$$
$$\bar{y}_1 = \tfrac{1}{3} \times 3'' = 1'' \qquad \bar{y}_3 = \tfrac{1}{3} \times 3'' = 1''$$
$$\bar{y}_2 = \tfrac{1}{2} \times 3'' = 1.5'' \qquad \bar{y}_4 = 1.5''$$

Substituting the above values into Eqs. 5.4 will give the values of $\bar{x}$ and $\bar{y}$ for the composite area, as follows:

$$\bar{x} = \frac{A_1x_1 + A_2x_2 + A_3x_3 - A_4x_4}{A_1 + A_2 + A_3 - A_4}$$
$$= \frac{(3 \times 1.33) + (12 \times 4) + (6 \times 7.33) - (0.785 \times 4)}{3 + 12 + 6 - 0.785} = 4.59''$$
$$\bar{y} = \frac{A_1y_1 + A_2y_2 + A_3y_3 - A_4y_4}{A_1 + A_2 + A_3 - A_4}$$
$$= \frac{(3 \times 1) + (12 \times 1.5) + (6 \times 1) - (0.785 \times 1.5)}{3 + 12 + 6 - 0.785} = 1.28'' \tag{5.5}$$

Alternatively, this problem can be solved with fewer calculations. Since the centroids of areas $A_2$ and $A_4$ coincide, choice of a coordinate system that passes through the center of the circle will make $\bar{x}_2$, $\bar{y}_2$ and

$\bar{x}_4$, $\bar{y}_4$ all equal to zero, and the numerators for $\bar{x}$ and $\bar{y}$ in Eqs. 5.5 will have only two terms. Notice that the $\bar{x}$ and $\bar{y}$ obtained this way will differ from those previously obtained because they refer to two different coordinate systems; in both cases, however, they represent the same point (the calculations are left to the student as a homework problem).

## 5.6 CENTROID OF STRUCTURAL SHAPES

The treatment given here for determining the center of gravity is a simplified one. A more general procedure involves the use of integrals, which is avoided in this book. Since structural shapes used in architecture are usually limited to a handful of specified and standard shapes, however, the use of direct integration is ordinarily not necessary. Some examples of standard cross-sectional areas used in architecture are shown in Fig. 5.11. To locate the centroids of the areas shown in these figures, notice that each of the three to the left can be decomposed into three rectangular areas and then the centroid of the entire composite area found as discussed in Example 3. The area to the far right is even simpler since it is comprised of only two rectangles.

In reality, the cross-sectional areas of structural members purchased from steel companies do not have the shapes shown in Fig. 5.11. To make structural shapes stronger as well as more economical, their sharp corners are eliminated as are portions of those areas that practically do not carry any load. The four cross-sectional areas in Fig. 5.11 are idealized ones; their actual cross-sectional profiles are shown in Fig. 5.12.

In the process of modifying areas from Figs. 5.11 to 5.12, obviously the sizes as well as the weight distributions of the areas are altered, and the centroids are relocated. Determining the exact location of the centroids of modified areas would have to take into account the exact profile configuration, which would be a tedious job. Fortunately, the manufacturers of structural members (such as steel companies) provide their customers with tables containing exact area sizes as well as the exact location of centroids and other related and useful information. Table 5.2

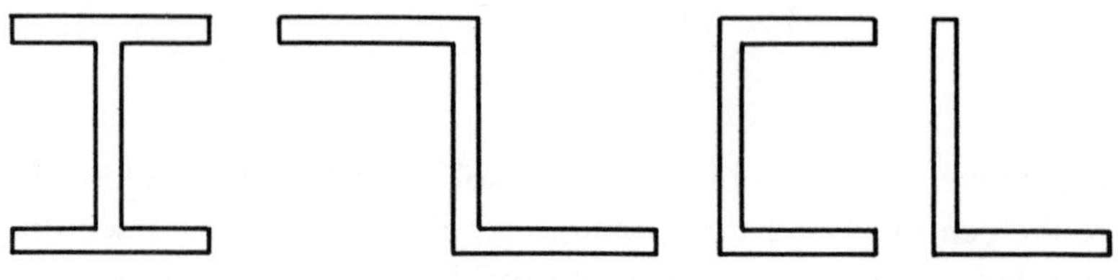

Figure 5.11

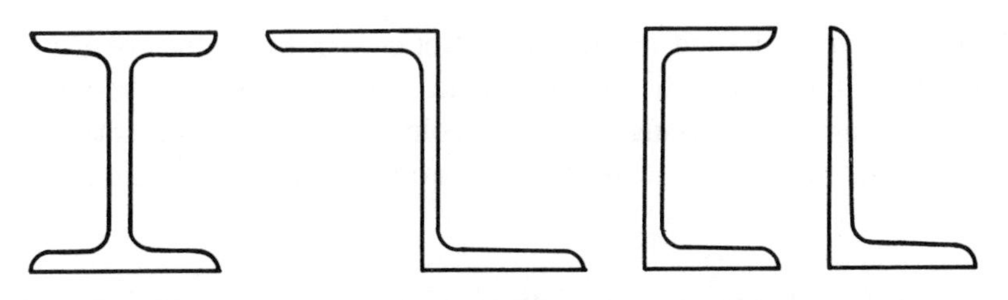

Figure 5.12

## TABLE 5.2. Centroids of Structural Shapes.

| Section | Weight Per Ft (lb) | Area $A$ (in.)$^2$ | $\bar{x}$ (in.) | $\bar{y}$ (in.) |
|---|---|---|---|---|
| W36×300 | 300 | 88.3 | 0 | 0 |
| W24×104 | 104 | 30.6 | 0 | 0 |
| W12×14 | 14 | 4.16 | 0 | 0 |
| C15×50.0 | 50.00 | 14.7 | 0.799 | 0 |
| C8×18.75 | 18.75 | 5.51 | 0.565 | 0 |
| C3×4.1 | 4.1 | 1.21 | 0.437 | 0 |
| WT18×150 | 150.0 | 44.1 | 0 | 4.13 |
| WT12×34 | 34.0 | 10.0 | 0 | 3.06 |
| WT4×20 | 20.0 | 5.87 | 0 | 0.735 |
| L9×4×5/8 | 26.3 | 7.73 | 1.04 | 3.36 |
| L5×3×1/4 | 6.6 | 1.94 | 0.657 | 1.66 |
| L3×2×3/8 | 5.9 | 1.73 | 0.539 | 1.04 |

shows a sample of such information. Notice that the values of $\bar{x}$ and $\bar{y}$ in the table are related to the coordinate system shown in the accompanying figures. In solving various problems, if one uses a coordinate system other than that shown, the calculations must be modified accordingly.

EXAMPLE 5: Determine the centroid of the area shown in Fig. 5.13. The composite area is composed of two standard *C*-channels welded to a rectangular (1″ × 10″) plate. Both *C*-channels are 8″ × 18.75 lb. (The letter "*C*" describes the shape of the channel; "8" is the nominal depth, in inches, of the channel; and "18.75" is the weight per foot in pounds.)

SOLUTION: Since the composite area is symmetric about the *y* axis, $\bar{x} = 0$, and only $\bar{y}$ has to be determined. The area of the *C*-channel can be found from Table 5.2 ($A = 5.51$ in.$^2$), and since it is symmetric about the horizontal axis, $\bar{y}$ for each channel is 4″ above the *x* axis as shown in Fig. 5.13. The centroid of the rectangle is along the *y* axis and 8.5″ above the *x* axis. By using Eqs. 5.3, we get

$$\bar{y} = \frac{A_1\bar{y}_1 + A_2\bar{y}_2 + A_3\bar{y}_3}{A_1 + A_2 + A_3}$$

$$= \frac{[(1 \times 10) \times 8.5] + 2[(5.51) \times 4]}{(1 \times 10) + (2 \times 5.51)} = 4.57''$$

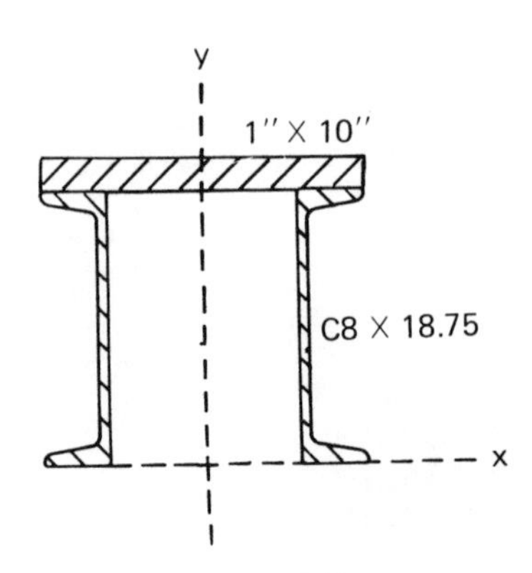


Figure 5.13

(The factor 2 in front of the second term in the numerator and denominator represents the two $C$-channels.) Thus,

$$\bar{x} = 0 \text{ and } \bar{y} = 4.57''$$

EXAMPLE 6: Determine the centroid of the cross-sectional area of the beam shown in Fig. 5.14. The beam is fabricated by welding an angle ($8'' \times 4'' \times 21.9$ lb) to a $1'' \times 20''$ plate.

SOLUTION: First, a coordinate system is chosen such as that shown in Fig. 5.14. The centroid of the rectangular plate is located at $\bar{x}_1 = 0.5''$ and $\bar{y}_1 = 10''$. Using the table in the Appendix (which is the complete version of Table 5.2), the following information is obtained for the angle: $A_2 = 6.43$ in.$^2$, $\bar{x}_2 = 0.88''$, and $\bar{y}_2 = 2.88''$.

Note that, as shown in the diagram of the channel on page 343 of the Appendix, $\bar{x}$ and $\bar{y}$ for the angle are measured from the bottom and left-side edges of the angle. These values must be modified with respect to the coordinate system used in Fig. 5.14, as follows:

$$\bar{x}_2 = 1'' + 0.88'' = 1.88''$$

$$\bar{y}_2 = 2.88'' \text{ (unchanged)}$$

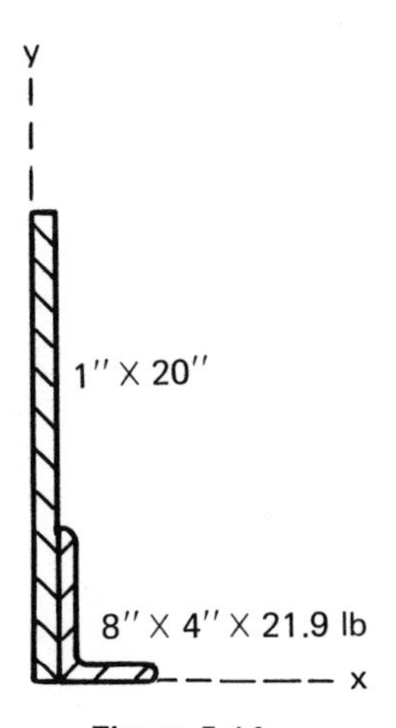


Figure 5.14

Now, using Eqs. 5.3,

$$\bar{x} = \frac{A_1x_1 + A_2x_2}{A_1 + A_2} = \frac{[(1 \times 20) \times 0.5] + [(6.43) \times 1.88]}{(1 \times 20) + 6.43} = 0.84''$$

$$\bar{y} = \frac{A_1y_1 + A_2y_2}{A_1 + A_2} = \frac{[(1 \times 20) \times 10] + [(6.43) \times 2.88]}{(1 \times 20) + 6.43} = 8.27''$$

## PROBLEMS

**5.1.** The known weights shown in Fig. P5.1 are distributed along a straight line. Find the location of the center of gravity of this system.

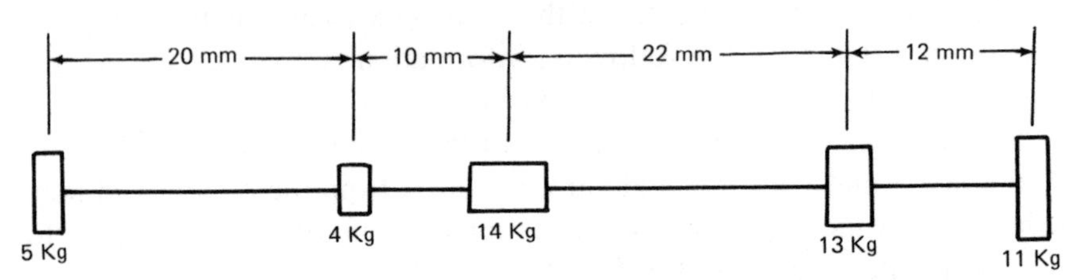


**Figure P5.1**

**5.2.** Three known weights are distributed along the $x$ and $y$ axes as shown in Fig. P5.2. Find the center of gravity of this system.

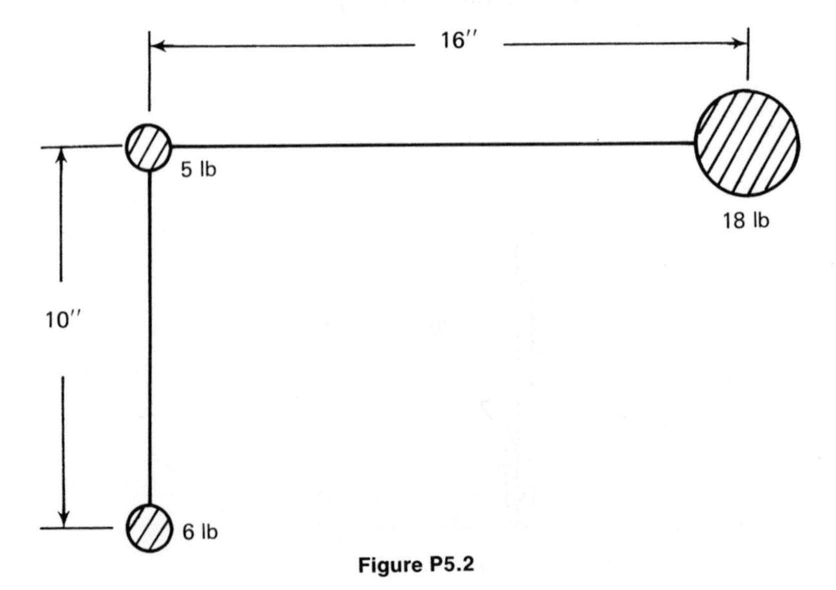


**Figure P5.2**

**5.3.** Three weights are located at the vertices of a triangle as shown in Fig. P5.3. Locate the center of gravity of the system.

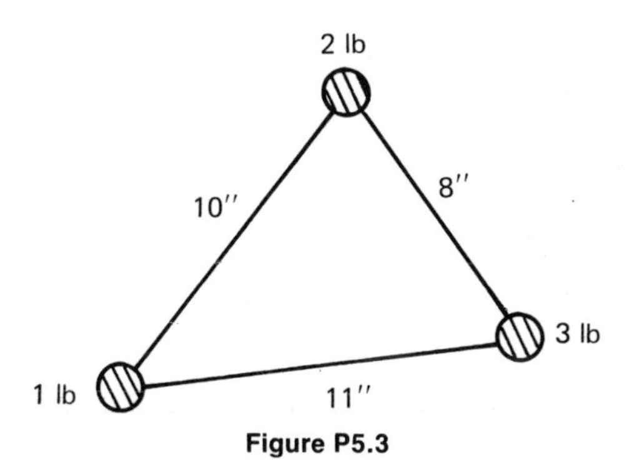


**Figure P5.3**

**5.4.** Determine the placement of a 10-kg weight in the rectangular system shown in Fig. P5.4 so that the center of gravity of the entire system will coincide with the geometric center of the rectangle.

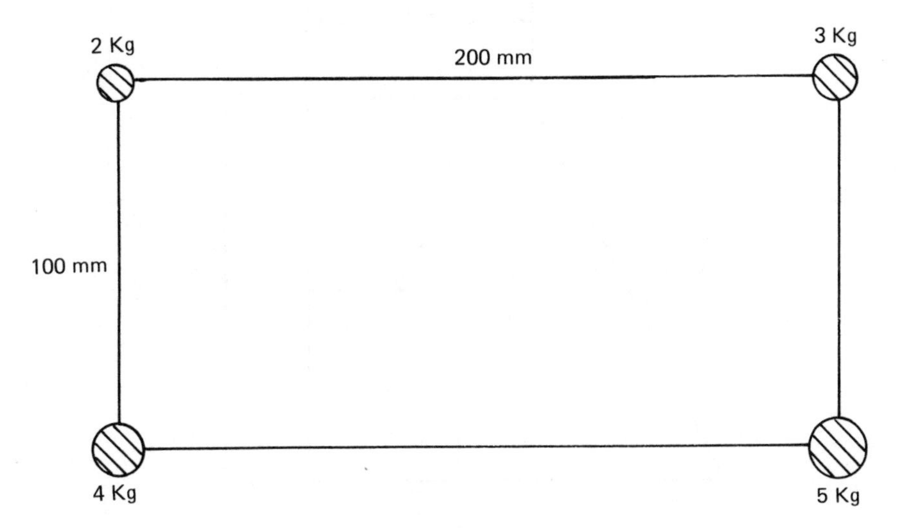


**Figure P5.4**

**5.5.–5.10.** Find the centroids of the plane geometric figures shown in Figs. P5.5, P5.6, P5.7, P5.8, P5.9, and P5.10.

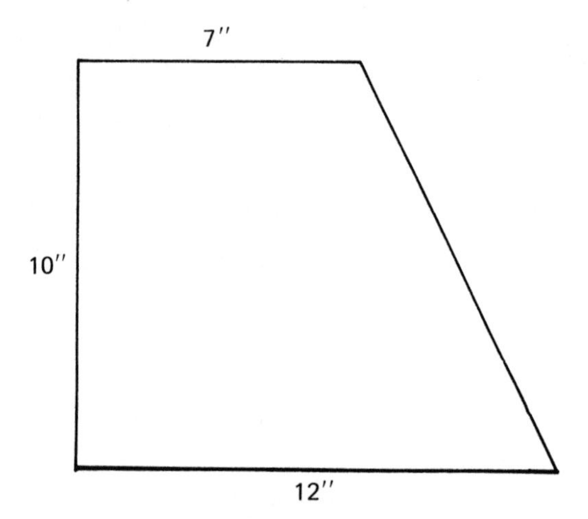


Figure P5.5

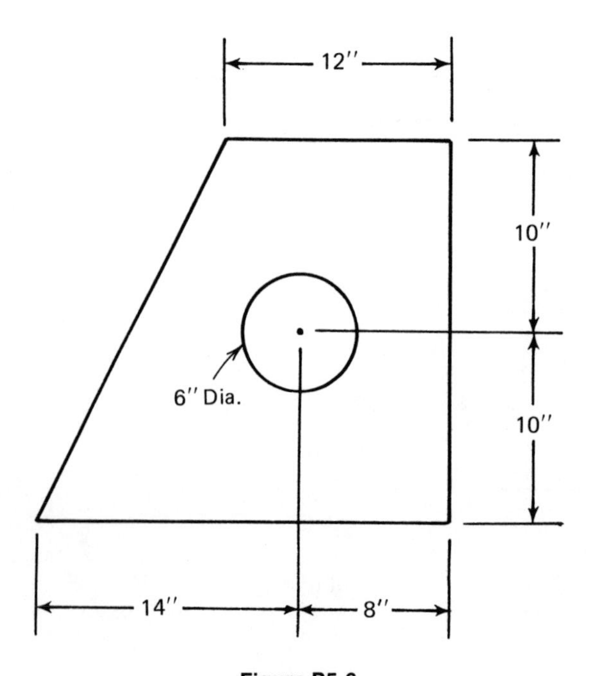


Figure P5.6

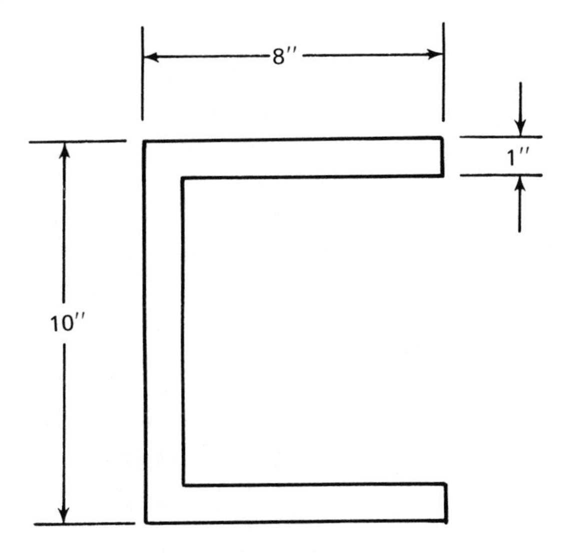


**Figure P5.7**

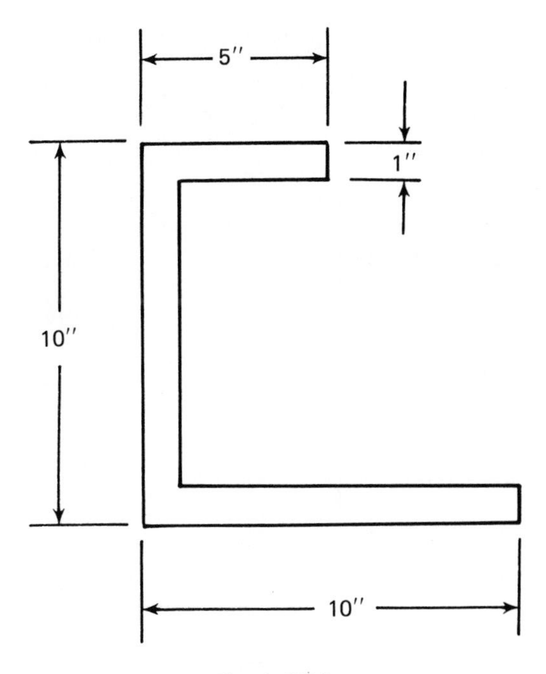


**Figure P5.8**

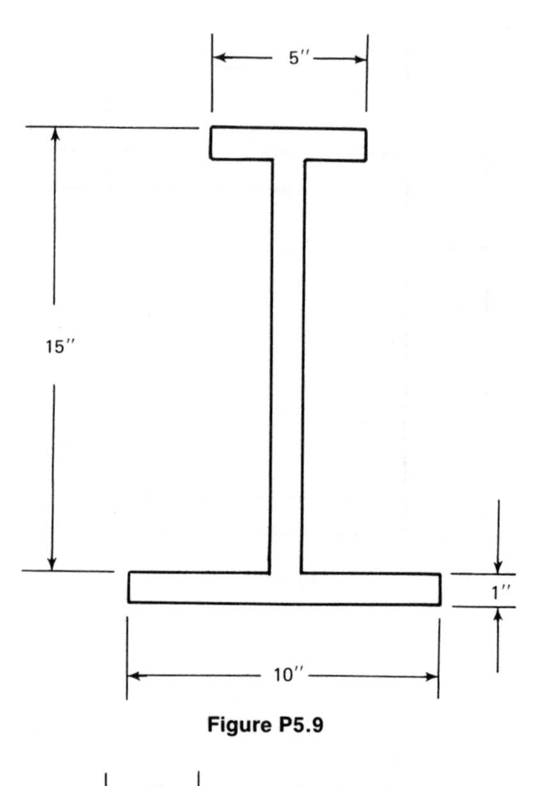


Figure P5.9

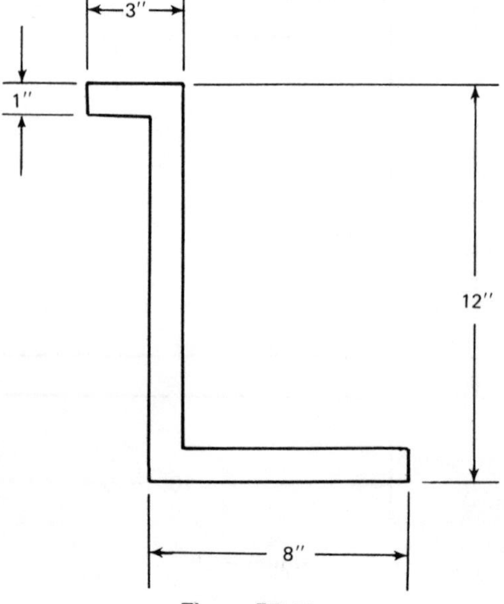


Figure P5.10

**5.11.–5.14.** Beams are fabricated by using various structural shapes. Determine the coordinates of the centroids of each shape shown in Figs. P5.11, P5.12, P5.13, and P5.14.

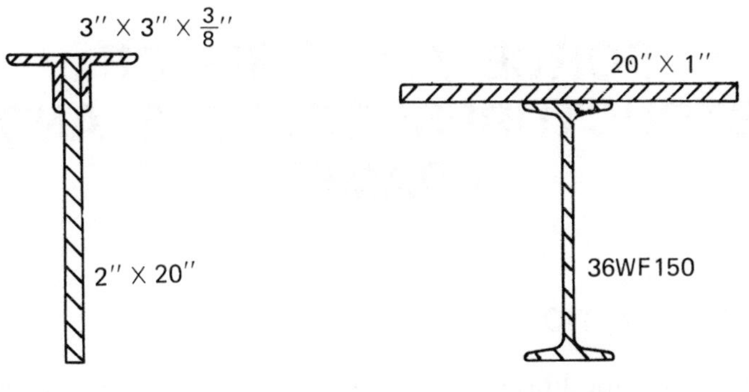


**Figure P5.11** **Figure P5.12**

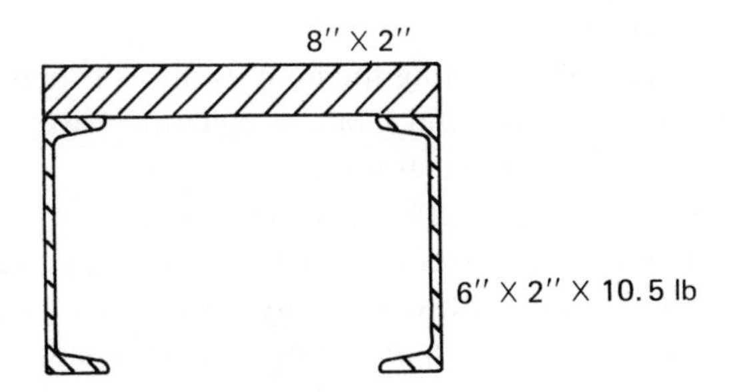


**Figure P5.13**

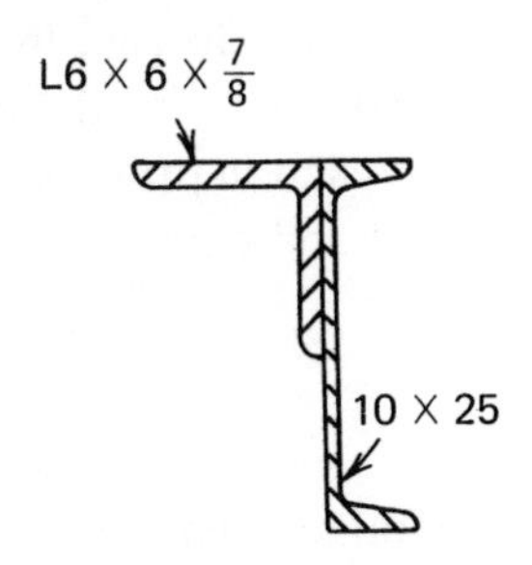


**Figure P5.14**

# 6

# FORCE ANALYSIS OF STRUCTURES: TRUSSES AND FRAMES

## 6.1 INTRODUCTION

In Chap. 4, the equilibrium of a single rigid body, or an assembly of connected members taken as a single rigid body, was studied. The treatment involved drawing the free-body diagram of the single body showing all the external forces that originally acted on it as well as the reaction forces developed by isolating the body from its supports. Next, the equilibrium equations (force and moment equations) were brought into play to obtain the unknown reactions.

In this chapter, we will apply the same procedure to structures and structural members. An engineering structure is a system built to support and/or transfer forces and withstand safely the applied loads. Generally speaking, such a structure is an assembly or combination of members. In Chap. 4, we analyzed the external forces on a body (i.e., externally applied forces and reaction forces that are external to the body). In this chapter, we will be more interested in internal forces. *Internal forces* are those forces, other than external forces, that act on each individual member or on portions of a structure. In such an analysis, we will use Newton's third law (or action–reaction principle), which states that every action is countered by an equal and opposite reaction.

## 6.2 TRUSSES

A framework composed of straight members connected at their ends to form a rigid structure, such as a bridge, roof panel, derrick, etc, is known as a *truss*. If an entire truss lies in a single plane, it is called a *plane truss*. When three straight members are connected to each other at their ends, they form a triangle that is a structurally rigid unit. This triangular unit is the basic component of the truss. In this chapter, we will analyze only *statically determinate trusses*, that is, those trusses that do not have more supports and members than are necessary to maintain equilibrium. As we learned in Chap. 4, the equilibrium equations are sufficient to analyze statically determinate problems.

## Analysis of Trusses

As just implied, trusses can be constructed by connecting straight members to each other. Common straight structural members used to build a truss include bars, channels, *I*-shapes, angles, and specially shaped members. Usually these members are connected to each other at their ends by welding, bolting, or riveting. Often a heavy wide plate, called a *gusset*, is used to reinforce the joints.

Plane trusses, such as bridges or roof trusses, are usually constructed in pairs, with one truss placed on each side of a structure and connected by cross members to support the roadway or the entire roof. Plane trusses are constructed in various configurations as shown in Fig. 6.1.

Three bars connected to one another by using pins at their ends constitute a rigid frame, as shown in Fig. 6.2(a), whereas four-member connections, as shown in Fig. 6.2(b), or polygons with more than four bars connected constitute a nonrigid frame.

A nonrigid frame such as that shown in Fig. 6.2(b) can be made rigid by adding a diagonal bar across joints $A$ and $C$ or across joints $B$ and $D$. Notice that by adding a single diagonal member to Fig. 6.2(b), two rigid triangular units are constructed. A rigid unit is referred to as a *noncollapsible unit*. Although adding one bar, say $AC$, to the frame of Fig. 6.2(b) will provide rigidity, adding a second, $BD$, while preserving the rigidity, will make the frame *statically indeterminate*. In general, when more members are present than are needed to prevent its collapse, a truss is a statically indeterminate truss. The extra member not necessary for rigidity, such as $BD$, is called a *redundant member*.

One of the first steps in designing a truss includes determination of the forces that will be developed in various members of the truss under the action of external loads. In this chapter, these forces will be calculated using two different methods: the *method of joints* and the *method of sections*. Before undertaking this force analysis, however, let us list the assumptions made in solving truss problems, as follows:

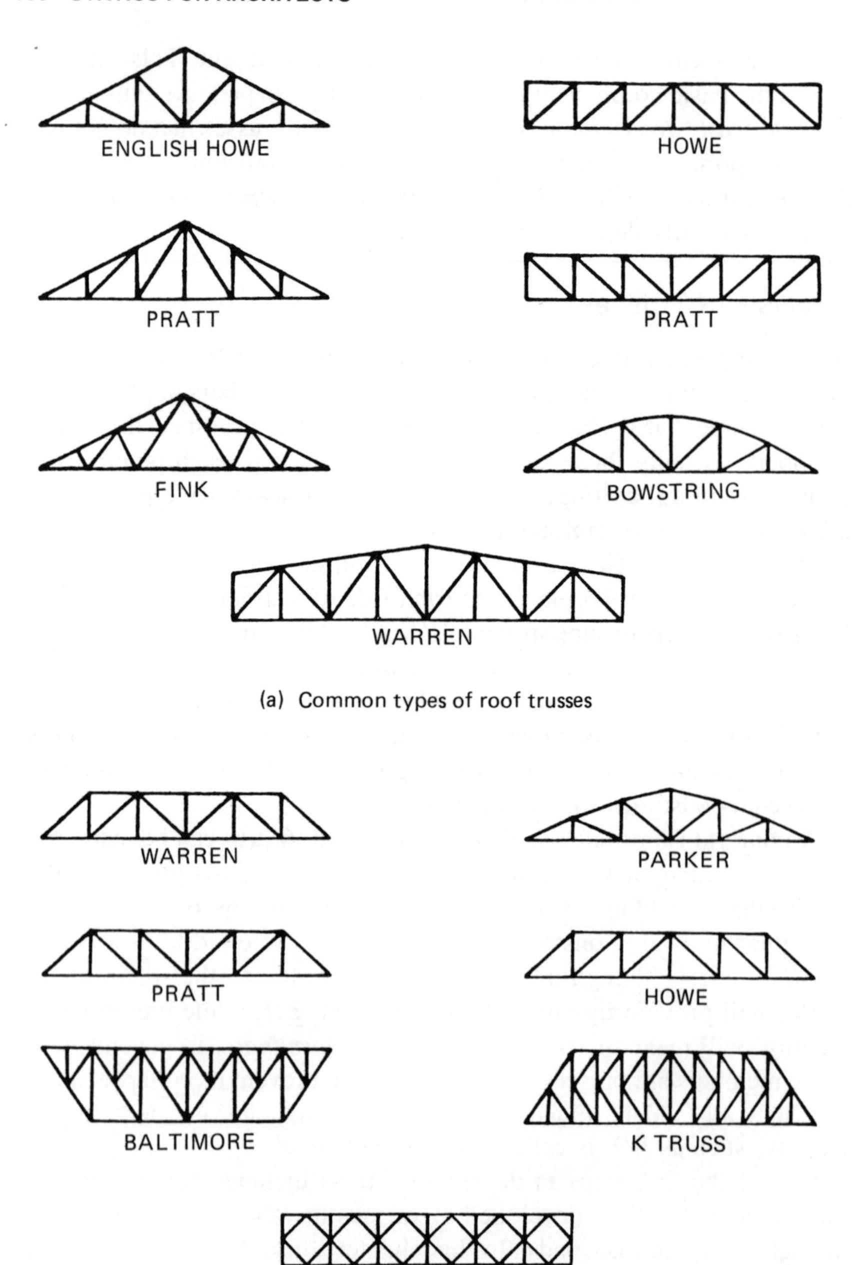


(a) Common types of roof trusses

(b) Common types of bridge trusses

**Figure 6.1**

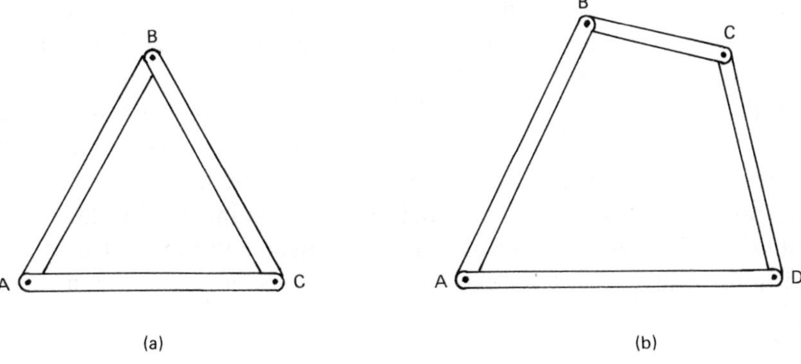


Figure 6.2

1. All members are pin-connected even though they may be welded to each other.
2. All external loads are applied only at the joints.
3. All members are two-force members.
4. The weights of the members are negligible compared to the applied external loads.

The consequence of the third assumption is that each individual straight member is in tension or compression, as shown in Fig. 6.3. As pointed out in Chap. 4, the two applied forces in a two-force member are of equal magnitudes and opposite directions and are aligned with

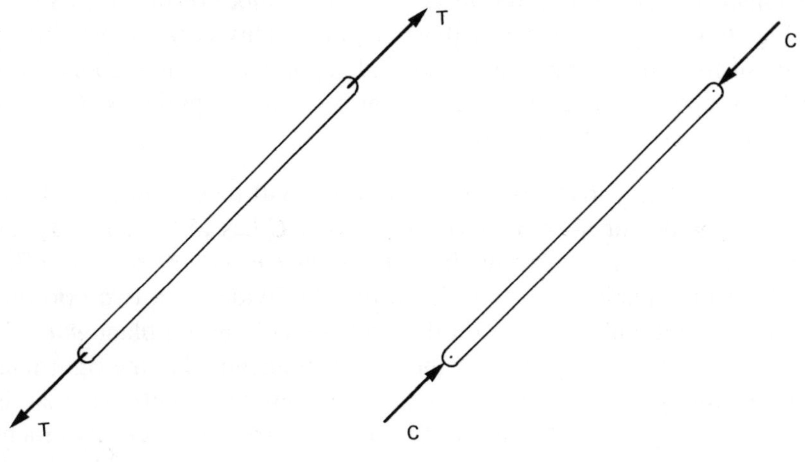


Figure 6.3

each other. In this case, since the individual member is a straight member, the forces are also aligned with the member itself; otherwise, the member would not be in equilibrium.

With regard to the fourth assumption, if the truss is small, the weight of the member, and indeed of the entire truss, when compared to the applied loads is negligible. For a very large truss, however, such as a long bridge, the weight of each member must be included in the calculations. In this case, the weight of the member will be divided by two and applied at each of its ends so that the second assumption is not violated.

## Method of Joints

Since the entire truss is in equilibrium, all its individual members (such as joints) must also be in equilibrium. Isolation of a joint will disturb this equilibrium unless we add the effect of the rest of the truss on this joint. This effect is nothing other than the forces exerted on the joint through the connecting members that meet at the joint. Since the truss members (bars) are two-force members, the direction of the force on each member (bar) is aligned with that member. Now, since the member (bar) is connected to the joint, there is an interacting force between the bar and the joint, and, according to Newton's third law, these forces have equal magnitudes but opposite directions. The following example demonstrates the force analysis of a truss using the method of joints.

EXAMPLE 1: Determine the force in each member of the truss shown in Fig. 6.4(a) by using the method of joints. This very simple truss is made of five equal-length members and supported at pin support *A* and rocker support *B*. The external load of 1000 lb is applied at *C* and is normal to member *BC* as shown.

SOLUTION: Let's start with joint *C* (later it will be explained why one selects a particular joint to start with). Joint *C* has to be isolated from the rest of the truss by introducing a fictitious cut, *mn*, as shown in Fig. 6.4(b). Consequently, the truss is artificially divided into two portions, one on the left and the other on the right side of cutting plane *mn*. The effect of the left-hand portion of the truss is transmitted to the right-hand portion through the members connecting these two portions, that is, members *DC* and *BC*. Isolation of joint *C* requires that these two members be cut, resulting in the appearance of internal forces at the cutting

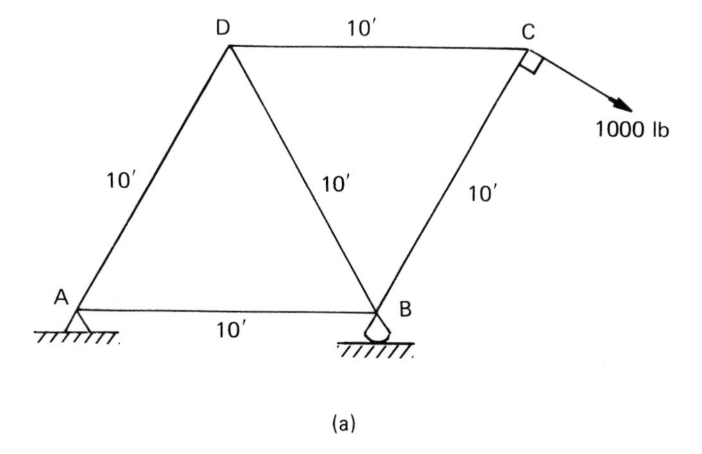


(a)

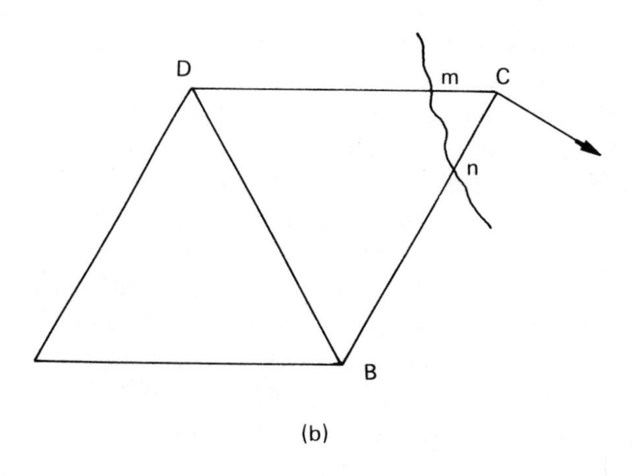


(b)

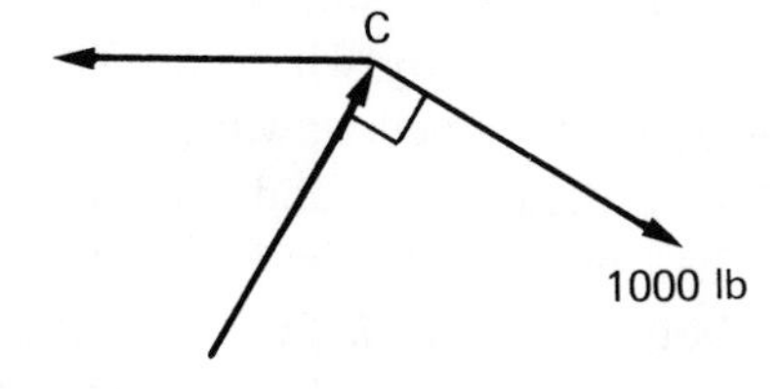


(c)

**Figure 6.4**

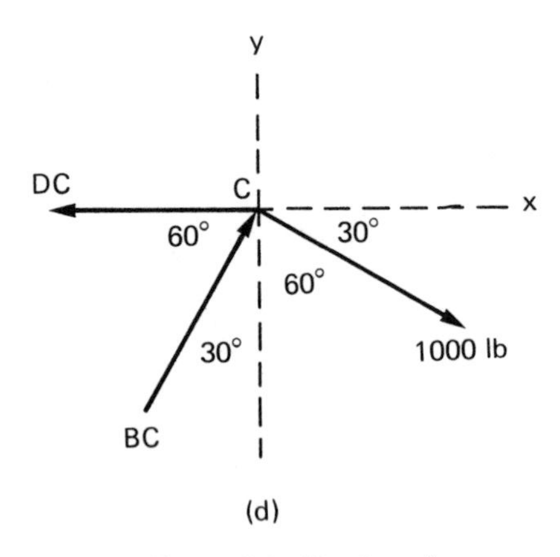


**Figure 6.4.** *(Continued)*

points. These forces are the forces that already existed in the members internally before they were cut at points *m* and *n*. (In actuality, the cutting plane passes through point *C*, and points *m* and *n* are immediately next to *C*.) Once member *DC* is split at *m*, the portions of the truss on both sides of *m* move against each other, thereby disturbing the equilibrium. In order to reestablish it, the internal force initially present in member *DC* now has to be applied externally.

In isolating a portion of a truss from the main body, therefore, we may have to cut some members; by doing so, the internal force within each member appears as a new force. This force is the exact force that that particular member was already carrying under the action of the external forces on the truss. To solve a truss is really to do no more than find these internal forces to which each member of the truss is inherently subject.

The free-body diagram of joint *C* is shown in Fig. 6.4(c). The horizontal force to the left represents the interacting force between member *DC* and joint *C*; this is the force by which joint *C* is being pulled by member *DC*. Similarly, the inclined upward force to the right is the interacting force between member *BC* and joint *C*, the force by which the latter is being pushed by the former. Note that we have arbitrarily assigned directions to both these forces, although, in general, the correct direction of such forces cannot be predicted ahead of time. As will be shown later, it is not necessary to predict the correct direction at this stage. Usually, we assume a certain direction to be the correct one. If our assumption proves to be correct, we will end up with a positive value for that force; a negative value, on the other hand, will indicate that the assumed direction was incorrect.

For bookkeeping purposes, we should name these unknown forces for easy identification. Since the horizontal force is caused by member *DC* and the inclined force by member *BC*, we shall therefore label them as *DC* and *BC*, respectively, as shown in Fig. 6.4(d). Notice that since an external force of 1000 lb was applied at joint *C*, it should also appear in the free-body diagram of joint *C*.

Now, the problem is to find unknown forces *DC* and *BC*. Because we know joint *C* to be in equilibrium, we can use the equilibrium equations. Notice three forces are involved in all (one known force and two unknowns), and these forces are concurrent forces (that is, they all pass through a common point). Since the force equilibrium equations of Chap. 4 ($\Sigma F_x = 0$ and $\Sigma F_y = 0$) are sufficient to solve such a problem, we should first choose a coordinate system. The presence of two unknowns means that we will need two equations. Although we can choose practically any coordinate system we might desire, certain choices might simplify the calculations. A little reflection will reveal that taking one of the axes (such as the $y$ axis) normal to one of the two unknown forces (such as *DC*) and using the equation, $\Sigma F_y = 0$, will offer us a single equation with only one unknown (since *DC* is normal to the $y$ axis and thus has no $y$ component). Since the angle between *DC* and *BC* is known from the truss as shown in Fig. 6.4(a)—triangle *DCB* is obviously an equilateral triangle—we have

$$+\uparrow \Sigma F_y = 0$$

$$BC \cos 30^\circ - 1000 \cos 60^\circ = 0$$

$$BC = 1000 \frac{\cos 60^\circ}{\cos 30^\circ}$$

$$= 577 \text{ lb}$$

To find *DC*, a new coordinate axis can be chosen normal to *BC*, or $\Sigma F_x$ in Fig. 6.4(d) can be used, as follows:

$$\overset{+}{\rightarrow} \Sigma F_x = 0$$

$$-DC + BC \cos 60^\circ + 1000 \cos 30^\circ = 0$$

Substituting for $BC$ from above, we have

$$-DC + 577 \cos 60^\circ + 1000 \cos 30^\circ = 0$$

$$DC = 577 \cos 60^\circ + 1000 \cos 30^\circ$$

$$= 1155 \text{ lb}$$

Since this solution results in positive values for both $BC$ (+577) and $DC$ (+1155), it implies that the initial assumptions for the directions of the unknown forces were correct ones.

These being the forces applied to joint $C$ by members $DC$ and $BC$, we may be sure that joint $C$ will apply equal forces of opposite direction to both $DC$ and $BC$ because of the action–reaction principle of Newton's third law. This can be shown by isolating these two members, as shown in Fig. 6.5. Since $DC$ is a two-force member, once it is pulled at $C$ by a 1155-lb force, it also has to be pulled at $D$ with the same amount of force in order to be in equilibrium. Similar treatment should be applied to member $BC$. Figure 6.5 shows the free-body diagram of joint $C$ as well as its joining members $DC$ and $BC$.

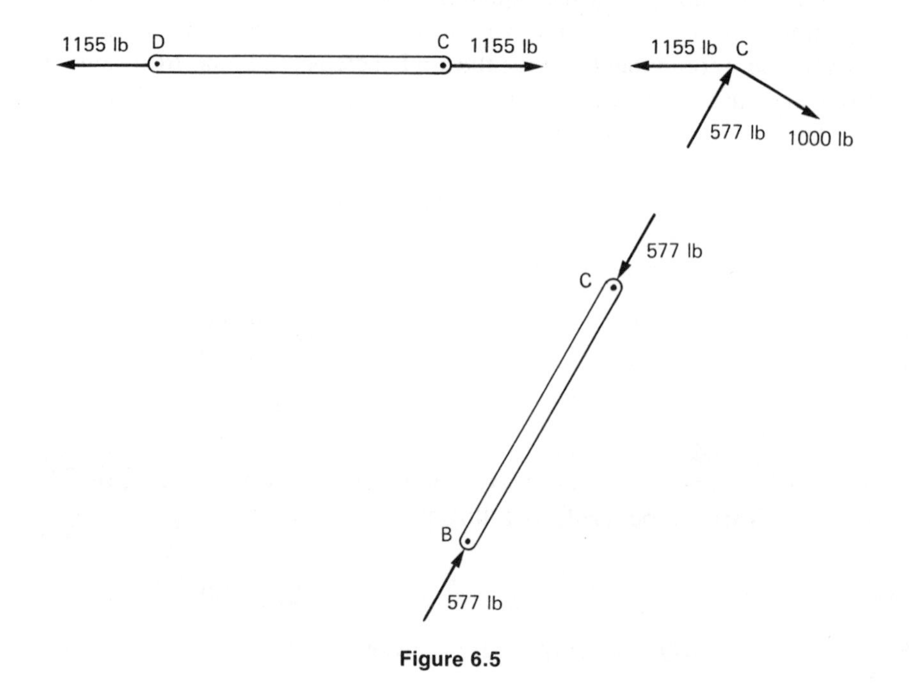


Figure 6.5

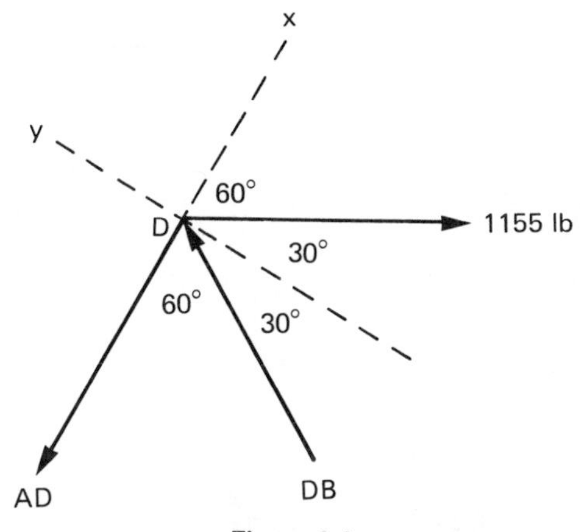


Figure 6.6

The next step is to solve joint $D$ in a similar manner. Figure 6.6 shows the free-body diagram of joint $D$. Since three members join at $D$, there will be three forces, one for each member and aligned with it. Forces $AD$ and $DB$ are unknown forces. The horizontal force should be equal to 1155 lb and its direction should be to the right because it should compensate for the force shown at $D$ in Fig. 6.5.

We choose the $y$-coordinate axis to be normal to $AD$. Then,

$$+\uparrow \Sigma F_y = 0$$

$$DB \cos 30° - 1155 \cos 30° = 0$$

$$DB = 1155 \text{ lb}$$

Moreover,

$$\overset{+}{\rightarrow} \Sigma F_x = 0$$

$$-AD + DB \cos 60° + 1155 \cos 60° = 0$$

Substituting $DB$ from above,

$$-AD + 1155 \cos 60° + 1155 \cos 60° = 0$$

$$AD = 1155 \text{ lb}$$

Again, both assumed directions have proved to be correct.

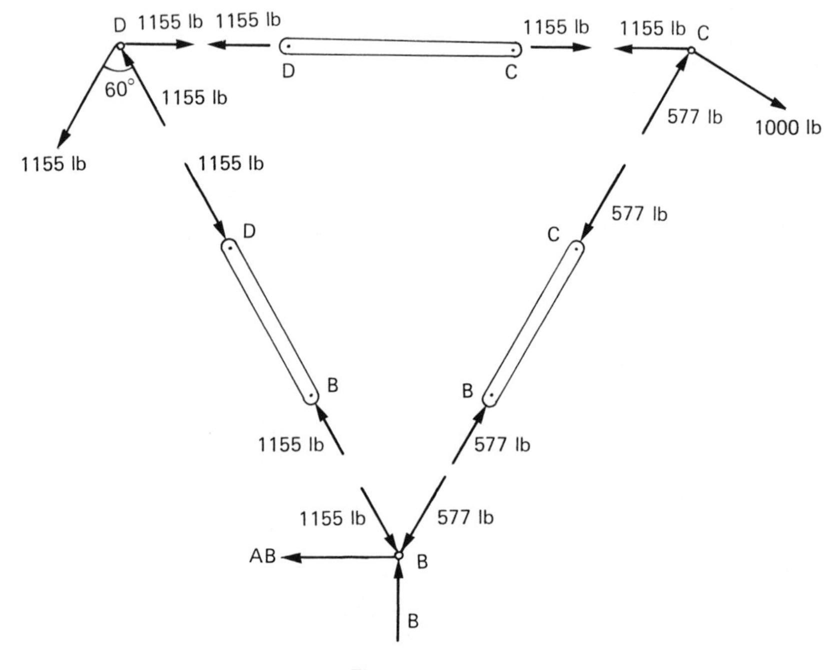


Figure 6.7

Next, we will examine joint $B$. Notice that joint $A$ cannot be solved at this stage because it has three unknowns (one because of member $AB$ and two because of the pin-reaction components at $A$). To draw the free-body diagram of joint $B$, we need to transfer those forces resulting from members $DB$ and $CB$ that were found earlier to joint $B$. This process is shown in Fig. 6.7. The two unknown forces at $B$ are the horizontal force, $AB$, resulting from member $AB$, and the vertical force, $B$, which is the reaction of the truss at $B$. Since support $B$ has a rocker, this reaction has only a vertical component. The free-body diagram of joint $B$ is given in Fig. 6.8(a).

Summing $F_x$, we have

$$\overset{+}{\rightarrow} \Sigma F_x = 0$$

$$-AB + 1155 \cos 60° - 577 \cos 60° = 0$$

$$AB = 289 \text{ lb}$$

Summing $F_y$, we have

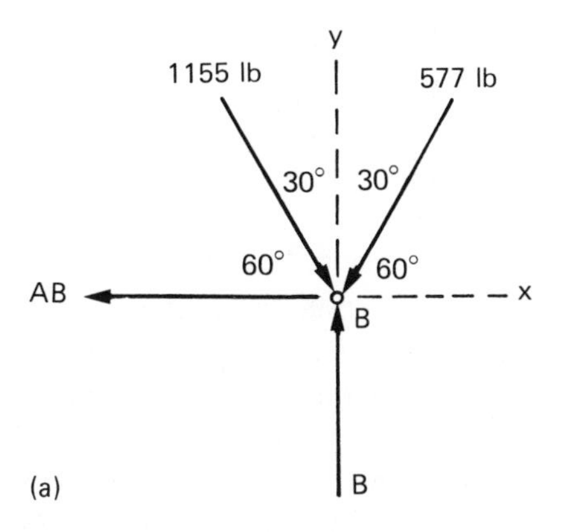


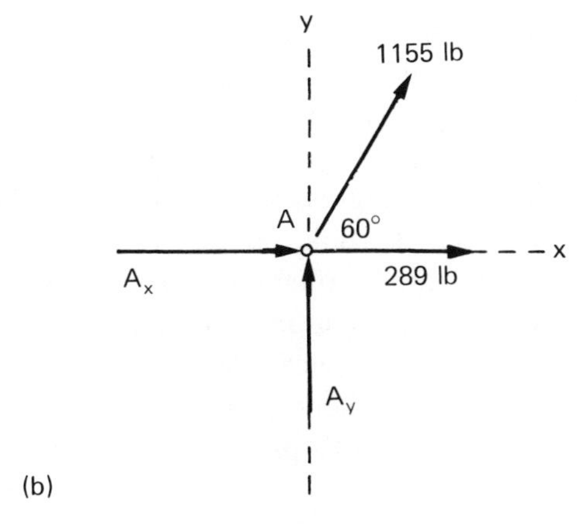


**Figure 6.8**

$$+\uparrow \Sigma F_y = 0$$

$$B - 1155 \cos 30° - 577 \cos 30° = 0$$

$$B = 1500 \text{ lb}$$

Similarly, the free-body diagram of joint $A$ can be drawn (Fig. 6.8b) and the unknown forces calculated, as follows:

$$\overset{+}{\rightarrow} \Sigma F_x = 0$$

$$A_x + 1155 \cos 60° + 289 = 0$$

$$A_x = -866 \text{ lb}$$

The vertical component of the reaction at $A$ can be calculated as:

$$+\uparrow \Sigma F_y = 0$$

$$1155 \cos 30° + A_y = 0$$

$$A_y = -1000 \text{ lb}$$

The fact that these calculations result in negative values for $A_x$ and $A_y$ indicates that the assumed directions were not correct.

As noticed from previous calculations, all internal forces resulted in positive values. This means that the directions for all forces except $A_x$ and $A_y$ were correctly assumed prior to calculations. Obviously, this selection was made intentionally. For some simple problems, the correct direction of the internal forces can be predicted merely from the basic physics of the problem. For example, for the truss shown in Fig. 6.4(a), it is fairly obvious that member $DC$ will be under tension and member $BC$ under compression because of the applied load shown. Similarly, since support $B$ (a rocker) has no resistance in the horizontal direction, the applied load at $C$ will compress members $AB$ and $CB$, but member $AD$ will be under tension. It would be nice if the correct directions of internal forces could always be predicted, of course, but the fact that arbitrary directions can be chosen and corrected later by means of the sign determined is still a decided advantage.

Figure 6.9 shows all the members of the truss with their corresponding tensile or compressive members as previously obtained.

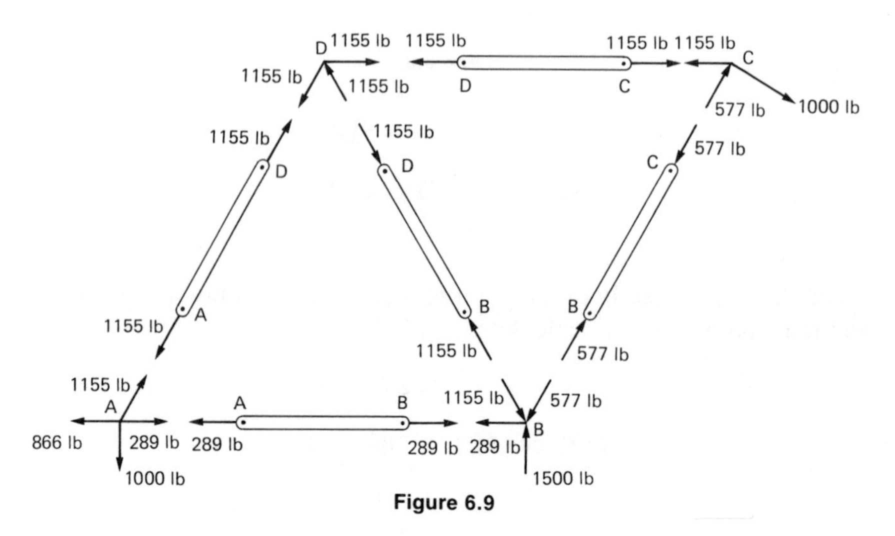


**Figure 6.9**

This problem has demonstrated the method of joints in detail. It was solved at length for purposes of demonstration. In the following example, another problem will be solved using the method of joints, but most of the intermediate steps will be left out.

EXAMPLE 2: Determine the force in each member of the truss shown in Fig. 6.10(a) using the method of joints.

SOLUTION: Inspection of each joint separately will establish the absence of any joint with only one or two unknowns; in fact, joint $A$ has four unknowns and the other joints have three unknowns each. In problems like this, the support reactions must be calculated first. Figure 6.10(b) shows the free-body diagram of the entire truss. Since joint $A$ is a hinge support, it has two reaction components, and since joint $B$ is a roller support, it has only a vertical reacting force. Thus the summation of forces, $\Sigma F_x = 0$, will result in $A_x = 0$. Taking the moment about $A$ will result in the following:

$$+ \circlearrowright \Sigma M_A = 0$$

$$(1000 \times 5) - (C \times 10) = 0$$

$$C = 500 \text{ N}$$

Since $\Sigma F_y = 0$,

$$A_y - 1000 + C = 0$$

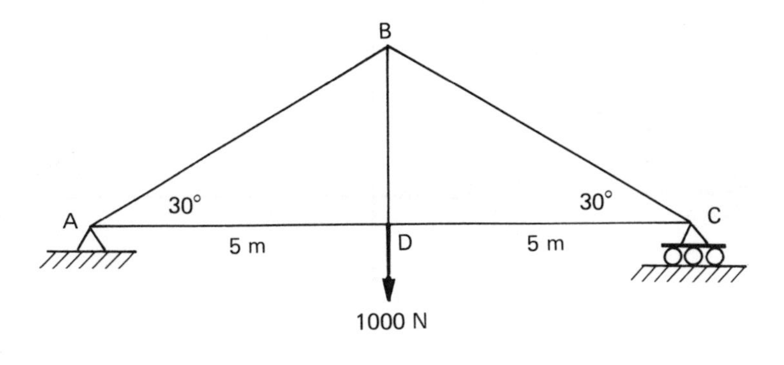


(a)

**Figure 6.10**

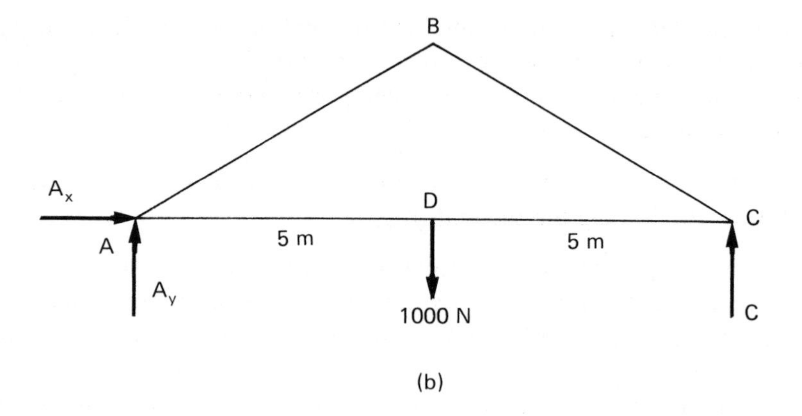


(b)

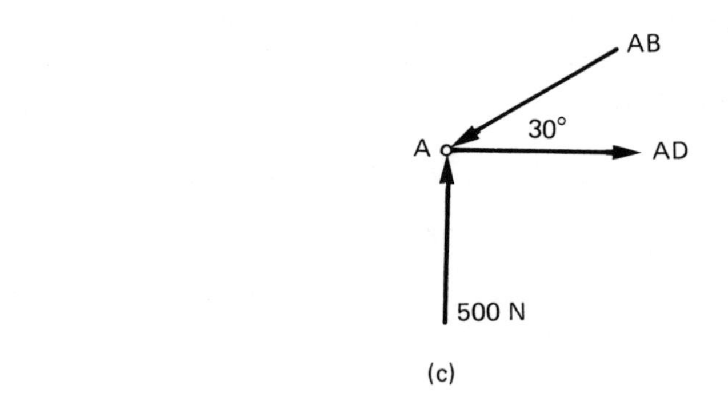


(c)

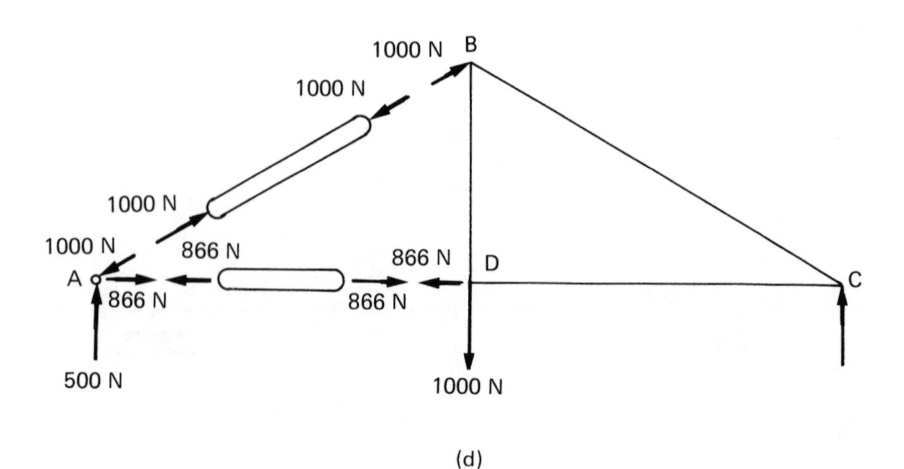


(d)

**Figure 6.10.** (*Continued*)

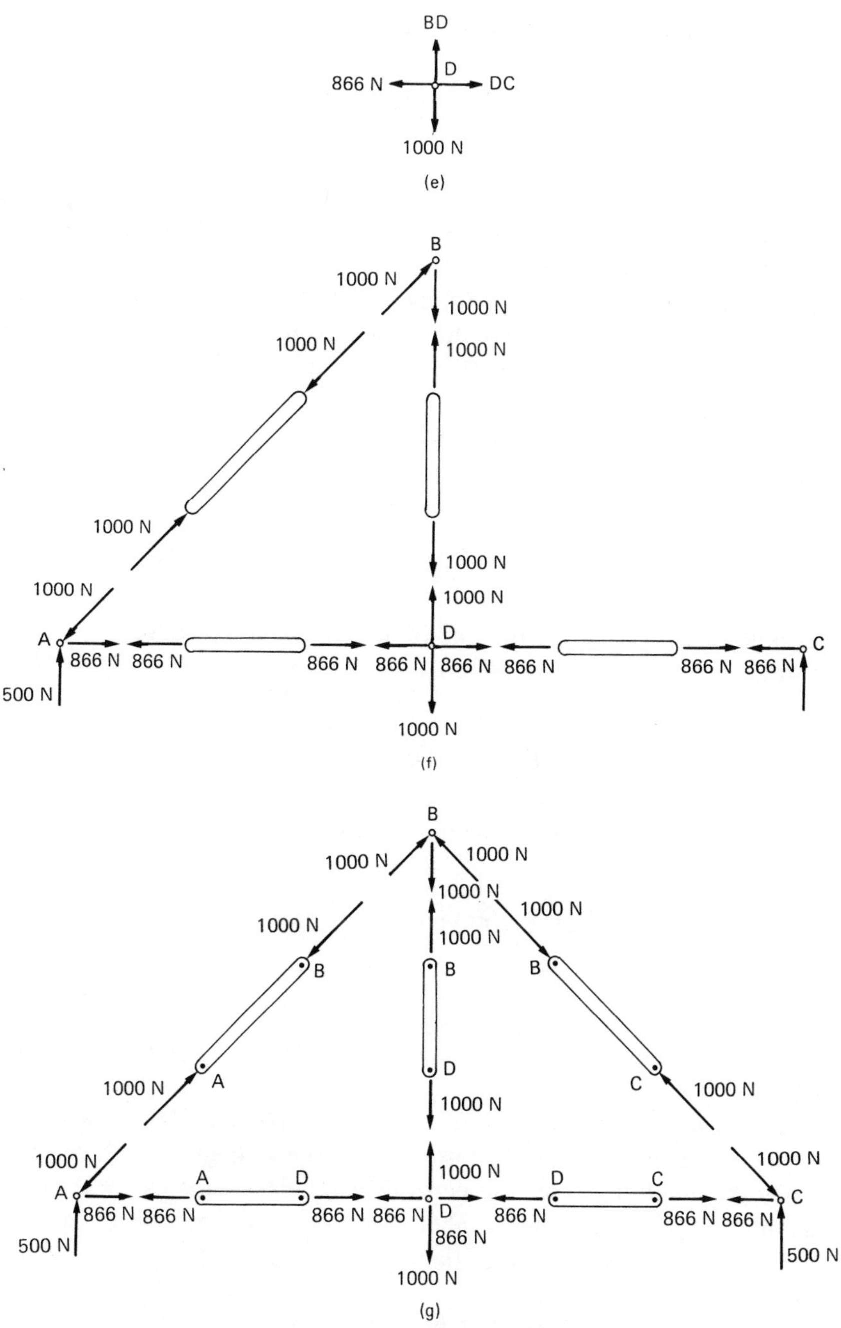


**Figure 6.10.** (*Continued*)

Using the value of $C$ obtained above,

$$A_y - 1000 + 500 = 0$$
$$A_y = 500 \text{ N}$$

(Notice that these values could have been obtained by considering only the symmetry of the truss.) At this point, we can attack the problem at either joint $A$ or $C$. Figure 6.10(c) shows the free-body diagram of joint $A$. Making use of $\Sigma F_y$ will give the value of force $AB$ (the $y$ axis is taken along the line of action of the 500-N force) as follows:

$$+\uparrow \Sigma F_y = 0$$
$$500 - AB \cos 60^\circ = 0$$
$$AB = 1000 \text{ N}$$

Taking the $x$ axis along the line of action of $AD$, we have

$$\Sigma F_x = 0$$
$$AD - AB \cos 30^\circ = 0$$

Substituting the value of $AB$ obtained above,

$$AD - 1000 \cos 30^\circ = 0$$
$$AD = 866 \text{ N}$$

The forces applied by members $AB$ and $AD$ to joint $A$ are thus 1000 N and 866 N, respectively. Joint $A$ consequently will react with the same forces but in opposite directions. Since both $AB$ and $AD$ are two-force members, equal but opposite forces will exist at their ends in order to satisfy the equilibrium requirements of each. These forces are the ones that joints $B$ and $D$ apply to members $AB$ and $AD$. In turn, these members apply forces to joints $B$ and $D$ that are opposite in direction, as shown in Fig. 6.10(d).

Figure 6.10(e) shows the free-body diagram of joint $D$, which can be analyzed as follows:

$$\overset{+}{\rightarrow}\Sigma F_x = 0$$

$$DC - 866 = 0$$

$$DC = 866 \text{ N}$$

Likewise,

$$+\uparrow\Sigma F_y = 0$$

$$DB - 1000 = 0$$

$$DB = 1000 \text{ N}$$

Similarly, the forces applied to member *DB* and *DC* will be opposite to the forces applied to joint *D*, as shown in Fig. 6.10(f). Notice that the forces on member *DC* will be the same as the forces on member *AD* because of symmetry. In like manner, member *BC* will be under the same loading as member *AB*.

The complete solution of this truss is illustrated in Fig. 6.10(g). Note that members *AB* and *BC* are compressive members, whereas other members are tensile members.

## Method of Sections

Forces on truss members can also be determined by using the method of sections, a method particularly suitable when only forces on selected members of the truss are needed. In the method of joints, as seen in the previous section, the force on a truss member located in the middle of a truss, say, can usually be determined only after the forces on many members have been determined. This can become a tedious and very lengthy procedure, especially for large trusses. In particular, checking calculations for certain forces by using the method of joints may almost require doing the problem over, taking about the same amount of time as the solution of the problem itself. The method of sections represents a shortcut method for solving or for checking and inspecting a previously solved problem. Occasionally both methods are used because, for some members, the method of joints, and, for other members, the method of sections is the faster and easier one.

In the method of sections, the truss is ''sectioned'' (that is, cut) by means of a fictitious cutting plane into two portions. The equilibrium of only one portion, or section, will be considered, and the cutting plane itself is chosen so as to cut the members of interest. In this method, however, the number of unknown forces should not exceed three, which is equal to the number of independent equilibrium equations.

EXAMPLE 3: Determine the force in members *CE*, *CF*, and *DF* of the truss shown in Fig. 6.11(a) using the method of sections.

SOLUTION: Since we are interested in members *CE*, *CF*, and *DF*, our cutting plane should cut all three members if possible. Such a cut is shown in Fig. 6.11(b), which produces two partial sectioned trusses, one to the right of section a–a and one to the left. Either portion can be used to solve the problem. The first step is to draw the free-body diagram of the portion selected. We will choose the portion to the right of section a–a—although it is the larger of the two—and for the following reason. Since the truss is supported at both *A* and *B*, a free-body diagram of the portion to the left of section a–a would include *seven* unknowns, four of them for the reaction components, whereas a free-body diagram of the portion to the right will have only three unknowns, these being the very forces in which we are interested.

Each member of the truss carries a certain amount of load that is called *internal load*. Upon sectioning, the equilibrium is disturbed, and a force equal to the internal force on each member appears at the cutting point as an external force to keep the section in equilibrium. In this example, three such forces appear for members *CE*, *CF*, and *DF*, as shown in Fig. 6.11(c). These three forces collectively constitute the entire effect of the portion of the truss to the left of a–a on the portion to the right. Notice that these forces are aligned along the members and that their directions are assumed as before, whether tensile or compressive. (An incorrect assumption will result in a negative value for the force involved.)

Now, consider the free-body diagram of the portion to the right of cut a–a in Fig. 6.11(c). To find the three unknown forces—*CE*, *CF*, and *DF*—three equations of equilibrium are needed, as follows:

$$+\uparrow \Sigma F_y = 0$$

$$CF \cos 45^\circ - 1000 - 1000 = 0$$

$$CF = 2828 \text{ lb}$$

$$+\circlearrowright \Sigma M_F = 0$$

$$(1000 \times 36) + (1000 \times 24) - (CE \times 12) = 0$$

$$CE = 5000 \text{ lb}$$

$$\Sigma F_x = 0$$

$$DF - CF \cos 45^\circ - CE = 0$$

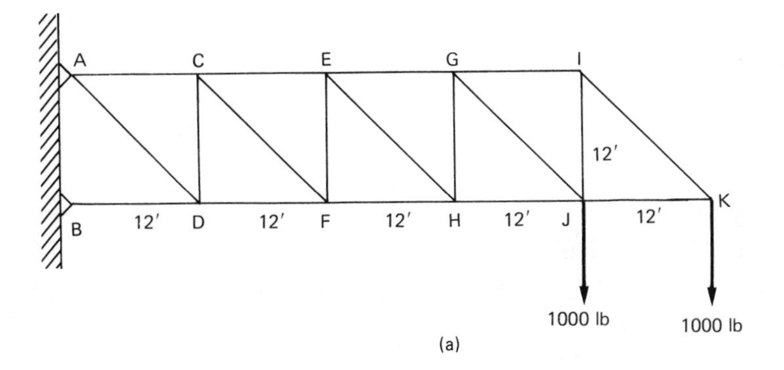

(a)

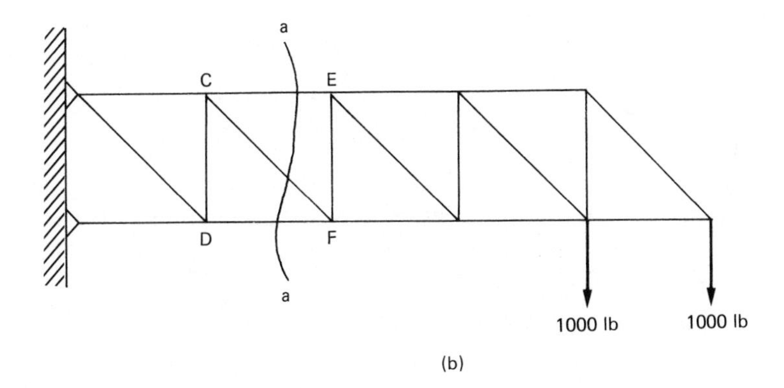

(b)

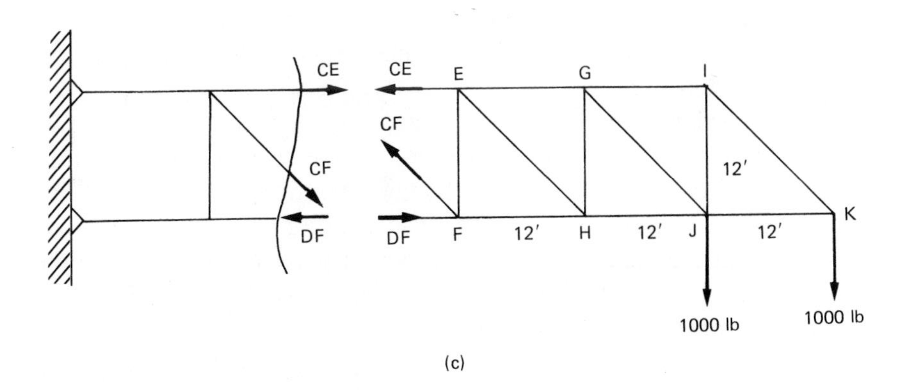

(c)

**Figure 6.11**

Substituting the known values,

$$DF - 2828 \cos 45° - 5000 = 0$$

$$DF = 7000 \text{ lb}$$

If this problem were to be solved using the method of joints, it would involve many more calculations. In fact, one would have to start with joint $K$ and march joint by joint toward joints $E$ and $F$ in order to be able to solve the latter since all the in-between joints would have to be analyzed as well. The superiority of the method of sections in a problem like this is obvious.

Sometimes it is not possible to determine all the required unknowns with a single cut, as illustrated in the following example.

EXAMPLE 4: Determine the force in members $BC$, $CD$, and $CF$ of the truss shown in Fig. 6.12(a) using the method of sections.

SOLUTION: The reactions at $A$ and $E$ are needed. They can be calculated using the free-body diagram of the entire truss and taking moments about $A$ and $E$, from which $A = 5$ kips and $E = 25$ kips, as shown in Fig. 6.12b.

In this case, a single cut will not be able to determine the forces in all three members, $BC$, $CD$, and $CF$. Notice that although it is possible to cut all three members with one cut, such as a–a, it will not be possible to solve for three unknowns because the necessity of cutting a fourth member, $CG$, as well will result in a total of four unknowns. Therefore, two independent cuts have to be introduced, as shown in Fig. 6.12(b) by cuts b–b and c–c.

Now consider the free-body diagram of the portion to the left of cut b–b, as shown in Fig. 6.12(c). Since b–b cuts only three members, there will be three unknowns, but we are interested in $BC$ only, and, if possible, we should solve for $BC$ without solving for $CG$ and $GF$. This can be easily done by taking the moment about point $G$. Since both other unknown forces pass through point $G$, unknowns $CG$ and $GF$ will not enter the moment equation. Thus,

$$+\circlearrowleft \Sigma M_G = 0$$

$$(5 \times 10) + [BC \times (10 \sin 60°)] = 0$$

$$BC = -5.77 \text{ kips}$$

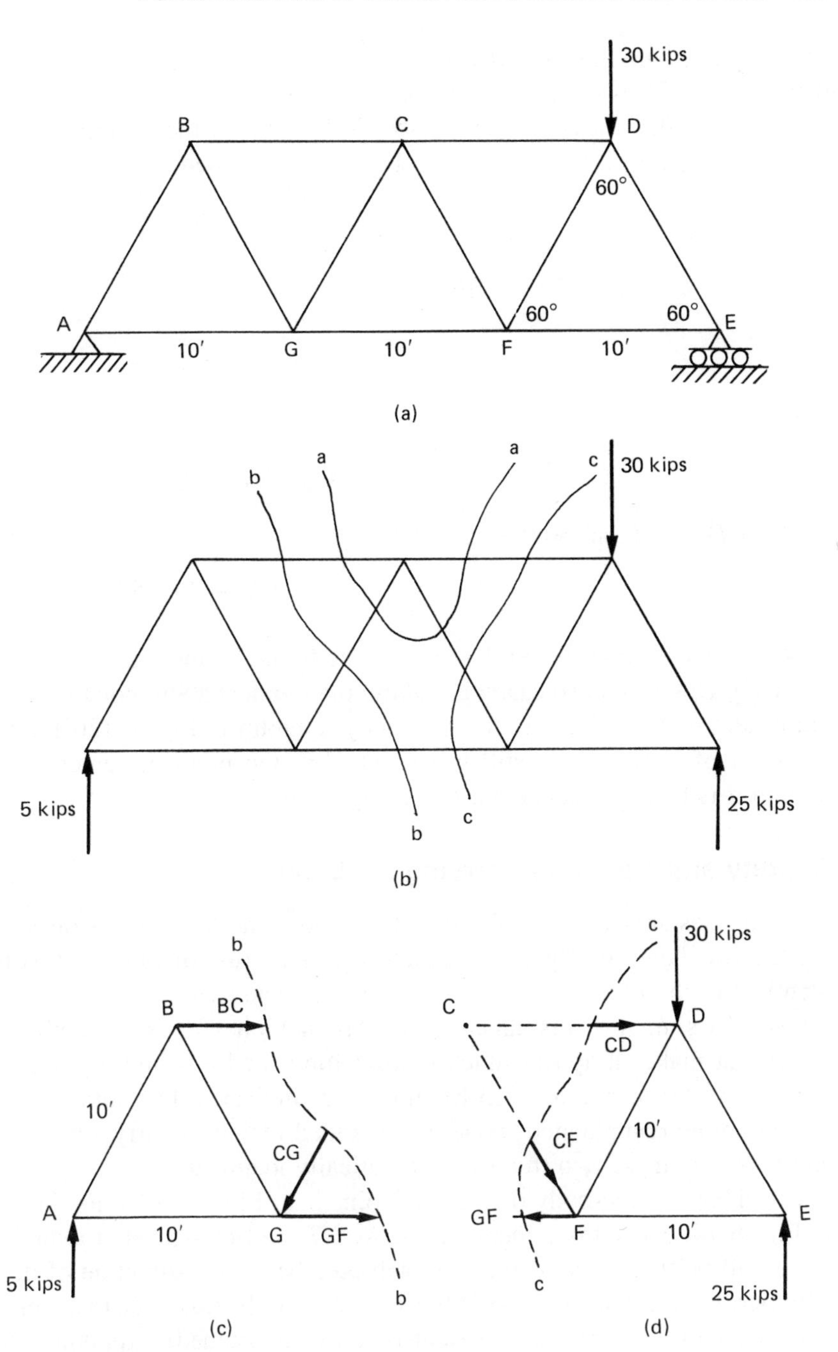
30 kips
B
C
D
60°
A
60°
60°
E
10′
G
10′
F
10′
(a)
b
a
a
c
30 kips
5 kips
25 kips
b
c
(b)
b
B
BC
10′
CG
A
10′
G
GF
5 kips
b
(c)
c
30 kips
C
D
CD
10′
CF
GF
F
10′
E
c
25 kips
(d)

Figure 6.12

The negative sign indicates that the assumed direction was incorrect and that member $BC$ is really a compressive member.

Now consider cut c–c. The free-body diagram of the portion of the truss to the right of c–c is shown in Fig. 6.12(d). Then,

$$+\uparrow\Sigma F_y = 0$$

$$25 - 30 - CF \cos 30° = 0$$

$$CF = -5.77 \text{ kips}$$

and

$$+ \circlearrowleft \Sigma M_F = 0$$

$$CD \times (10 \sin 60°) - 25 \times 10 + 30 \times 5 = 0$$

$$CD = 11.55 \text{ kips}$$

Now consider that force $GF$ must also be found (without knowing $CD$ and $CF$). One can do so easily by taking the moment about point $C$. The point here is that, although the free-body diagram in Fig. 6.12(d) is to the right of c–c, we can still use point $C$ as the moment center even though it is located outside the free-body diagram.

## Rigidity and Members Carrying No Load

Often trusses are much simpler than they appear at the outset. Consider the trusses shown in Figs. 6.13(a) and (b). The truss of Fig. 6.13(a) is identical to the truss analyzed in Example 3. The truss of Fig. 6.13(b) is basically similar to this same truss except that it has 10 extra members, a fact that makes it appear much more complicated. Careful consideration of the truss in Fig. 6.13(b), however, indicates that, although it appears more complicated, none of the added members carry any load, and therefore these two trusses are practically identical.

Consider the free-body diagram of joint $U$ in Fig. 6.13(b), as shown in Fig. 6.14. Since three members—$UK$, $UT$, and $UJ$—meet at joint $U$, there will be three forces, one for each member in the direction of that member. Notice that, because of the geometry of the truss, the two members, $UJ$ and $UK$, are colinear (that is, they are aligned). Applying the force equilibrium equation, $\Sigma F_y = 0$, will result in force $UT$ being zero,

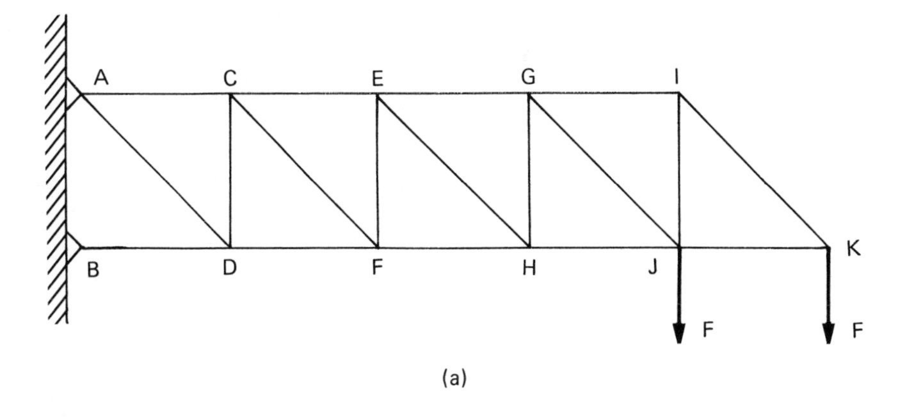


(a)

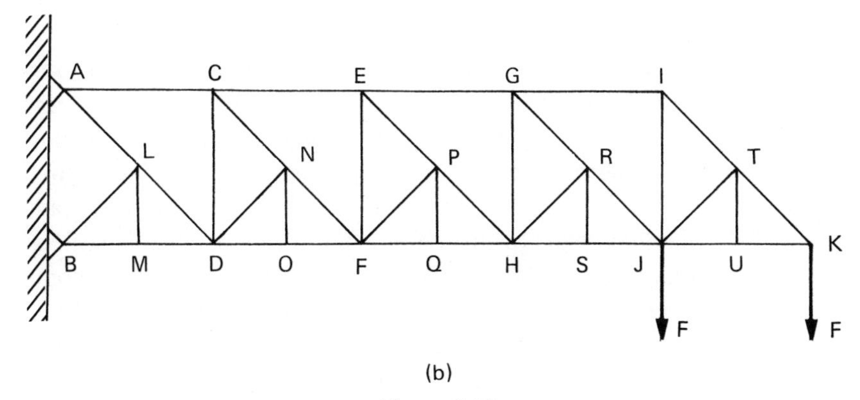


(b)

**Figure 6.13**

which indicates that member *UT* carries no load and, for the purpose of analysis, can be removed.

Now consider the free-body diagram of joint *T* in Fig. 6.13(b), as shown in Fig. 6.15 after zero-force member *UT* has been removed. Let's take the coordinate axes so that the *x* axis coincides with colinear forces *TI* and *TK* and the *y* axis passes through *T* and is, of course, perpendic-

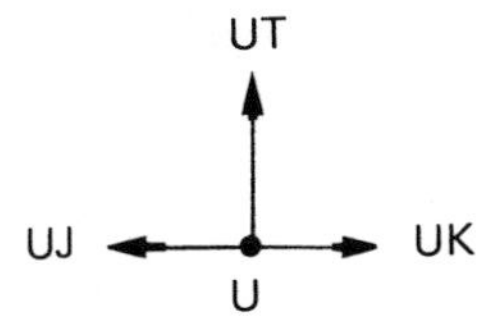


**Figure 6.14**

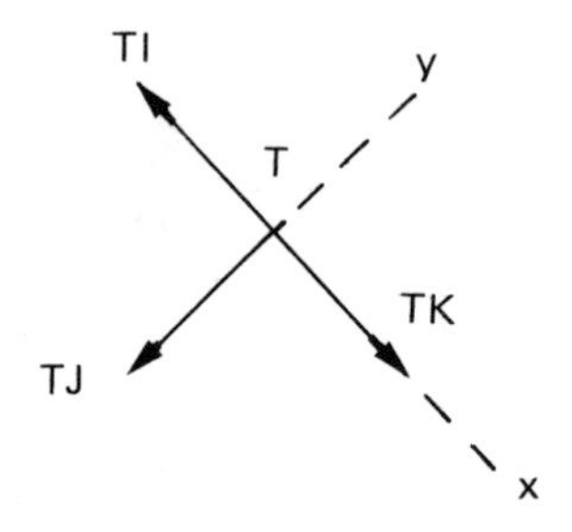

Figure 6.15

ular to the $x$ axis. Applying the equilibrium equation, $\Sigma F_y = 0$, will result in a force $TJ$ of 0. This process can be extended to joints $S$, $R$, $Q$, $P$, $O$, $N$, $M$, and $L$ to show that all ten added members are zero-force members.

It is not really necessary to draw a free-body diagram for each joint and then apply a force equilibrium equation to identify zero-force members. This process can be done quickly by *inspection* only. Identifying zero-force members before the analysis of a truss will considerably simplify that analysis when the truss has some members that carry no load.

Zero-force members make a truss more stable and also act as stiffeners. The addition of such stiffeners may also contribute to more economically designed trusses.

To identify zero-force members easily, inspect each joint with only three joining members and no external force upon it. If two of the members are colinear (or aligned) the third will be a zero-force member, as shown in Fig. 6.16.

Consider the truss shown in Fig. 6.17(a). The free-body diagrams of joints $B$, $D$, and $F$ are shown in Figs. 6.17(b), (c), and (d). In Fig. 6.17(b), the two members that meet at corner joint $B$ are both zero-force members because $BC$ is the only force in the horizontal direction and $BA$ is the only force in the vertical direction and therefore, to maintain equi-

Figure 6.16

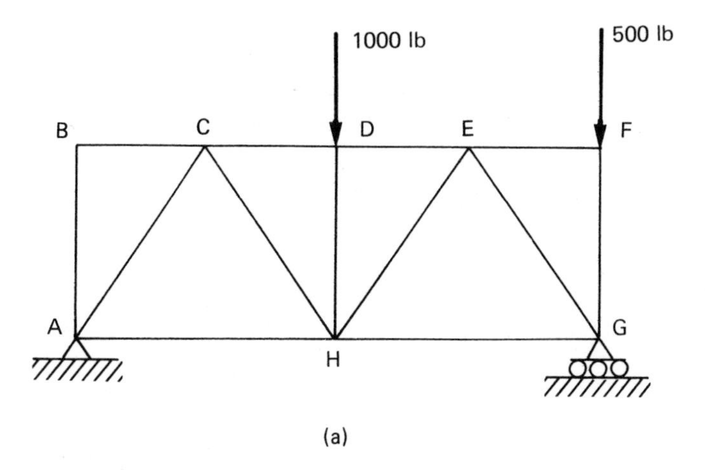


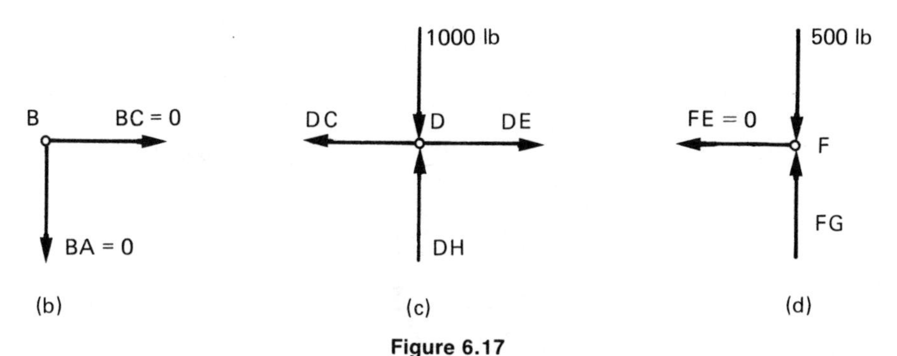


**Figure 6.17**

librium, both forces must be equal to zero. Notice that even if the angle between *BC* and *BA* were not a right angle the same results would be obtained, and both members would still be zero-force members.

Now consider joint *D* as shown in Fig. 6.17(c). Because of the applied 1000-lb force at *D*, *DH* is not a zero-force member. On the other hand, although there is an external force (500 lb) on joint *F* in Fig. 6.17(d), member *FE is* a zero-force member because the external force is aligned with member *FG*. Notice that if the external 500-lb force at joint *F* were applied horizontally, *FG* would be the zero-force member, not *FE*. Finally, if the external force at *F* were applied in an inclined manner, neither one of the members, *FE* or *FG*, would be a zero-force member.

EXAMPLE 5: Identify the zero-force members in the truss shown in Fig. 6.18(a) by inspection.

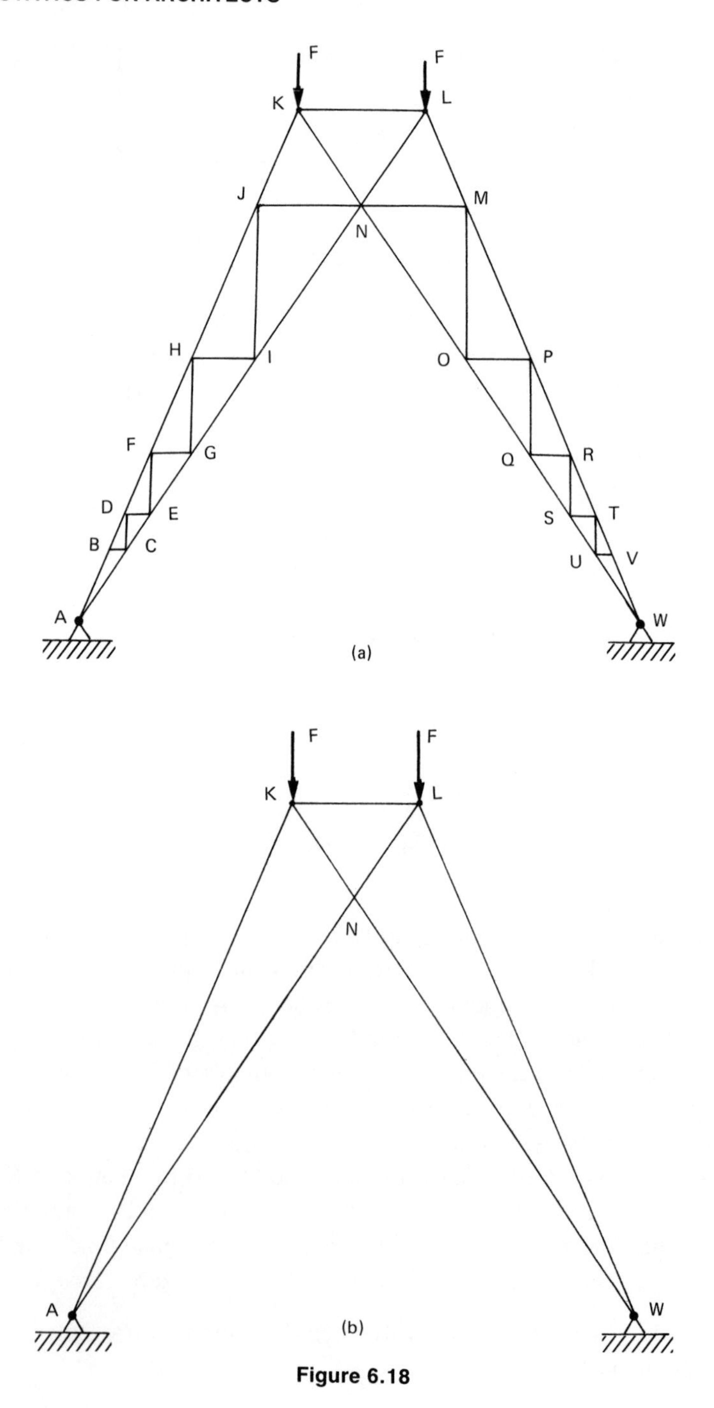


Figure 6.18

SOLUTION: Inspection of joint $B$ first identifies zero-force member $BC$. Then, inspection of joint $C$ with zero-force member $BC$ removed reveals that member $CD$ is a zero-force member, and so on. Figure 6.18(b) shows all the zero-force members removed.

## 6.3 FRAMES

Frames are another type of structure that is similar to a truss and capable of carrying external loads. The main difference between trusses and frames is the way that external loads are applied to them. In trusses, members are connected to each other at end joints only; in frames, they may not be. In trusses, external loads are applied at joints only; in frames, they may be applied at any point of any member. The consequence of these two differences is that not all members of frames are two-force members as are all truss members (the latter are either tensile or compressive). Because of this construction, frame members may be subjected to bending as well.

### Analysis of Frames

As it does in trusses, the analysis of frames involves the determination of the forces on each member. The method of solution is known as the *method of members* and involves drawing a free-body diagram for each member as well as for the entire frame so that we can apply the equations of equilibrium and the principle of action–reaction to determine all forces and support reactions.

EXAMPLE 6: Determine the forces in each member of the simple frame shown in Fig. 6.19(a). The three members of this frame are pinned to one another at $B$, $C$, and $E$.

SOLUTION: Support reactions are determined by using the free-body diagram of the entire frame shown in Fig. 6.19(b), as follows:

$$\Sigma F_x = 0$$

$$A_x = 0$$

$$+\circlearrowleft \Sigma M_A = 0$$

$$(4000 \times 17) - (F \times 10) = 0$$

$$F = 6800 \text{ lb}$$

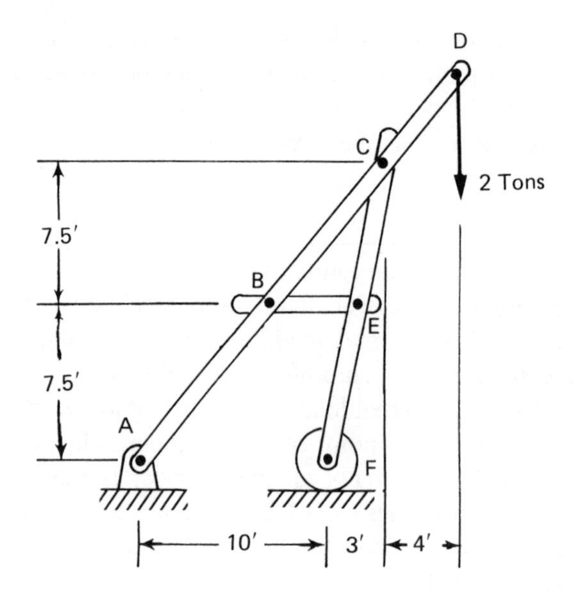


(a)

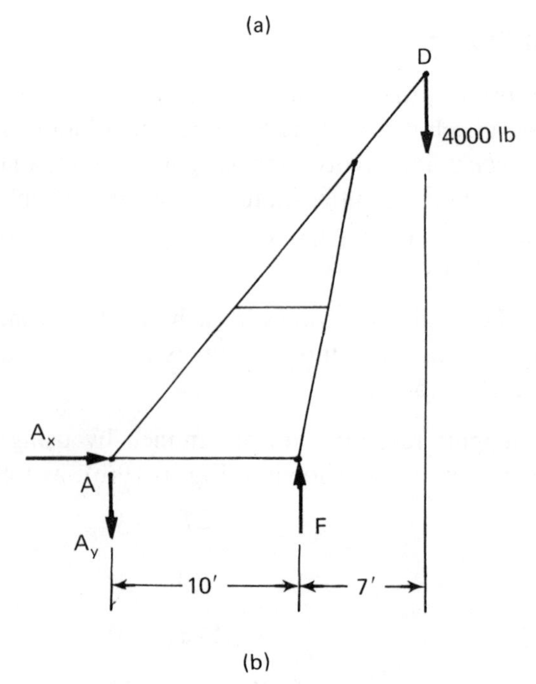


(b)

**Figure 6.19**

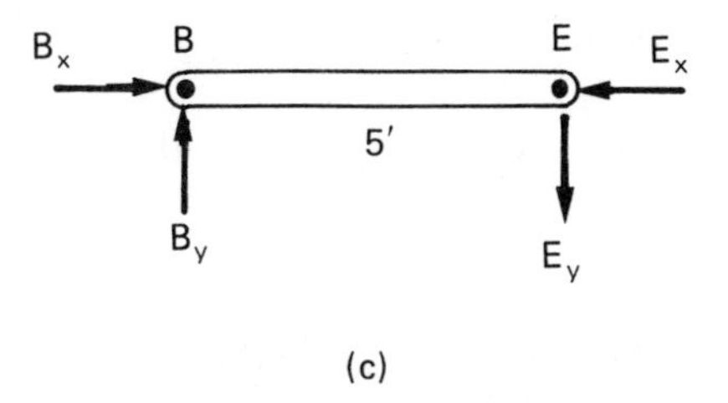


(c)

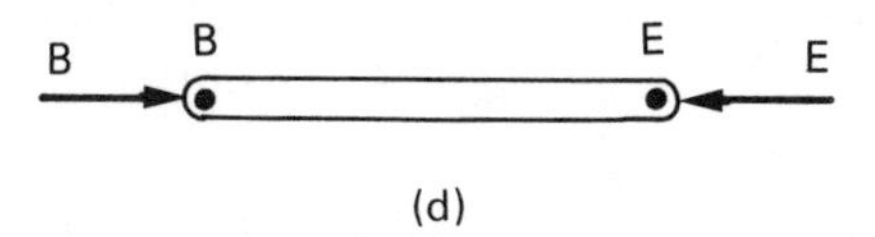


(d)

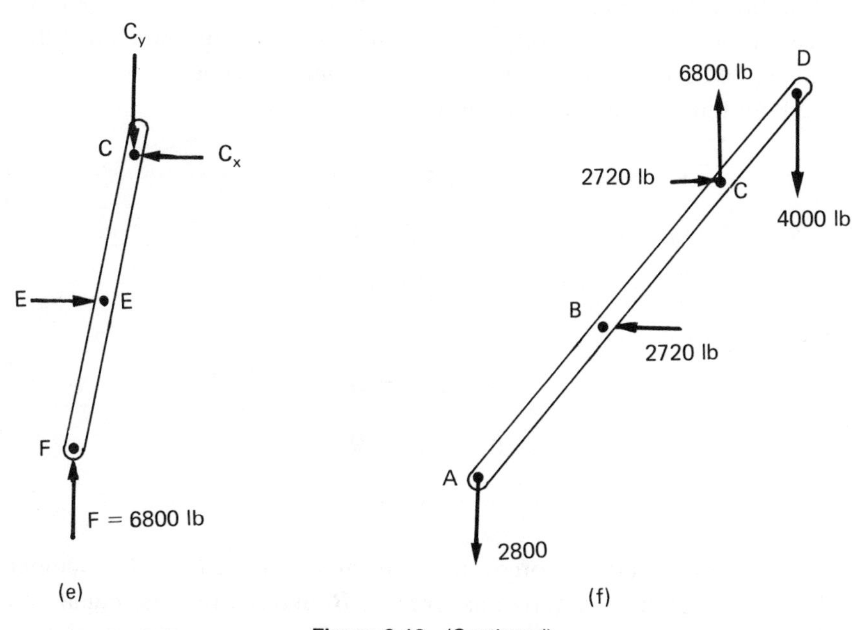


(e)

(f)

**Figure 6.19.** *(Continued)*

Moreover,

$$+\uparrow \Sigma F_y = 0$$

$$-A_y + F - 4000 = 0$$

Substituting for $F$ from the preceding,

$$-A_y + 6800 - 4000 = 0$$

$$A_y = 2800 \text{ lb}$$

The next step is to draw the free-body diagram of each member. Let's start with the simplest of the members, $BE$, as shown in Fig. 6.19(c). This member is connected to member $AD$ at $B$, and the interacting force between them at this point has a vertical as well as horizontal component (because the members are *pinned* at $B$). Similarly, there are two interacting force components at $E$ because member $BE$ is pinned to member $FC$ at $E$. The length of $BE$ is calculated to be 5′ (from the similarity of the two triangles).

For the time being, we cannot determine all four of the unknown forces. Usually in such cases, we calculate the forces that we can at this stage and determine the rest later by using other members.

Taking moment about $B$, we have

$$+\circlearrowleft \Sigma M_B = 0$$

$$E_y \times 5 = 0$$

$$E_y = 0$$

Moreover,

$$+\uparrow \Sigma F_y = 0$$

$$B_y - E_y = 0$$

$$B_y = 0$$

Therefore, both vertical forces are zero, and, since $\Sigma F_x = 0$, member $BE$ is subjected to two horizontal forces, $B_x$ and $E_x$, that are equal and opposite to one another. Notice that these calculations were unnecessary because it is obvious from Fig. 6.19(a) that member $BE$ is a two-force member (first, because it is connected to two other members at two points only and, second, because no external force is applied to it). The final free-body diagram of member $BE$ is redrawn in Fig. 6.19(d), the subscript $x$ having been dropped from forces $B_x$ and $E_x$.

Next consider member $CF$. Its free-body diagram is shown in Fig.

6.19(e). Notice that it is redrawn with the same orientation it had in Fig. 6.19(a). The reaction at point *F* has already been calculated at 6800 lb. Since this member is pin-connected to member *AD* at *C*, two reacting components, $C_x$ and $C_y$, are shown at point *C*. Point *E* is held in common with member *BE*, and, because of the principle of action-reaction, the forces at *E* on both *BE* and *FC* must be the same although in opposite directions.

To crystallize this situation, consider that members *BE* and *CF* were originally pin-connected at *E*. At this point, there existed an interacting force between these two members. Force *E* in Fig. 6.19(d) is the force applied by member *CF* to member *BE*. Because of the action–reaction principle, we should add an equal and opposite force at *E* for the member *CF* shown in Fig. 6.19(e); that's why the same symbol *E* is used for the forces on both members. Another way to look at this is the following: Let's think of bringing member *BE* and pinning it back to member *CF* at *E*; the two forces, *E* (to the left) on *BE* and *E* (to the right) on *CF*, now cancel each other out. Notice that, at this stage, we cannot assume an arbitrary direction for force *E* on member *FC*. Since we have already assumed this direction in Fig. 6.19(d), the direction of *E* in Fig. 6.19(e) must match that of Fig. 6.19(d).

The free-body diagram of member *CF* has only three unknowns, and therefore all can be determined by using the equilibrium equations, as follows:

$$+\uparrow \Sigma F_y = 0$$

$$6800 - C_y = 0$$

$$C_y = 6800 \text{ lb}$$

$$+\circlearrowleft \Sigma M_C = 0$$

$$(6800 \times 3) - (E \times 7.5) = 0$$

$$E = 2720 \text{ lb}$$

$$\Sigma F_x = 0$$

$$E - C_x = 0$$

$$2720 - C_x = 0$$

$$C_x = 2720 \text{ lb}$$

Once $E$ is known, the value of $B$ in Fig. 6.19(d), previously incalculable, can now be evaluated:

$$\Sigma F_x = 0$$
$$B - E = 0$$
$$B - 2720 = 0$$
$$B = 2720 \text{ lb}$$

Our next step is to draw the free-body diagram of member $AD$, as shown in Fig. 6.19(f). The reactions at point $A$ found earlier are indicated there. The forces at $B$ and $C$ used earlier are transferred to $B$ and $C$ merely by changing their directions (because of the principle of action–reaction).

After adding all the known common forces acting on $AD$ from other members [see Fig. 6.19(f)], no further calculation is needed. Use of the equilibrium equations to verify this free-body diagram, however, will be a way to validate our previous calculations, as follows:

$$\Sigma F_x = 0$$
$$2720 - 2720 = 0$$
$$\Sigma F_y = 0$$
$$-2800 + 6800 - 4000 = 0$$
$$+ \quad \Sigma M_A = 0$$
$$(4000 \times 17) - (6800 \times 13) + (2720 \times 15) - (2720 \times 7.5) = 0$$

The fact that all three equilibrium equations are verified indicates that all calculations have been performed correctly.

EXAMPLE 7: Determine the forces in each member of the frame shown in Fig. 6.20(a). All members are pin-connected. The two concentric pulleys are rigidly connected to each other.

SOLUTION: The frame has a pin support at $A$ and a roller support at $B$. Figure 6.20(b) shows the free-body diagram of the entire frame. Deter-

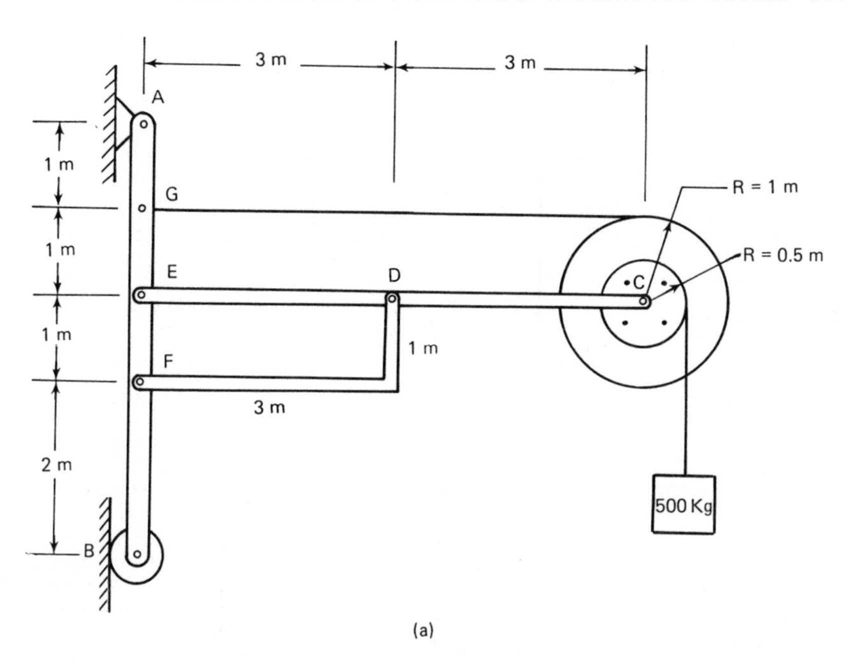


(a)

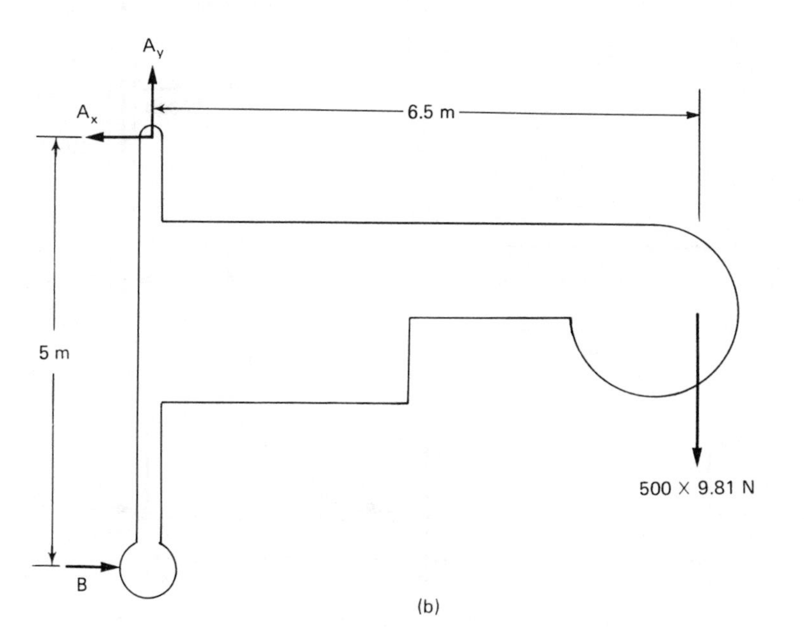


(b)

**Figure 6.20**

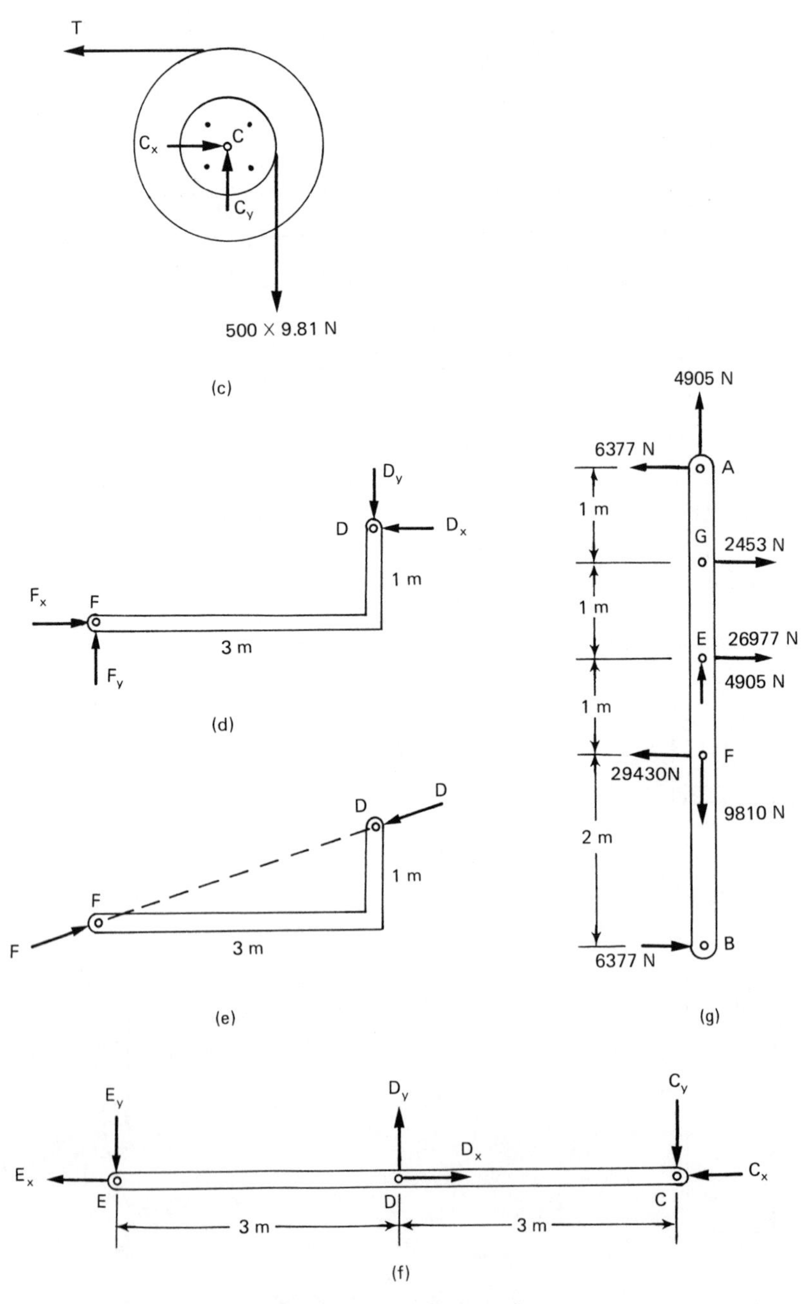


**Figure 6.20.** *(Continued)*

mination of support reactions is shown below:

$$+\uparrow \Sigma F_y = 0$$

$$A_y - (500 \times 9.81) = 0$$

$$A_y = 4905 \text{ N}$$

$$+ \circlearrowleft \Sigma M_B = 0$$

$$(500 \times 9.81 \times 6.5) - (A_x \times 5) = 0$$

$$A_x = 6377 \text{ N}$$

$$\Sigma F_x = 0$$

$$B - A_x = 0$$

$$B - 6377 = 0$$

$$B = 6377 \text{ N}$$

Now consider the free-body diagram of the pulley system shown in Fig. 6.20(c). The two pulleys are rigidly connected to each other and both pinned to member *EC* at *C*. Since *C* is a pin support, there will be two reaction components at *C*. Taking moment about *C* will result in the value of tension force, *T*, as follows:

$$+ \circlearrowleft \Sigma M_C = 0$$

$$(500 \times 9.81 \times 0.5) - (T \times 1) = 0$$

$$T = 2453 \text{ N}$$

$$\Sigma F_x = 0$$

$$C_x - T = 0$$

$$C_x - 2453 = 0$$

$$C_x = 2453 \text{ N}$$

$$\Sigma F_y = 0$$

$$C_y - (500 \times 9.81) = 0$$

$$C_y = 4905 \text{ N}$$

Next, let's solve for member *FD*. The free-body diagram of this member is shown in Fig. 6.20(d). Although at this point there appear to be four unknowns, member *FD* is really a two-force member, and the two forces at *F* and *D* must therefore be aligned to satisfy equilibrium, as shown in Fig. 6.20(e).

Notice that all the two-force members we have used before have been straight members. In fact, a two-force member does not have to be a straight member; it just has to be connected to other members at only two points (i.e., loaded at only two points). Now, by connecting points *F* and *D*, the directions of the forces at *F* and *D* can be determined. The slope of line *FD* is $\frac{1}{3}$ because of the geometry of member *FD* in Fig. 6.20(a). At this point, forces *F* and *D* cannot be determined by what we know to be the horizontal and vertical components at *F* and *D* in Fig. 6.20(d). Since the slopes of forces *F* and *D* are equal to $\frac{1}{3}$ (this slope can also be determined by taking moment about point *F* or *D*), we therefore have

$$\frac{D_y}{D_x} = \frac{1}{3} \quad \text{and} \quad \frac{F_y}{F_x} = \frac{1}{3}$$

or

$$D_x = 3D_y \text{ and } F_x = 3F_y$$

Although the number of unknowns is hereby reduced from four to two, their values cannot be determined by using member *FD* at this stage.

Now, let's consider the free-body diagram of member *EC* as shown in Fig. 6.20(f). Forces at *C* are known from calculations related to the pulley (they must be equal and opposite to those shown in Fig. 6.20(c)). By taking moment about *E*, $D_y$ can be evaluated, as follows:

$$+\circlearrowright \Sigma M_E = 0$$

$$(4905 \times 6) - (D_y \times 3) = 0$$

$$D_y = 9810 \text{ N}$$

Since we know from member *FD* that $D_x = 3D_y$, therefore $D_x = 3 \times 9810$, or

$$D_x = 29430 \text{ N}$$

$$\Sigma F_x = 0$$

$$-E_x + D_x - 2453 = 0$$

$$-E_x + 29430 - 2453 = 0$$

$$E_x = 26977 \text{ N}$$

$$+\uparrow\Sigma F_y = 0$$

$$-E_y + D_y - 4905 = 0$$

$$-E_y + 9810 - 4905 = 0$$

$$E_y = 4905 \text{ N}$$

Since $D_x$ and $D_y$ are now known, forces $F_x$ and $F_y$ can be calculated by using member *FD* again:

$$\Sigma F_x = 0$$

$$F_x - D_x = 0$$

$$F_x = D_x = 29{,}430 \text{ N}$$

$$\Sigma F_y = 0$$

$$F_y - D_y = 0$$

$$F_y - 9810 = 0$$

$$F_y = 9810 \text{ N}$$

A free-body diagram of member *AB* is shown in Fig. 6.20(g). Notice that all the forces applied to this member have already been calculated using other members. As a check, since this member should also be in equilibrium, the following equations should verify this fact:

$$\Sigma F_x = 0$$

$$-6377 + 2453 + 26977 - 29{,}430 + 6377 = 0$$

$$+\uparrow\Sigma F_y = 0$$

$$4905 + 4905 - 9810 = 0$$

$$+\circlearrowright\Sigma M_B = 0$$

$$(-6377 \times 5) + (2453 \times 4) + (26{,}977 \times 3) - (29{,}430 \times 2) = -2$$

As can be seen, the first two equations are equal to zero, and the third equation is equal to $-2$, a value that, compared to the magnitudes of the other forces, is practically negligible (equal to zero). We arrived at the value of $-2$ instead of zero as a result of approximating decimal fraction values of some forces that were not integers.

## PROBLEMS

**6.1.–6.6.** In Figs. P6.1 through P6.6, determine the force in each member of the truss by using the method of joints.

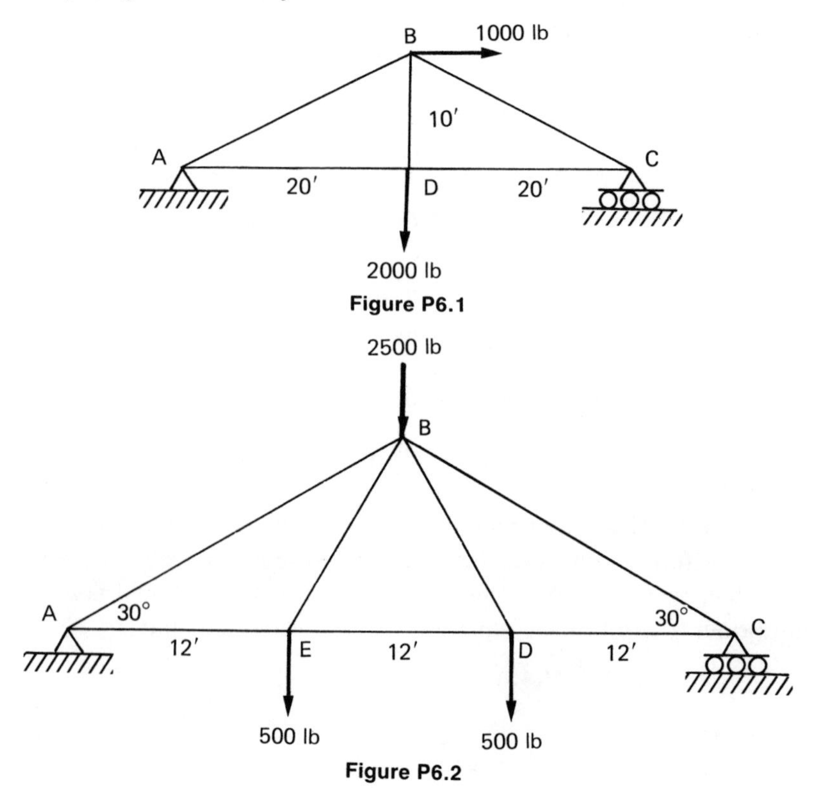


**Figure P6.1**



**Figure P6.2**

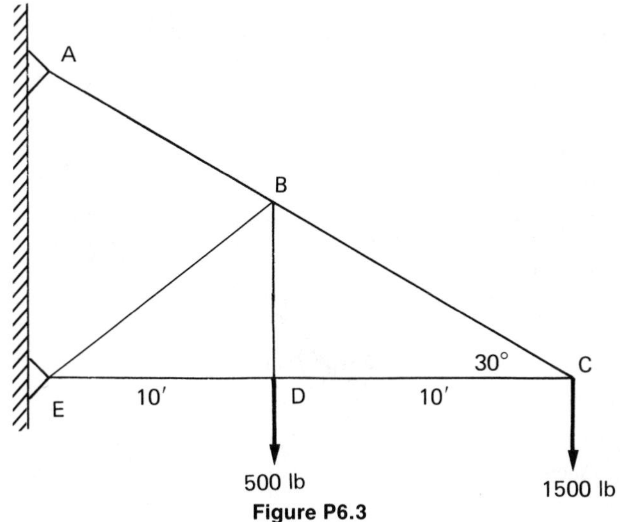

**Figure P6.3**

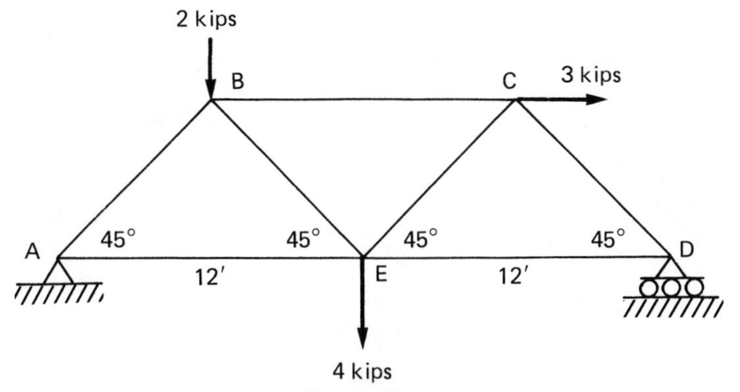

**Figure P6.4**

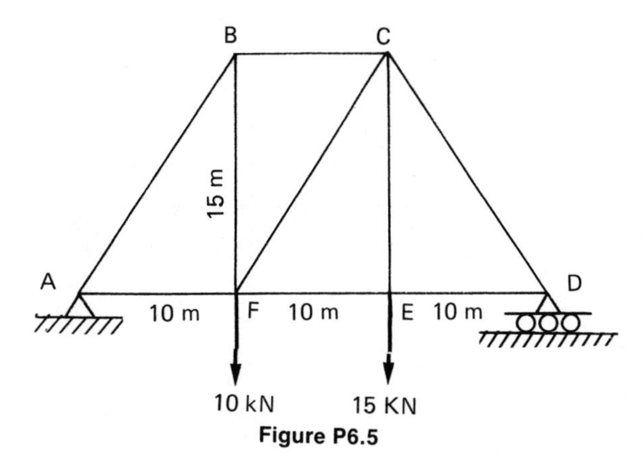

**Figure P6.5**

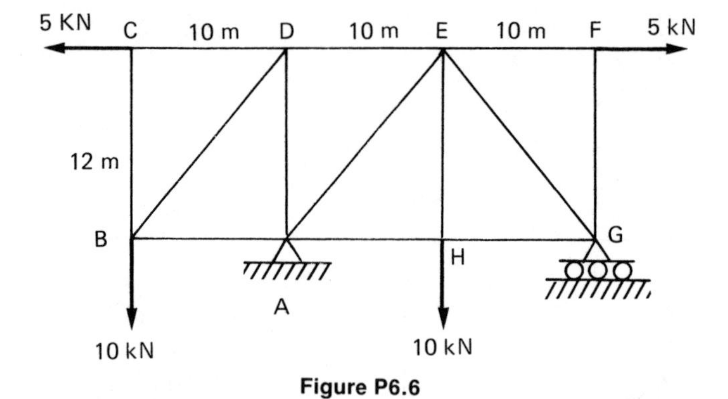


**Figure P6.6**

**6.7.** Using the method of joints, determine the force in members *BC*, *CH*, and *CG* in the truss shown in Fig. P6.7.

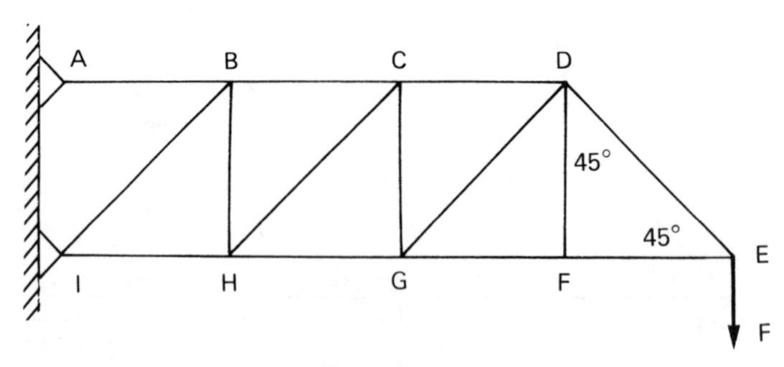


**Figure P6.7**

**6.8.** Using the method of sections, determine the force in members *BC*, *CE*, and *ED* in the truss shown in Fig. P6.8.

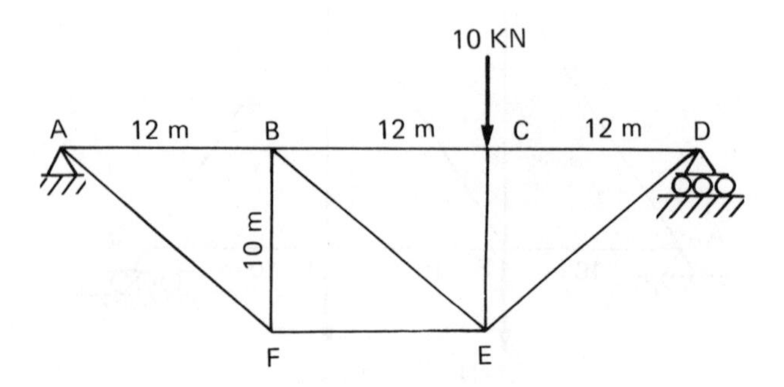

**6.9.** Using the method of sections, determine the force in members *BJ*, *CI*, and *ID* in the truss shown in Fig. P6.9.

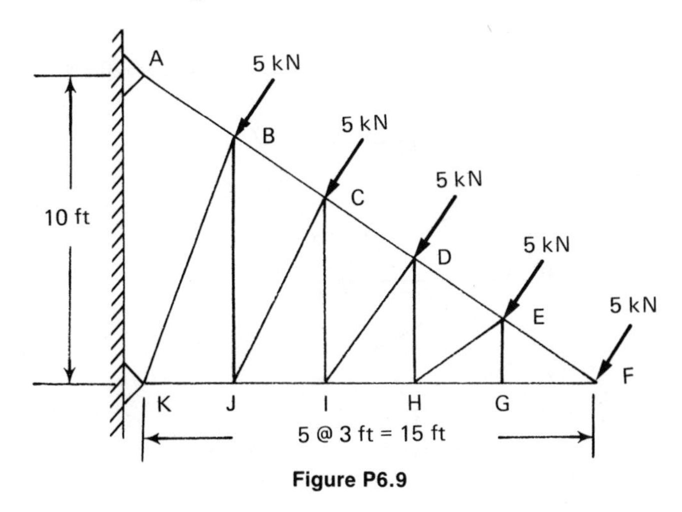


**Figure P6.9**

**6.10.** Using the method of sections, determine the force in members *AB*, *BC*, and *CD* in the truss shown in Fig. P6.10.

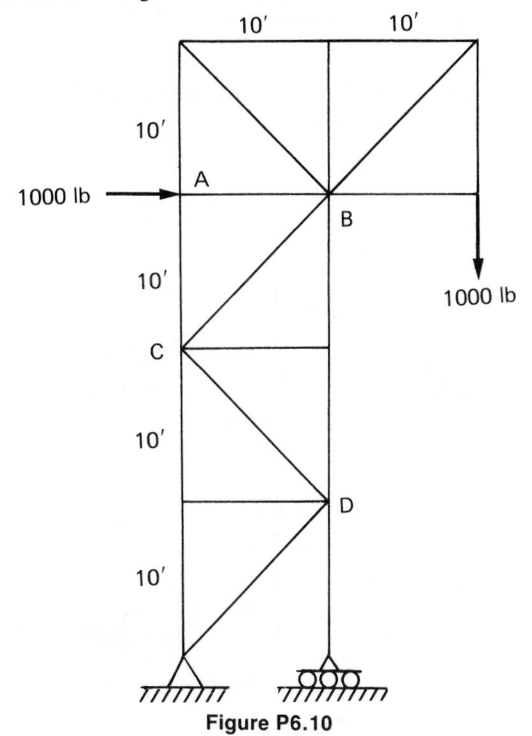


**Figure P6.10**

**6.11.** Using the method of sections, determine the force in member *AB*, *BC*, and *CD* in the truss shown in Fig. P6.11.

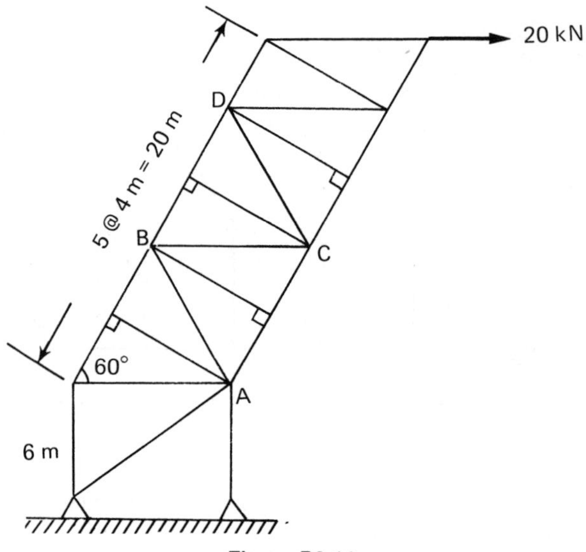


**Figure P6.11**

**6.12.–6.17.** In the trusses shown in Figs. P6.12 through P6.17, determine the zero-force members by inspection alone.

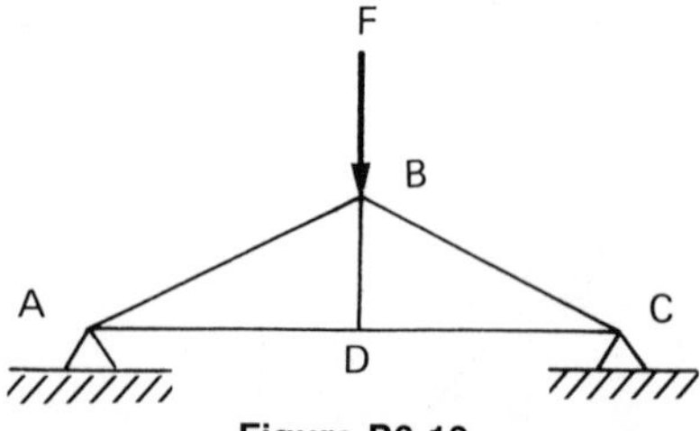


**Figure P6.12**

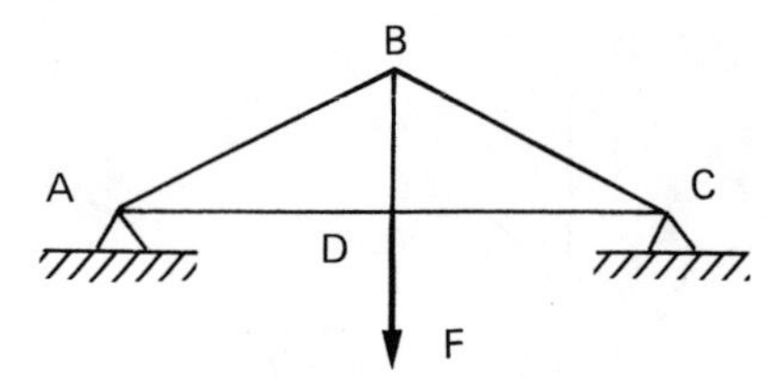


**Figure P6.13**

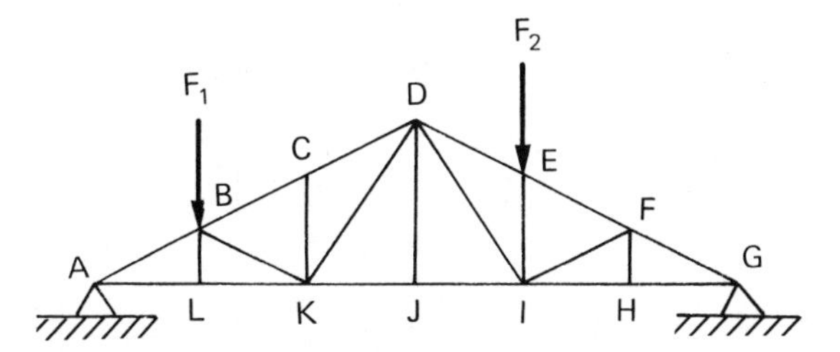


**Figure P6.14**

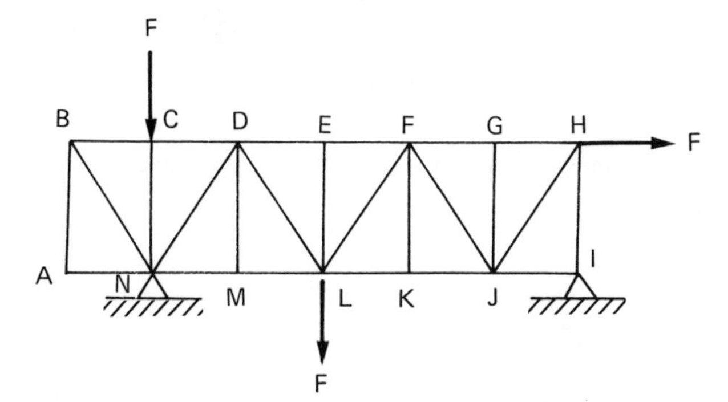


**Figure P6.15**

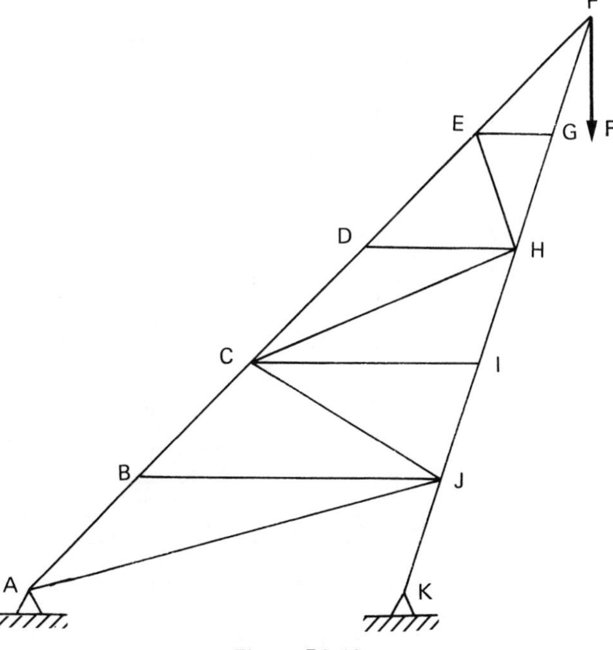


**Figure P6.16**

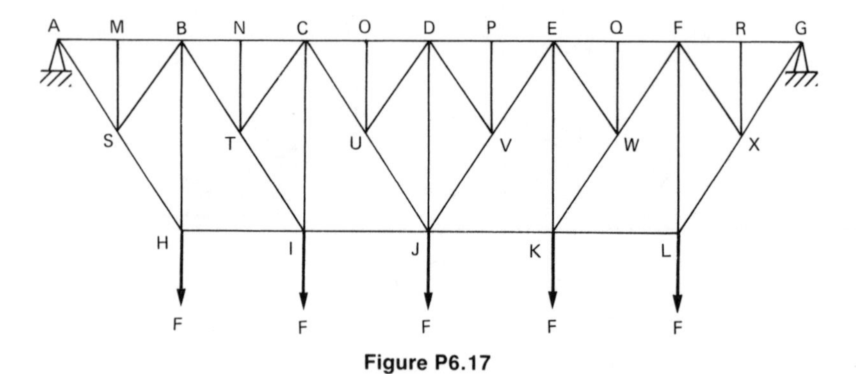

**Figure P6.17**

**6.18.–6.20.** Determine the forces in each member of the frame in Figs. P6.18 through P6.20.

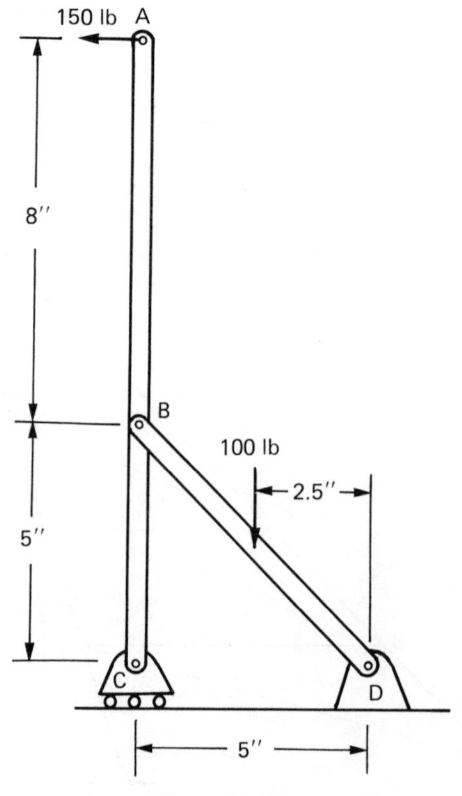

**Figure P6.18**

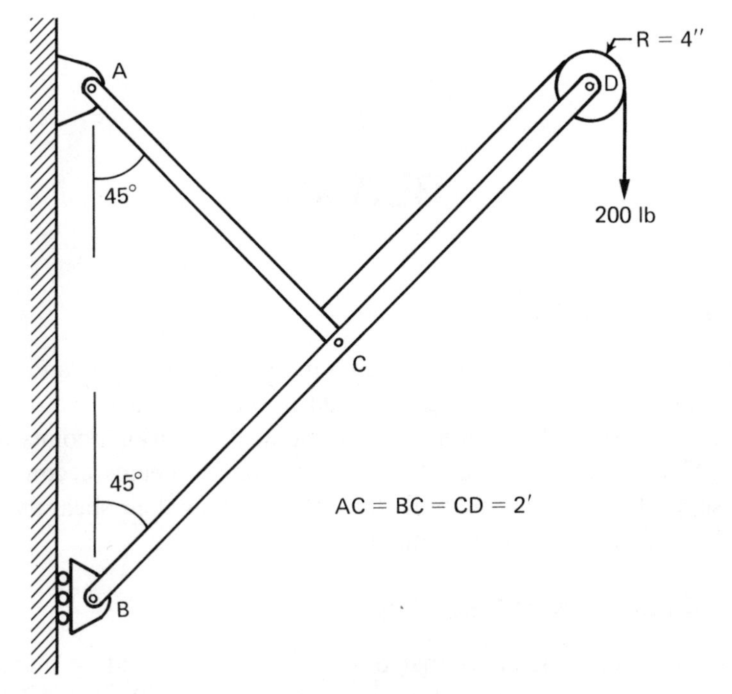


Figure P6.19

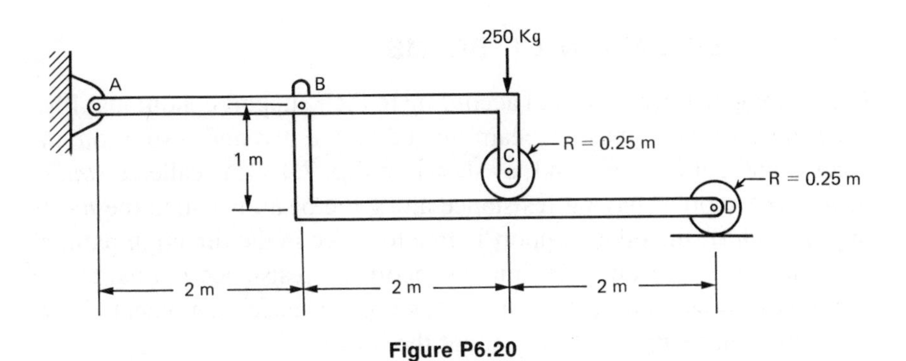


Figure P6.20

# 7

# BEAMS

## 7.1 INTRODUCTION

Beams are perhaps the most important and widely used structural members. Among the various steps involved in their design is the determination of the shear force and bending moment at various points along them. This chapter will classify different types of beams according to their support conditions and will show how to draw their shear and moment diagrams using different methods.

## 7.2 DEFINITION OF BEAMS

Slender structural members that offer resistance to bending under the action of applied loads are called *beams*. They are usually long prismatic bars with constant cross sections. The loads they are meant to sustain are usually applied normally to their axes.

## 7.3 CLASSIFICATION OF BEAMS

Beams are generally classified according to their support conditions; Fig. 7.1 shows various types. A beam hinged at one end and resting on another support at the other end, as shown in Fig. 7.1(a), is called a *simple beam*. The support having resistance along the beam is called the *hinge* or *pin support*; the other support is free to move in the direction parallel to the axis of the beam. The hinge support has resistance normal to the beam as well as along the beam; it has no resistance to moment, however, and can easily be rotated about the hinge.

Usually a beam is free to move along its axis at one of its supports. The reason for this is to allow the beam to compensate for length changes as a result of load or environmental temperature changes.

A *cantilever beam*, on the other hand, is fixed rigidly at one end, as shown in Fig. 7.1(b), and will thus show resistance to rotation as well. This kind of support is also called a *built-in support*. An *overhanging*

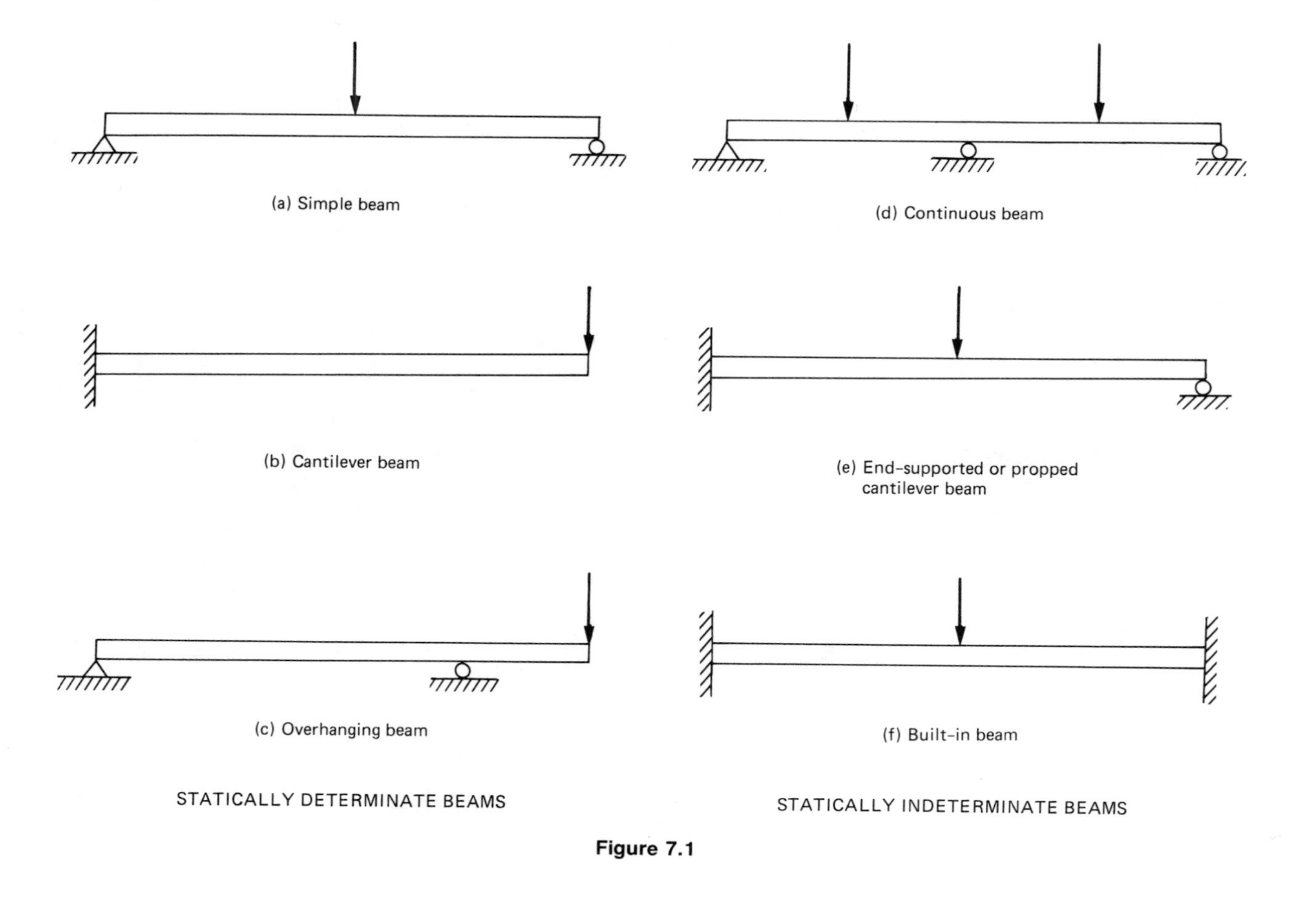
(a) Simple beam
(d) Continuous beam
(b) Cantilever beam
(e) End-supported or propped cantilever beam
(c) Overhanging beam
(f) Built-in beam
STATICALLY DETERMINATE BEAMS
STATICALLY INDETERMINATE BEAMS

Figure 7.1

*beam* is a beam a part of which extends beyond the support; one such beam is shown in Fig. 7.1(c).

If the unknown support reactions can be determined by using equations of static equilibrium alone ($\Sigma F_x = \Sigma F_y = 0$; $\Sigma M = 0$), the beams are classified as *statically determinate beams*; otherwise, they are known as *statically indeterminate beams*. A simple beam with more than two supports—the *continuous beam* shown in Fig. 7.1(d)—and the end-supported cantilever beam shown in Fig. 7.1(e) are examples of statically indeterminate beams, as is the *built-in beam* of Fig. 7.1(f). In this book we will consider only statically determinate beams, and therefore all support reactions can be determined by using equilibrium equations.

## 7.4 CLASSIFICATION OF LOADS

Loads applied to beams can take five different forms: (1) A *concentrated load* is a load applied at a point, as shown in Fig. 7.2(a). Ideally, it is applied over an area that is considered to be very small. (2) A *distributed load* is one that is distributed over a rather larger area; it can have a *uniform*, as shown in Fig. 7.2(b), or *nonuniform distribution*, as shown in Figs. 7.2(c) and (d), the latter being a special case. (3) A pure *moment load* is a load so applied as to place the beam under pure moment, as shown in Fig. 7.2(e). (4) A pure *torsional load* is a load that causes the beam to become twisted by application of a torque, as shown in Fig. 7.2(f). (5) Some combination of the above, as shown in Fig. 7.2(g). In this chapter, torsional load will not be considered.

All these loads can be of two distinct types: dead loads or live loads. The weight of all construction materials—beams, walls, etc.—is considered to be dead load (or weight), and this load is always present. On the other hand, live loads come and go. For example, loads applied by moving cars on a bridge or the weight of people going in and out of a building or the load applied to a building by a gust of wind (wind load) are all considered live loads.

Figure 7.3 nicely demonstrates the concepts of live and dead loads as well as of concentrated and distributed loads.

## 7.5 SHEAR, BENDING AND TORSIONAL LOADS

Beams can support any of the external loads that are applied to them, as shown in Fig. 7.4(a). These external loads are somehow carried by every particle of the beam and in some manner finally transmitted to the sup-

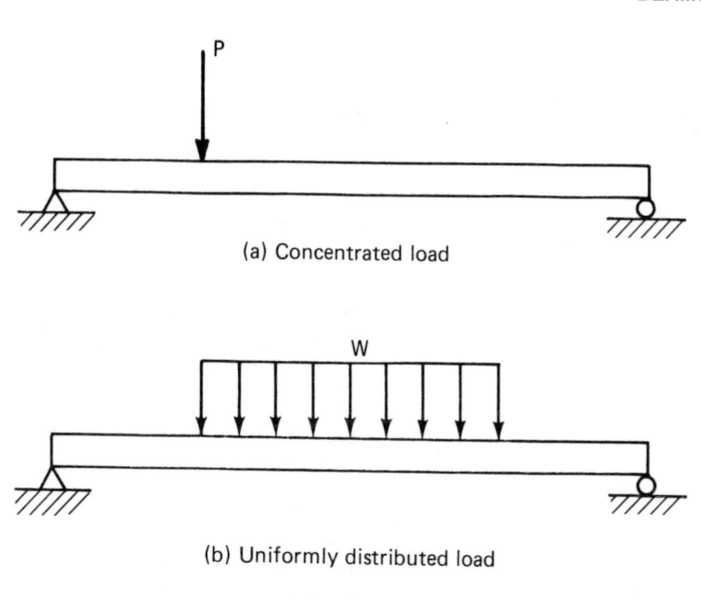


(a) Concentrated load

(b) Uniformly distributed load

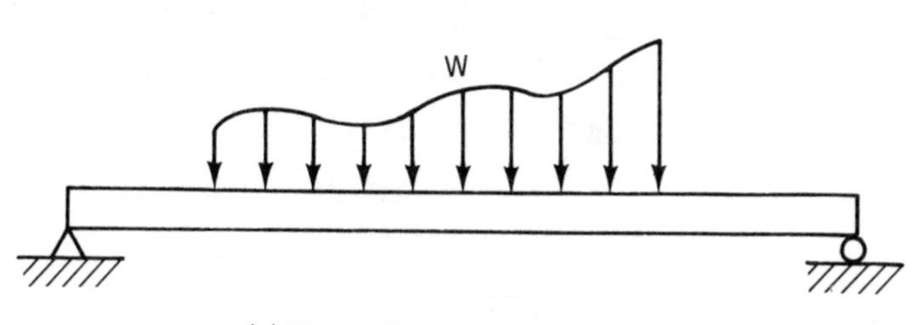


(c) Nonuniformly distributed load

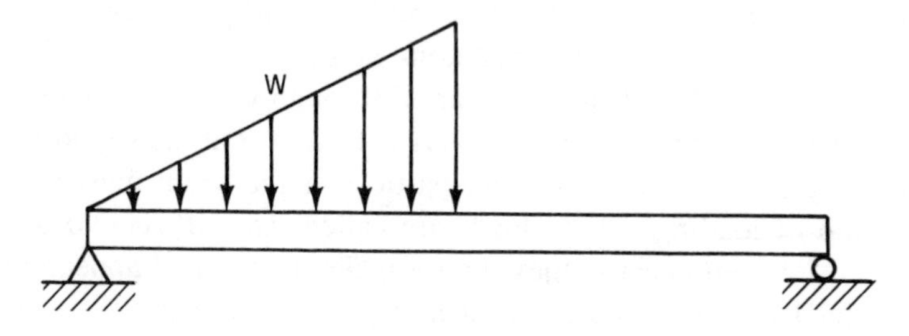


(d) Triangular load

**Figure 7.2**

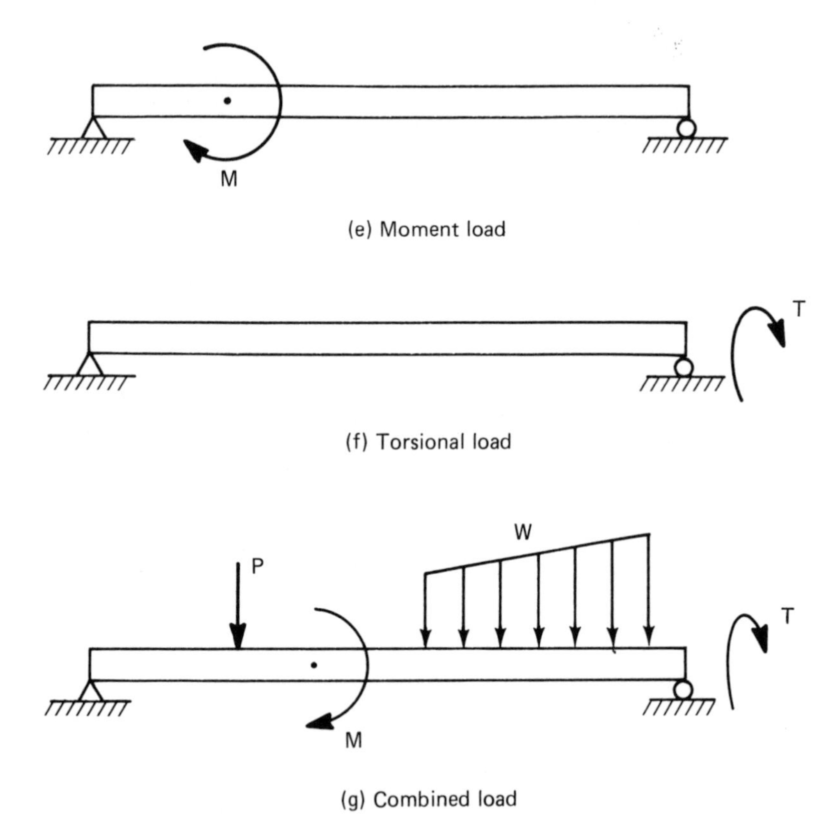


(e) Moment load

(f) Torsional load

(g) Combined load

**Figure 7.2.** *(Continued)*

ports. The support reactions thus represent the overall effect of the external loads at the supports, as shown in Fig. 7.4(b). Forces $A_x$, $A_y$, and $B$ are the components of the reactions at supports $A$ and $B$. In other words, the external loads are transferred through various particles of the beam and finally appear as reactions $A_x$, $A_y$, and $B$ at the supports.

Not only the supports but every particle of the beam must be under some kind of loading. These loads are called *internal reactions.* They are called *internal* because they do not make an external appearance. If we artificially section the beam at any point, however, these internal reactions will appear as external loads. Consider the beam shown in Fig. 7.5(a), which is sectioned by cutting plane *a–a*. Since the continuity of the beam is thus lost, as shown in Fig. 7.5(b), the effect of the portion of the beam to the right of section *a–a* should appear as external loads on the section to the left of *a–a*, and vice versa. In general, this effect

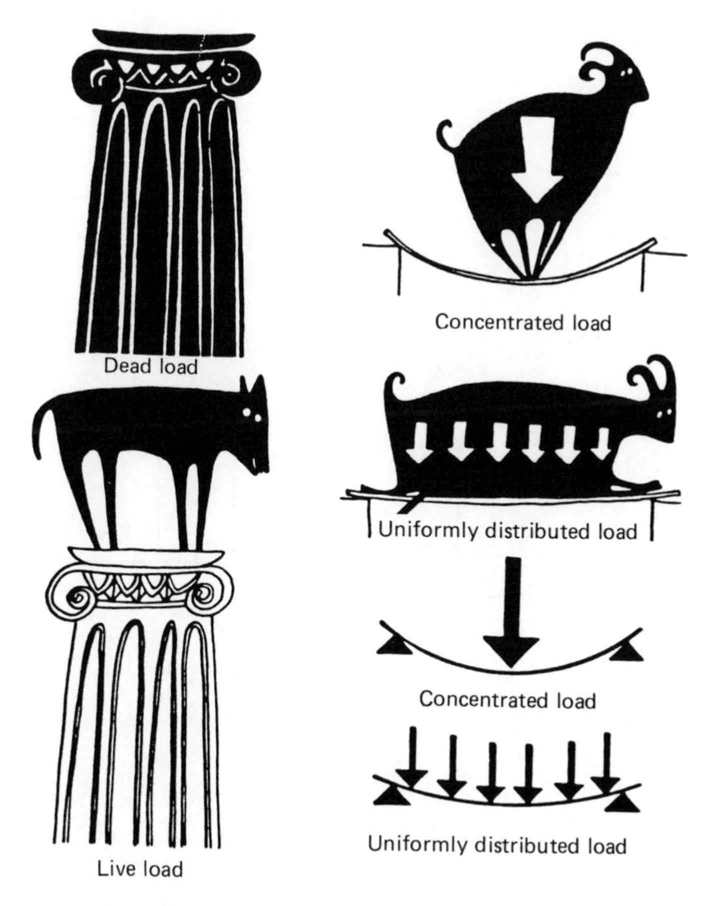


**Figure 7.3** (From *Structural Systems* by Cowan and Wilson, © Van Nostrand Reinhold, 1981)

can be thought of in terms of the three kinds of loading—moment, torque, and shear—shown in the detailed drawing in Fig. 7.5(c). The nature of these three types of internal loading on section *a–a* of the beam will now be discussed in detail.

## Shear Force

Consider a series of blocks that have equal cross-sectional surface areas and are set next to each other in the manner shown in Fig. 7.6(a). Although these blocks are not connected to each other, a strong enough person can apply a set of forces to the end blocks, as shown in Fig.

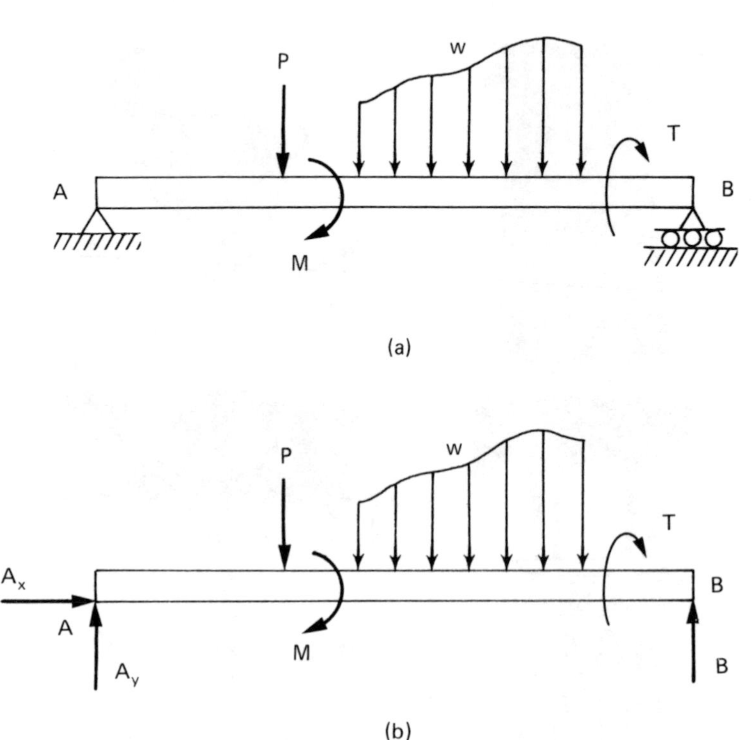


Figure 7.4

7.6(b), and lift all of the blocks as a single piece. Now consider a downward load, *P*, applied to one of the blocks at point *A*, as shown in Fig. 7.6(c). This load will try to slide the block at *A* downward. (Obviously, this presentation is overly simplified for demonstration purposes. Actually, sliding may occur among all blocks because of load *P*.) In doing so, the side surfaces of block *A* will apply some friction forces to the side surfaces of the adjacent blocks. The force produced between the two blocks because of this sliding is called *shear force*.

Now, imagine that all blocks in Fig. 7.6(a) are rigidly cemented to each other. Upon application of load *P* in this case, the blocks cannot physically slide against neighboring blocks, but the shear (friction) force just mentioned will be produced between the blocks on the cemented surfaces, and actually not only on the cemented surfaces but also on all layers normal to the axis of the blocks. This shear force can be thought of as an internal reaction, and once the beam is sectioned, this internal reaction will behave just like an externally applied load on that section.

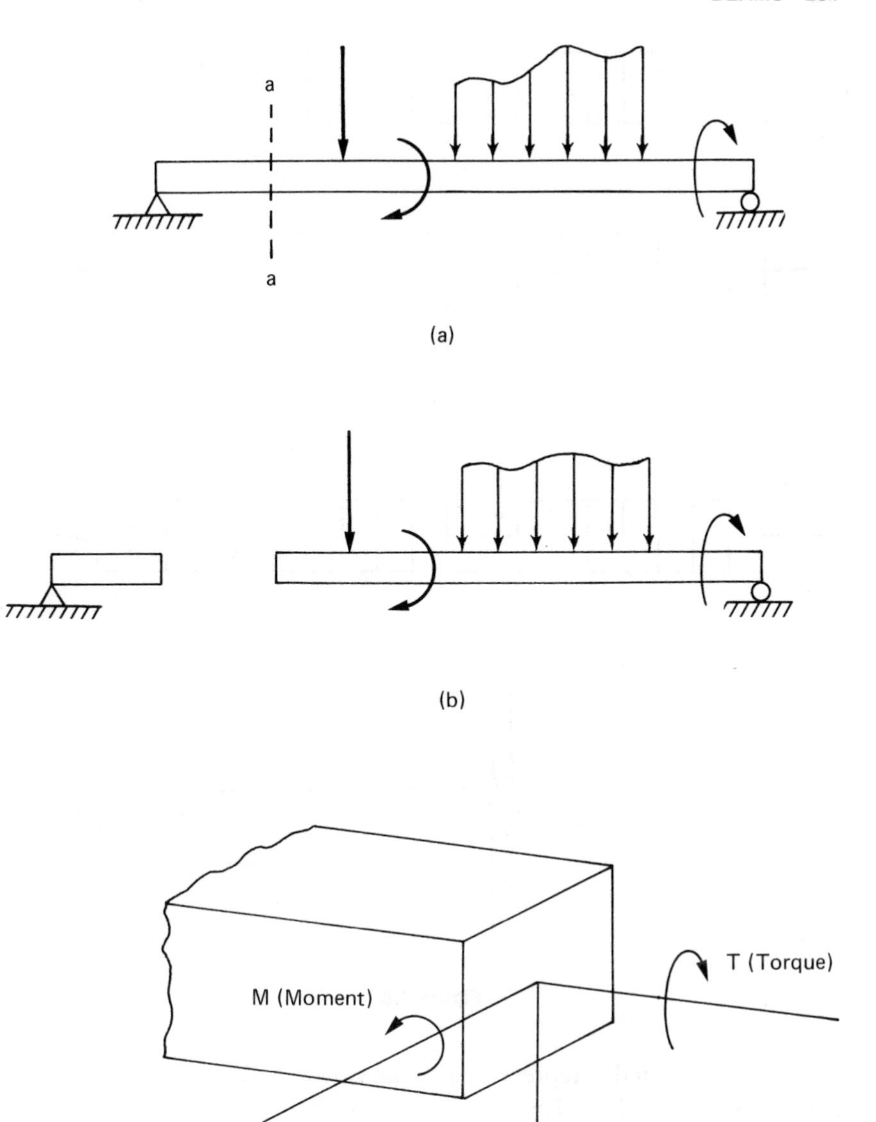


**Figure 7.5**

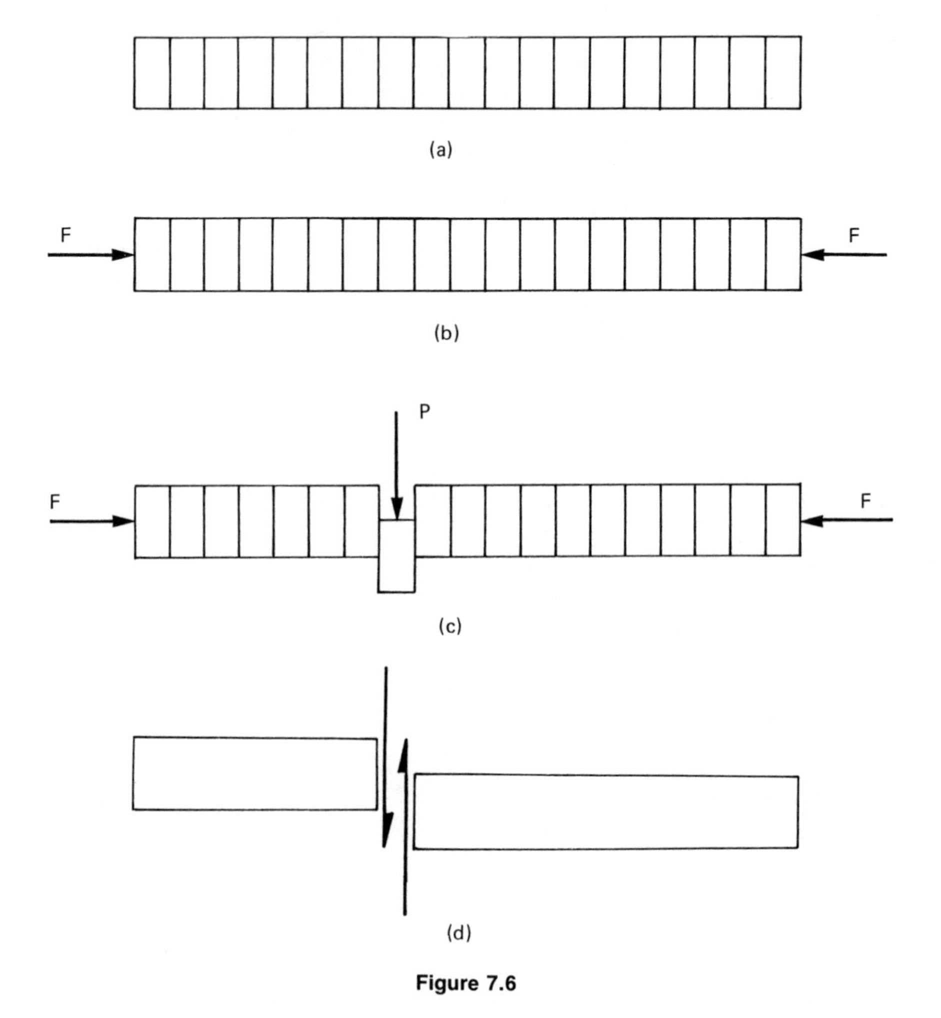


Figure 7.6

Shear force is usually represented by an arrow with only half an arrow head, as shown in Fig. 7.6(d).

## Bending Moment

Consider a point (or section) on a beam; the effect of loads applied at points (or sections) elsewhere on the beam will appear as a bending moment at this particular point (or section). Therefore, after sectioning, the effect of the other parts of the beam on this section should also include a bending moment as internal reaction, as shown in Fig. 7.7.

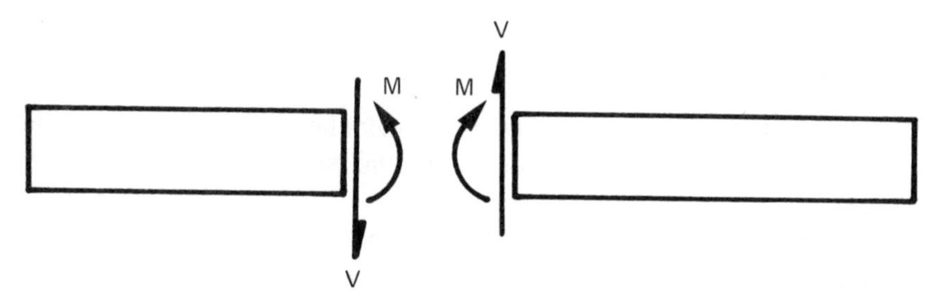


Figure 7.7

Notice the similarity here to the internal reactions studied in trusses and frames in Chap. 6. These internal reactions appear externally when the beam is sectioned. In Fig. 7.7, a downward shear force and a counterclockwise bending moment appear on the sectioned block on the left, whereas the shear force and bending moment on the block on the right have opposite directions. Another way to look at this phenomenon is to think of the two blocks as being cemented back together where previously sectioned, in which case these internal reactions should cancel each other out. Obviously in Fig. 7.7, the downward and upward shearing forces as well as the counterclockwise and clockwise bending moments do cancel each other out.

Design of a beam includes calculation of stresses and deflections. It can be shown that these stresses and deflections are functions of the internal reactions (shear force, $V$, and bending moment, $M$). The designer has to know the magnitudes and directions of these internal shear forces and bending moments at each point (section) of the beam, usually by making use of diagrams that show the distribution of shear forces and bending moments from one end of the beam to the other. These diagrams, entitled "Shear Force" and "Bending Moment" or simply Shear and Moment Diagrams, are the first step in the process of designing beams.

In the following sections, various methods for drawing shear and bending moment diagrams will be introduced and actual examples given.

## 7.6 SHEAR FORCE AND BENDING MOMENT DIAGRAMS

### Method of Sectioning and Equilibrium

This method will be illustrated in detail by using the following two examples.

EXAMPLE 1: Sketch the shear and moment diagrams for the beam shown in Fig. 7.8(a).

SOLUTION: The reactions at $A$ and $D$ are calculated simply by using the free-body diagram of the entire beam and the equilibrium equations discussed in Chap. 4. This will result in the values: $R_A = 2800$ lb and $R_D = 2200$ lb. Now imagine the beam to be sectioned by a transverse cutting plane, $a$–$a$, at a point between $A$ and $B$, say at a distance $x$ from $A$. The free-body diagram of the left portion of the sectioned beam is shown in Fig. 7.8(b). To keep this portion of the beam in equilibrium, we must add a shear force, $V$, and a bending moment, $M$, at the location where it is sectioned. This force and moment are the internal reactions of the beam at this particular section. In fact, they are the effect of the portion of the beam at the right of section $a$–$a$ on the portion at the left of the cutting plane. These internal reactions usually depend on the location of section $a$–$a$ (in this particular example, at least, the value of the moment depends on the axial location of cutting plane $a$–$a$, whereas the shear force is constant). The location of the cutting plane is always measured from the far left point of the beam, which, in this case, is $A$. If we assume section $a$–$a$ is located at a distance $x$ from point $A$, we can determine $V$ and $M$ by using the equations of equilibrium, as follows:

$$\Sigma F_y = 0$$

$$2800 - V = 0$$

$$V = 2800 \text{ lb}$$

Taking moments of the forces about a point located on section $a$–$a$ in Fig. 7.8(b) will give the value of $M$, as follows:

$$+\circlearrowright \Sigma M_a = 0$$

$$2800(x) - M = 0$$

$$M = 2800x$$

where subscript $a$ represents a point in cutting plane $a$–$a$. Notice that the equations will remain unchanged so long as cutting plane $a$–$a$ is introduced between points $A$ and $B$.

Next, the beam in Fig. 7.8(a) will be sectioned by cutting plane $b$–$b$ somewhere between $B$ and $C$; the free-body diagram of that portion of the beam to the left of section $b$–$b$ is shown in Fig. 7.8(c). Similarly, the values of internal reactions $V$ and $M$ at section $b$–$b$ can be calculated by using the equations of equilibrium, as follows:

$$\Sigma F_y = 0$$

$$2800 - 2000 - V = 0$$

$$V = 800 \text{ lb}$$

The moment of forces in Fig. 7.8(c) will be taken about point $b$ located on section $b$–$b$ at a distance $x$ from point $A$:

$$+\circlearrowright \Sigma M_b = 0$$

$$2800(x) - 2000(x - 2) - M = 0$$

$$M = 2800x - 2000(x - 2)$$

Although this equation can be simplified, as will be seen later, it is more convenient to keep it as shown here. Finally, the beam will be sectioned at a point between $C$ and $D$ by using cutting plane $c$–$c$. The free-body diagram of that portion of the beam to the left of section $c$–$c$ is shown in Fig. 7.8(d). The internal reactions at $c$–$c$ can be similarly determined by using the equilibrium equations, as follows:

$$\Sigma F_y = 0$$

$$2800 - 2000 - 3000 - V = 0$$

$$V = -2200 \text{ lb}$$

and

$$+\circlearrowright \Sigma M_c = 0$$

$$2800x - 2000(x - 2) - 3000(x - 6) - M = 0$$

$$M = 2800x - 2000(x - 2) - 3000(x - 6)$$

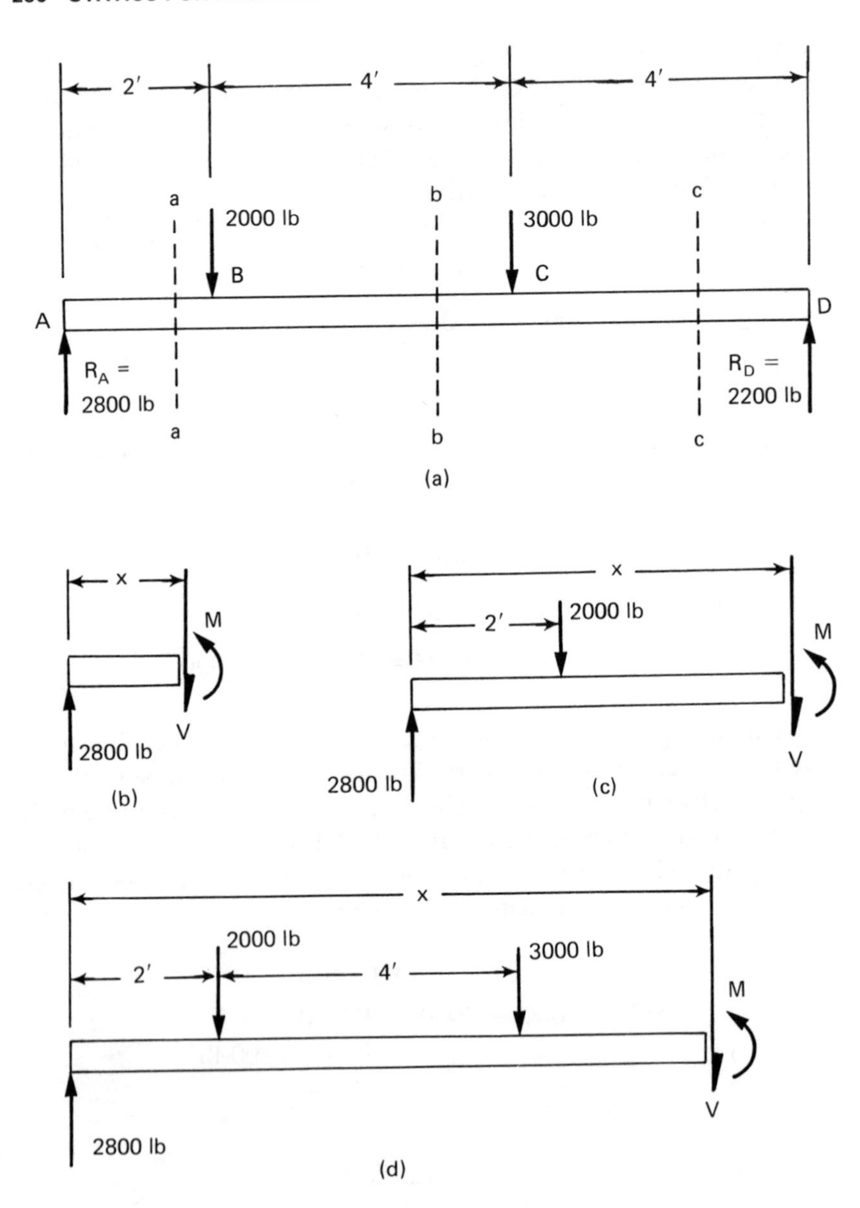
2′
4′
4′
a
b
c
2000 lb
3000 lb
B
C
A
D
$R_A$ =
2800 lb
$R_D$ =
2200 lb
a
b
c
(a)
x
M
V
2800 lb
(b)
x
2′
2000 lb
M
V
2800 lb
(c)
x
2000 lb
2′
4′
3000 lb
M
V
2800 lb
(d)

**Figure 7.8**

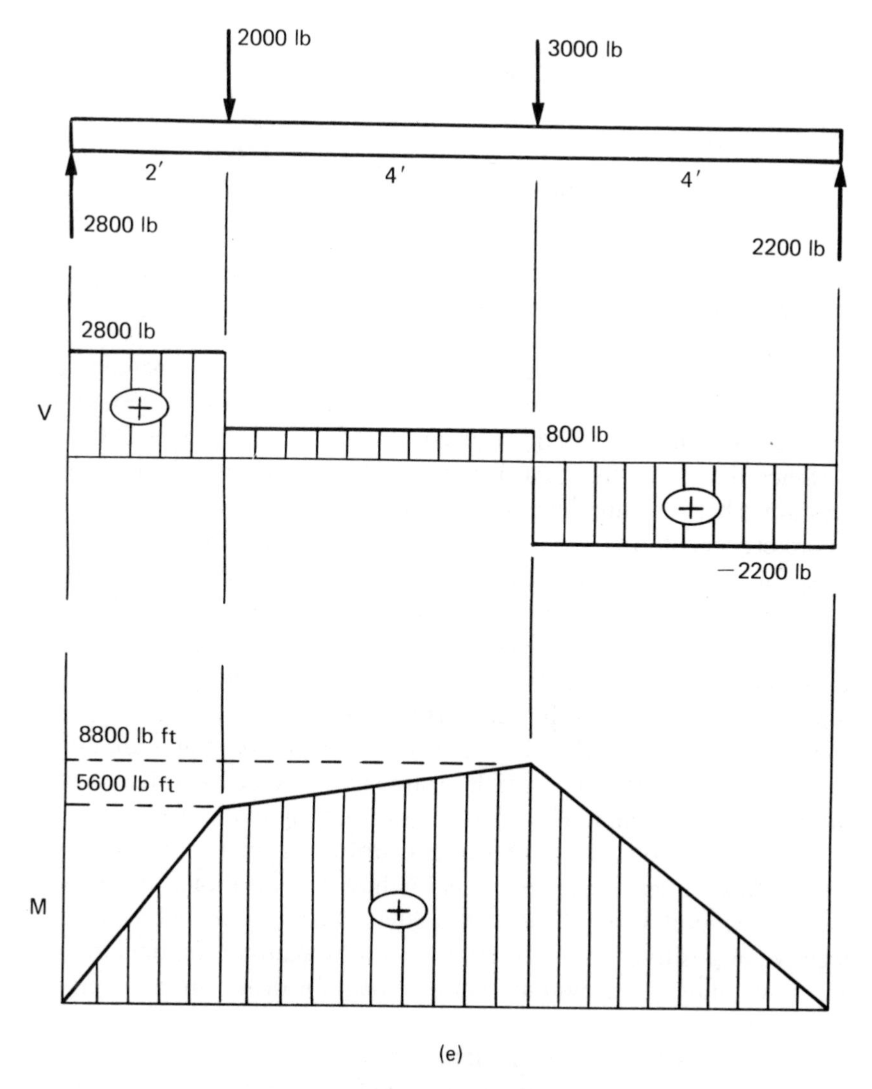


**Figure 7.8.** *(Continued)*

Notice that although the same symbols, $V$ and $M$, are used in all three figures, 7.8(b), (c), and (d), their values are as different as the preceding equations indicate, each equation being valid only for those particular values of $x$ that define each region, such as $AB$, $BC$, and $CD$ in this example. The equations for $V$ and $M$ here obtained are collectively listed below with their ranges of validity:

$$\left.\begin{aligned} V &= 2800 \text{ lb} \\ M &= 2800x \end{aligned}\right\} 0' \leq x \leq 2'$$

$$\left.\begin{aligned} V &= 800 \text{ lb} \\ M &= 2800x - 2000(x - 2) \end{aligned}\right\} 2' \leq x \leq 6'$$

$$\left.\begin{aligned} V &= -2200 \text{ lb} \\ M &= 2800x - 2000(x - 2) - 3000(x - 6) \end{aligned}\right\} 6' \leq x \leq 10'$$

These equations give the precise values of the shear and bending moment at any point along the beam but are valid in their own regions only. For example, in the second set of equations, $M = 2800x - 2000(x - 2)$. Since this is the moment distribution between $B$ and $C$, $x$ can have any values between and including $2'$ and $6'$. Although the equation itself is correct, substituting a value of $x$ different from these will give a wrong value for the moment. For example, if $x$ were set at 0 in this equation, the value of the moment would be 4000 lb ft, an answer that is obviously wrong. If one is interested in the value of moment at $x = 0$, the equation for moment from the first set, $M = 2800x$, should be used, which results in a zero moment at $x = 0$, which is the correct value.

In order to show the distribution of $V$ and $M$, two diagrams are usually plotted right beneath the beam; these are known as the *shear* and *moment diagrams* for the beam. The diagrams say nothing more than the preceding sets of equations do, but now the distributions of $V$ and $M$ are made visual. Also, these diagrams quickly show the positions of the beam under maximum $V$ or $M$, as well as those portions of the beam under negative or positive moments. All of this information is particularly useful, for example, in providing reinforcement for the beam in the process of design.

The shear and moment diagrams for this beam are plotted in Fig. 7.8(e). To construct these diagrams, fine lines are drawn vertical to the axis of the beam under supports and concentrated loads. Doing so divides the beam into three distinct regions; the $x$ coordinates that define each region are as follows:

Region I: $0' \leq x \leq 2'$

Region II: $2' \leq x \leq 6'$

Region III: $6' \leq x \leq 10'$

Next, two horizontal lines are drawn that serve as reference lines for $V$ and $M$. Above these reference lines, the values of $V$ and $M$ are considered positive; below them, negative. Using an appropriate scale for $V$ and $M$ respectively, the equations for shear and moment can be plotted for each region as shown in Fig. 7.8(e).

Notice that the values for shear obtained from each of the three equations are constant and that they can easily be plotted as horizontal lines corresponding to their values. The equations for moments are all equations of straight lines. To plot a straight line, one needs to locate only two points, say, the beginning and the end points of the region.

To determine the value of $V$ or $M$ at any location on the beam, simply draw a line normal to the axis of the beam at that point and extend this line to intersect the diagrams; then simply read the values of $V$ and $M$ off the diagrams that correspond to that point (using the same scale used to plot these diagrams). Notice that at point $A$ the value of shear is 2800 lb, which corresponds to the magnitude of the reaction at $A$ and represents the maximum shear at all points of Region I. The values of moments at $A$ and $D$ are both zero, which also checks with the fact that these points are hinged-end supports, and, as shown in the free-body diagram of Fig. 7.8(a), there is no moment reaction at $A$ and $D$. The maximum moment occurs right under the 3000-lb load, and its value is equal to 8800 lb ft.

EXAMPLE 2: Draw shear and bending moment diagrams for the beam shown in Fig. 7.9(a).

SOLUTION: The reactions are calculated by considering the symmetry of the beam. The total distributed load on the beam is equal to 250 lb/ft × 20 ft = 5000 lb; thus, $R_A = R_B = 5000/2 = 2500$ lb. The beam is sectioned by using a transverse cut $a$–$a$, and the free-body diagram of the portion of the beam to the left of cutting plane $a$–$a$ is shown in Fig. 7.9(b). By using the equations of equilibrium, $V$ and $M$ can be determined. To obtain the former, we have

$$\Sigma F_y = 0$$

$$2500 - 250(x) - V = 0$$

$$V = 2500 - 250x$$

To use the moment equilibrium equation, the distributed load in Fig. 7.9(b) has to be converted to its equivalent concentrated load. Since it is a uniformly distributed load of intensity 250 lb/ft that acts over a distance $x$, the total equivalent concentrated load will be equal to $250x$ lb; it will act at the midpoint, moreover, which is at a distance $x/2$ to the left of cutting plane $a$–$a$, as shown in Fig. 7.9(c). Now the moment equilibrium equation can be written by taking the moments of all the forces in Fig. 7.9(c) about a point on cutting plane $a$–$a$, as follows:

$$+\circlearrowright\ \Sigma M_a = 0$$

$$2500x - (250x)\frac{x}{2} - M = 0$$

$$M = 2500x - 125x^2$$

Both the preceding equations are valid for the entire beam because everything is uniform from $A$ to $B$. The first, the equation for shear, is

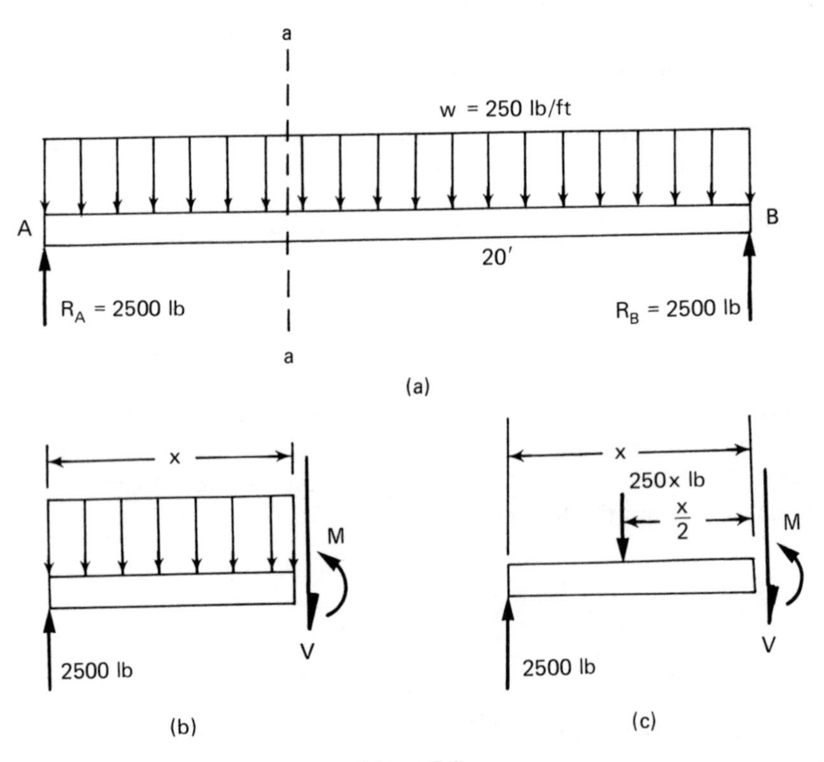


Figure 7.9

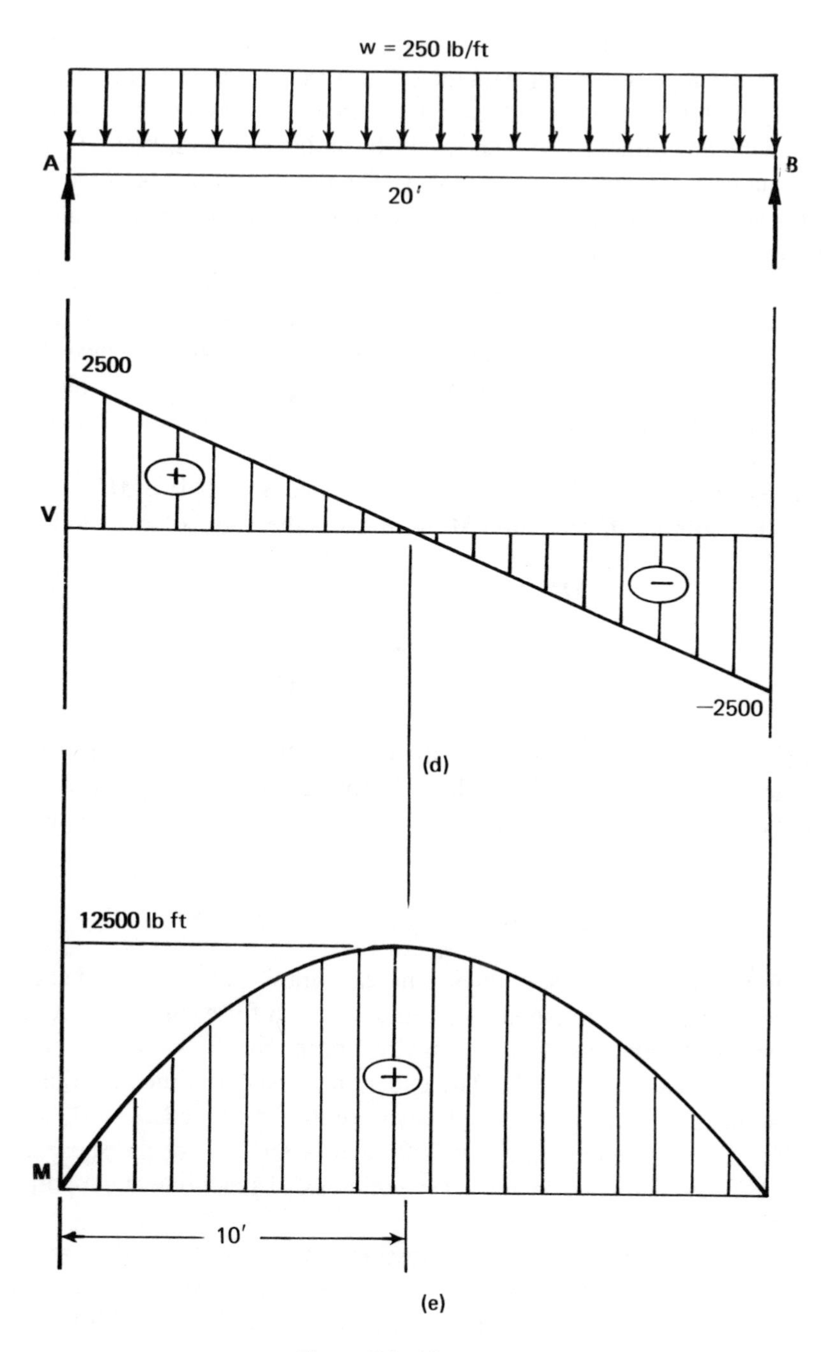


Figure 7.9. *(Continued)*

the equation for a straight line, whereas the second, the moment equation, is the equation for a parabola. Since the parabola has a standard shape, to plot it we need to locate only three of its points. Two of them can be the end points, $x = 0$ and $x = 20'$ in this problem, both of which give the value of zero for the moment. The third point can be some point in between the end points or a point corresponding to the maximum or minimum of the parabolic curve.

To locate where the maximum or minimum of the equation, $M = 2500x - 125x^2$, occurs, one has to find the derivative of this function and then set the derivative equal to zero and solve for $x$. Once the $x$ corresponding to the maximum or minimum is found, it can be plugged back into the equation for moment to evaluate the maximum or minimum value of the moment.

The derivative of $M = 2500x - 125x^2$ (see Sec. 1.4) is $M' = 2500 - 250x$. Solving the equation $M' = 0$ gives the value of $x$, as follows:

$$M' = 2500 - 250x = 0$$

$$250x = 2500$$

$$x = 10$$

This simply says that at $x = 10'$ the beam will have maximum or minimum moment and that the value of moment that corresponds to this point can be evaluated by substituting $x = 10'$ into the moment equation as follows:

$$M = 2500(10) - 125(10)^2 = 12500 \text{ lb ft}$$

After arriving at these values, one can draw the shear and bending moment diagrams corresponding to the equation for each. To do so in a way similar to that used in the previous problem, draw two horizontal reference lines for $V$ and $M$. The equation for shear is the equation for a straight line, and to plot it only two points are needed: $x = 0'$ and $x = 20'$. These provide values of 2500 and $-2500$ lb for the shear at $A$ and $B$. They also check with the magnitudes of reactions at supports $A$ and $B$.

To plot moment distribution, substitute the values $x = 0'$, $x = 20'$, and $x = 10'$ in the equation, $M = 2500x - 125x^2$. At $x = 0'$ and $20'$, the moments are both equal to zero, a value that checks with the fact that these end supports are pin supports and have no resistance to moment. At $x = 10'$, the moment has a value of 12500 lb ft. Draw a smooth parabolic-looking curve passing through these three points. Obviously, the value of the moment in this case is maximum at $x = 10'$ since the values of moment start from zero, increase to 12500 lb ft, and then decrease to zero so that the point corresponding to $x = 10$ must be the point of maximum moment. Notice also that the peak of the bending moment diagram ($x = 10'$, corresponding to the point of maximum moment) is located at a point corresponding to zero shear. This observation will be elaborated in the following section. The plots of shear and moment diagrams for this beam are shown in Fig. 7.9(d) and (e).

## Relationship between Applied Load, Shear Force, and Bending Moment

In this section, a different method will be introduced to construct the shear and moment diagrams. This is less time-consuming, and, particularly since it uses the relationship between load, shear, and moment, it can thus be used to check values of shear or moment at any point against each other and the applied load.

Consider the beam shown in Fig. 7.10(a), with an arbitrarily distributed load of variable intensity $w$, which can be specified as a function of $x$, where $x$ is measured from the furthest left point of the beam (Point $A$ in this case). Now consider a small segment $CD$ of the beam of length $\Delta x$ cut out so as to be isolated from the rest of the beam. Figure 7.10(b) shows the complete free-body diagram of this small segment of the beam. Since $\Delta x$ is very small, the distributed load on it can be considered uniform and then converted to its equivalent concentrated load, $w\Delta x$, which is applied at the midpoint of element $CD$, as shown in Fig. 7.10(c).

Isolation of this small section from the beam physically requires sectioning the beam with two transverse cutting planes at $C$ and $D$ infinitesimally close to each other ($\Delta x$ is very small). The effect of the portion of the beam to the left of the cut at $C$ will appear as internal reactions $V$ and $M$ on the left side of the element $CD$. Since the right side of the element is infinitesimally close to the left side, one does not expect the values of $V$ and $M$ to change appreciably from one side of the small element to the other (from $C$ to $D$). Therefore, if shear and moment on the left side are $V$ and $M$, the same quantities on the right side are changed by very small amounts, $\Delta V$ and $\Delta M$ ($\Delta$ stands for a very small change of a quantity), as shown in Fig. 7.10(b).

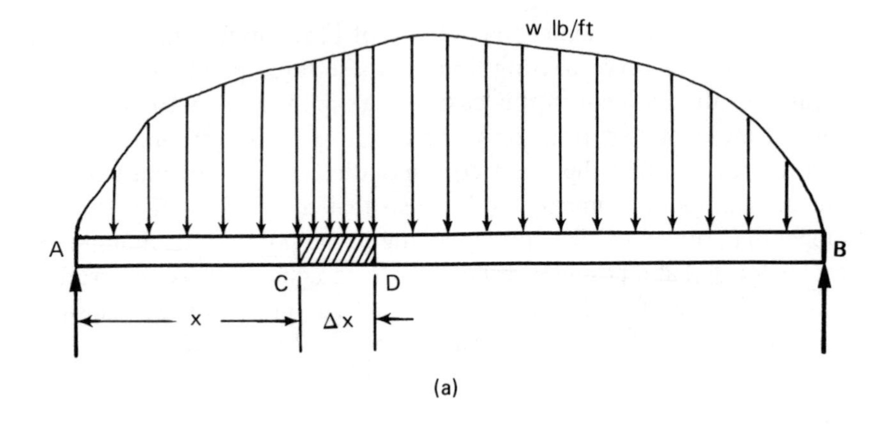


(a)

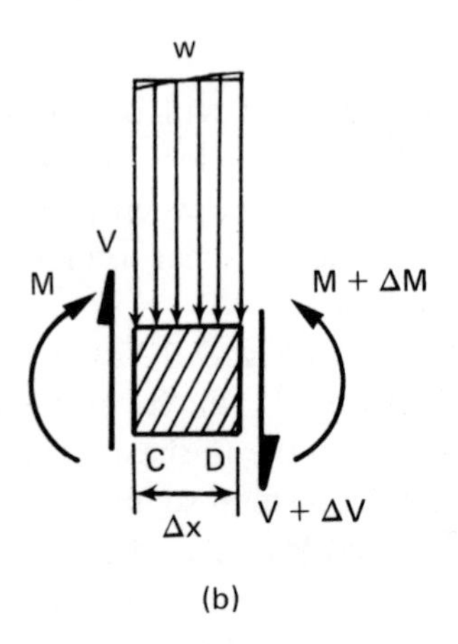


(b)

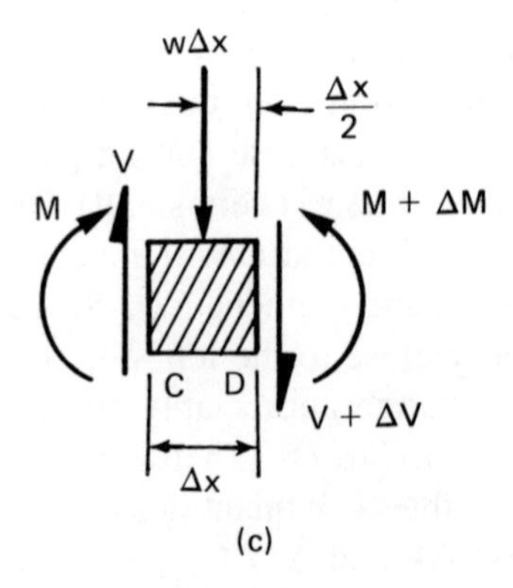


(c)

**Figure 7.10**

Figure 7.10(c) shows the same diagram as Fig. 7.10(b), but the downward uniformly distributed load is converted to its equivalent concentrated load, $w\Delta x$; as can be seen from the figure, it is located in the middle of $CD$. The equations of equilibrium can now be applied to the free-body diagram of Fig. 7.10(c) as follows:

$$\Sigma F_y = 0$$

$$V - w\Delta x - (V + \Delta V) = 0$$

$$\Delta V = -w\Delta x \qquad (7.1a)$$

or

$$\frac{\Delta V}{\Delta x} = -w \qquad (7.1b)$$

Likewise,

$$+\circlearrowright \Sigma M_D = 0$$

$$M + V\Delta x - w\Delta x(\Delta x/2) - (M + \Delta M) = 0$$

Since $\Delta x$ is small, $w\Delta x(\Delta x/2)$ will be very small compared to the other terms in this equation and therefore can be eliminated. The moment equation then reduces to

$$\Delta M = V\Delta x \qquad (7.2a)$$

or

$$\frac{\Delta M}{\Delta x} = V \qquad (7.2b)$$

Equation 7.1(a) can also be rewritten as

$$\Delta V = V_D - V_C = -w\Delta x$$

Remember that $\Delta V$ is the change in the value of shear from $C$ to $D$ and that the term, $V_D - V_C$ is the difference of shear values between the two sides of element $CD$ that are separated by the distance $\Delta x$. The term $w\Delta x$ is the area under the loading profile between points $C$ and $D$.

Now consider another infinitesimal element of the same sort next to element *CD*. Similar arguments can be applied to this element and to others as well. By accumulating these findings, we can state that the difference in the values of shear between any two points, 1 and 2, located along the length of the beam is equal to the value of the area under the loading profile between points 1 and 2:

$$V_2 - V_1 = -w\Delta x = \text{area under loading profile between points 1 and 2}$$

Applying similar reasoning to Eq. 7.2(a), one can state the following:

$$M_2 - M_1 = V\Delta x = \text{area under shear diagram between points 1 and 2}$$

Considering equations 7.1(b) and 7.2(b) and using the fact that $\Delta V/\Delta x$ is the slope of the shear diagram and $\Delta M/\Delta x$ the slope of the moment diagram, we can state the following important facts about the relationship between load, shear force, and bending moment:

1. The slope of a shear diagram at a point is equal to the value of the applied load at that point.
2. The slope of a moment diagram at a point is equal to the value of the shear at that point.

The preceding formulations introduce another method for drawing shear and moment diagrams for beams. Specifically, the use of this method will be better appreciated if one needs to calculate the magnitude of the maximum moment of a beam and its location. To do so requires the use of the well-known theorem in calculus that states that *a function is maximum (or minimum) when its derivative is equal to zero.* Since *the derivative of moment is equal to shear*, as expressed by Eq. 7.2(b), this means that at that point of the beam where shear is zero, the moment must have maximum (or minimum) value, or vice versa.

This new method will be demonstrated in the following examples.

EXAMPLE 3: Sketch the shear and moment diagrams for the cantilever beam shown in Fig. 7.11(a) using the relationship between load, shear, and moment.

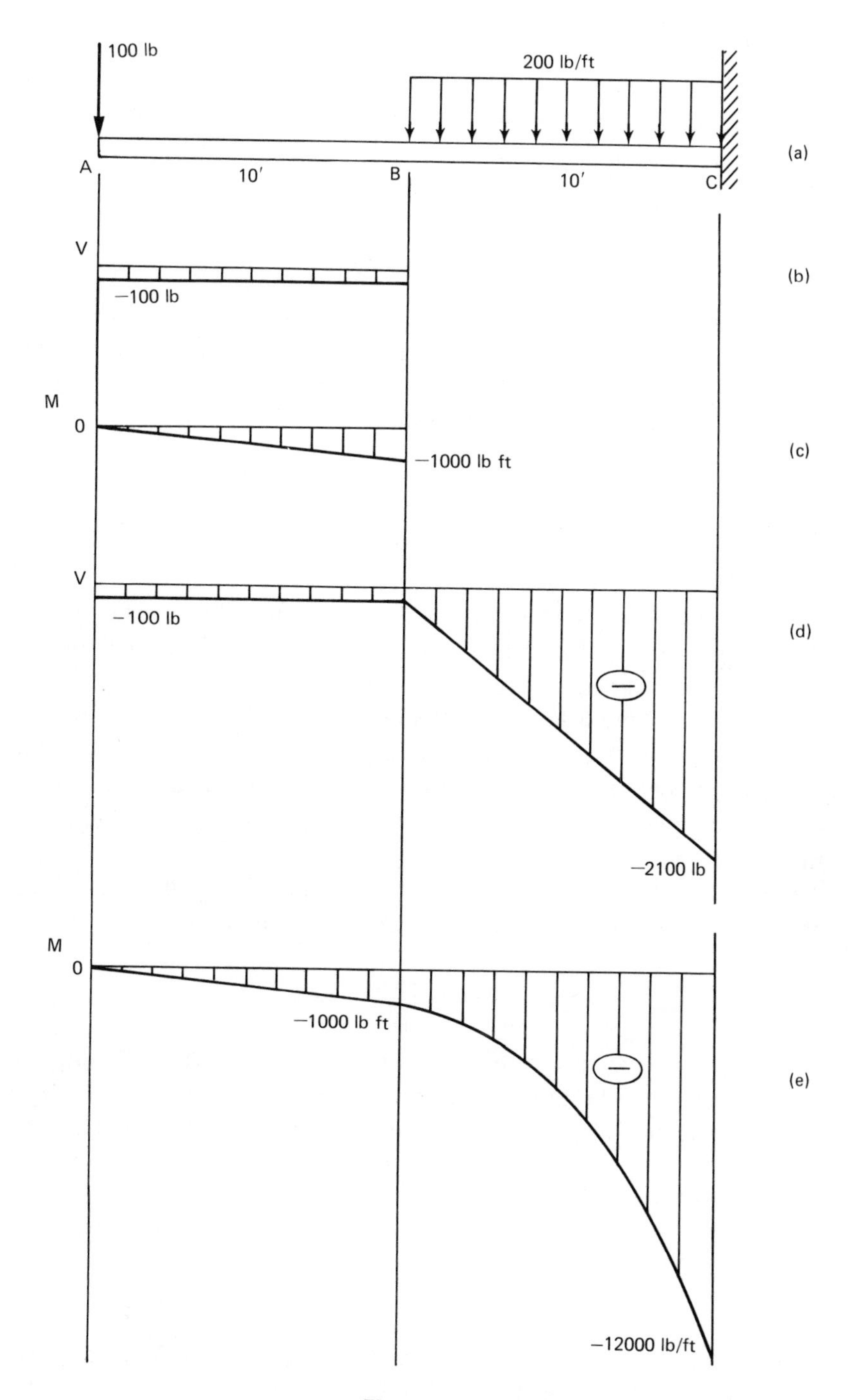
100 lb
200 lb/ft
(a)
A
10′
B
10′
C
V
(b)
−100 lb
M
0
(c)
−1000 lb ft
V
(d)
−100 lb
−2100 lb
M
0
−1000 lb ft
(e)
−12000 lb/ft

**Figure 7.11**

SOLUTION: The shear at point $A$ is equal to $-100$ lb (downward force at $A$), and the shear difference between points $B$ and $A$ is equal to the area of the loading diagram between these two points, as follows:

$$V_B - V_A = -[\text{area}_{BA}]_{\text{load}}$$

Since the area of the loading diagram between $B$ and $A$ is equal to zero (there is no distributed load from $A$ to $B$), the value of shear at $B$ can be calculated by substituting $-100$ for $V_A$ and zero for the value of the area in the above equation, as follows:

$$V_B - (-100) = 0$$

$$V_B = -100 \text{ lb}$$

Now that the values of shear at $A$ and $B$ are known, they can be plotted as shown in Fig. 7.11(b). Notice that we calculated shear at points $A$ and $B$ only and assumed in this plot that the shear does not vary from $A$ to $B$ (the result being a horizontal straight line). The reason for this is that since *the derivative of shear is equal to the applied load* on the beam (which is zero between $A$ and $B$ in this case), the only function whose derivative is equal to zero is a linear function with constant value (i.e., a horizontal straight line).

The moment difference between points $B$ and $A$ is equal to the value of the area of the shear diagram between these two points, or

$$M_B - M_A = [\text{area}_{BA}]_{\text{shear}}$$

Since $M_A = 0$ ($A$ being the free end) and the area between $B$ and $A$ of the shear diagram (the area of the rectangle) is equal to $-1000$ lb ft ($-100 \times 10$), the value of the moment at point $B$ can be calculated as follows:

$$M_B - 0 = -1000$$

$$M_B = -1000 \text{ lb ft}$$

Note that since the derivative of moment is equal to shear and since the shear is constant between $A$ and $B$, the moment must change linearly (the derivative of a linear function is constant), and, furthermore, since the value of this constant (shear) is negative, the moment must be a straight line with diminishing slope. The values of moment between $A$ and $B$ are therefore known and can now be plotted, as shown in Fig. 7.11(c).

The shear difference between $C$ and $B$ is equal to the area of the loading diagram between $C$ and $B$. The shear at $B$, $V_B$, has already been calculated to be $-100$ lb, and the area under the loading diagram between $B$ and $C$ is $200 \times 10$, or 2000. The shear at $C$ is then calculated as follows:

$$V_C - V_B = -[\text{area}_{CB}]_{\text{load}}$$

$$V_C - (-100) = -2000$$

$$V_C = -2100 \text{ lb}$$

Notice that since the loading is uniform (constant) and also equal to the derivative of the shear, the shear has to be distributed linearly (with negative, or diminishing, slope). The shear at $B$ and $C$ are now known and can be plotted as shown in Fig. 7.11(d).

The moment difference between $C$ and $B$ is equal to the area of the shear diagram between $C$ and $B$, or

$$M_C - M_B = [\text{area}_{CB}]_{\text{shear}}$$

Since $M_B = -1000$ lb ft from earlier calculations and since the area of the shear diagram between $C$ and $B$ (the area of the trapezoid) is equal to (1/2) (2100 + 100) (10) but has a negative value, we have

$$M_C - (-1000) = -(1/2)\,(2100 + 100)\,(10)$$

$$M_C = -12{,}000 \text{ lb ft}$$

Notice that the moment distribution between $C$ and $B$, as shown in Fig. 7.11(e), is parabolic since its derivative (shear) is linear with diminishing slope (shear being negative).

The complete shear and moment diagrams obtained from using load–shear–moment relationships are shown in Fig. 7.11(d) and (e).

EXAMPLE 4: Draw shear and moment diagrams for the beam shown in Fig. 7.12(a) using load–shear–moment relationships.

SOLUTION: Support reactions can be easily calculated since the beam is symmetric and $R_A = R_F = 2000$ lb. Shear at various points is calculated as shown below.

$$V_A = 2000 \text{ (the reaction at } A)$$

$$V_B - V_A = -[\text{area}_{BA}]_{\text{load}} = 0$$

$$V_B - 2000 = 0$$

$$V_B = 2000 \text{ lb}$$

At point $B$, shear drops by 1000 lb (because of the 1000-lb downward concentrated load at $B$), or $V_B = 2000 - 1000 = 1000$ lb. Then,

$$V_C - V_B = -[\text{area}_{CB}]_{\text{load}} = 0$$

$$V_C - 1000 = 0$$

$$V_C = 1000 \text{ lb}$$

$$V_D - V_C = -[\text{area}_{DC}]_{\text{load}}$$

$$V_D - 1000 = -200 \times 10 = -2000$$

$$V_D = -1000 \text{ lb}$$

The value of shear changes from $+1000$ lb at $C$ to $-1000$ lb at $D$ linearly because its derivative (load) is constant between $C$ and $D$ and only the derivative of a linear function is constant.

$$V_E - V_D = -[\text{area}_{ED}]_{\text{load}}$$

$$V_E - (-1000) = 0$$

$$V_E = -1000 \text{ lb}$$

Since the 1000-lb load at $E$ drops the shear by 1000 lb,

$$V_E = -1000 - 1000 = -2000 \text{ lb}$$

$$V_F - V_E = -[\text{area}_{FE}]_{\text{load}}$$

$$V_F - (-2000) = 0$$

$$V_F = -2000 \text{ lb}$$

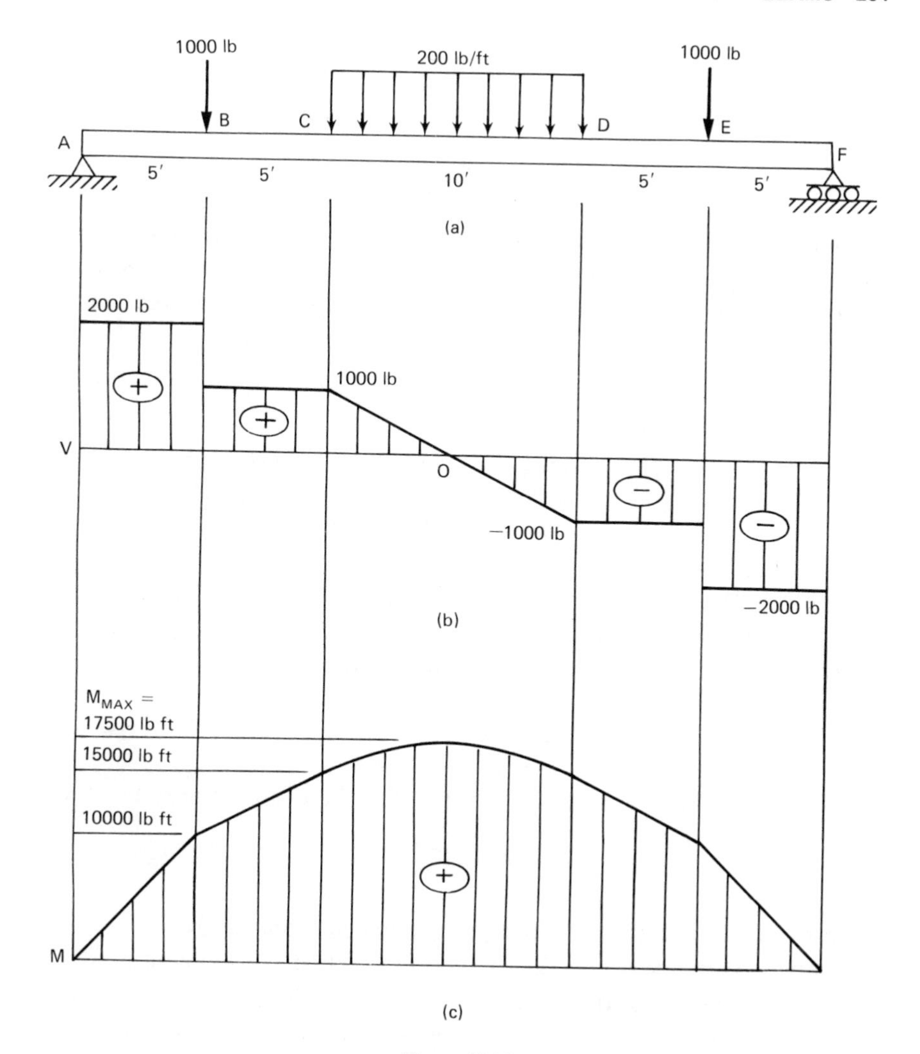


**Figure 7.12**

Now with the preceding information the shear diagram for the entire beam can be drawn, as shown in Fig. 7.12(b).

Similarly, the value of moment at each point can be evaluated, as follows:

$$M_A = 0 \text{ (hinged end support)}$$

$$M_B - M_A = [\text{area}_{BA}]_{\text{shear}}$$

$$M_B - 0 = 2000 \times 5 = 10{,}000$$

$$M_B = 10{,}000 \text{ lb ft}$$

Since shear has constant value between $A$ and $B$, moment should be distributed linearly between $A$ and $B$ (remember that *the derivative of moment is shear*, and that *only the derivative of a linear function is constant*).

$$M_C - M_B = [\text{area}_{CB}]_{\text{shear}}$$

$$M_C - 10{,}000 = 1000 \times 5 = 5000$$

$$M_C = 15{,}000 \text{ lb ft}$$

At point $O$ in the shear diagram, the value of shear is equal to zero, and since shear is the derivative of moment, the value of zero for the derivative indicates that moment assumes maximum or minimum value.

$$M_O - M_C = [\text{area}_{OC}]_{\text{shear}}$$

$$M_O - 15{,}000 = (1/2)(1000 \times 5)(\text{area of the triangle})$$

$$M_O = 17{,}500 \text{ lb ft}$$

Notice that point $O$ is the mid-point between $C$ and $D$. Now we know the values of moment at points $C$ and $O$ and that the moment between these two points is distributed parabolically with diminishing slope since its derivative is linear (only the derivative of a parabolic function, which is a second-order polynomial, is a straight line).

$$M_D - M_O = [\text{area}_{DO}]_{\text{shear}}$$

$$M_D - 17{,}500 = -(1/2)(1000 \times 5) = -2500$$

$$M_D = 15{,}000 \text{ lb ft}$$

Notice that the area appears with negative sign, since the value of shear at $D$ is negative or the area is under the reference axis for shear.

$$M_E - M_D = [\text{area}_{ED}]_{\text{shear}}$$

$$M_E - 15{,}000 = -1000 \times 5 = -5000$$

$$M_E = 10{,}000 \text{ lb ft}$$

$$M_F - M_E = [\text{area}_{FE}]_{\text{shear}}$$

$$M_F - 10{,}000 = -2000 \times 5 = -10{,}000$$

$$M_F = 0$$

The fact that the value of moment at $F$ turned out to be zero checks with the fact that support $F$ is a pin support that does not have resistance to moment ($M_F = 0$). The moment diagram for the entire beam is shown in Fig. 7.12(c).

## Method of Superposition

In determining the deflection of beams, which is the subject of strength of materials, one has to calculate the moment of the area under a moment diagram. If the beam is under several concentrated loads or a combination of various types of loading, such as concentrated load, distributed load, and pure moment, etc., the area under the moment diagram will have a complicated shape and calculating its moment will not be so simple. The use of the *superposition method* will considerably simplify such calculations. In this method, the moment diagram is drawn "in parts" by considering only one of the loads on the beam at a time and repeating the process for all the loads on the beam. The sum of these individual moment diagrams is equivalent to the moment diagram for the beam with all the loads acting on it.

EXAMPLE 5: Sketch the moment diagram for the cantilever beam shown in Fig. 7.13(a) using the method of superposition.

SOLUTION: Figure 7.13(b) shows the moment diagram of the beam using only the end load of 1000 lb alone. Figure 7.13(c) shows the moment diagram of the beam with only the 2000-lb load on it. Notice that since there is no load on the beam from $A$ to $B$ in this case, the moment is equal to zero, and the moment diagram starts from a point below point $B$. Similarly, the moment diagram of the beam with only the 3000-lb load on it can be drawn as shown in Fig. 7.13(d).

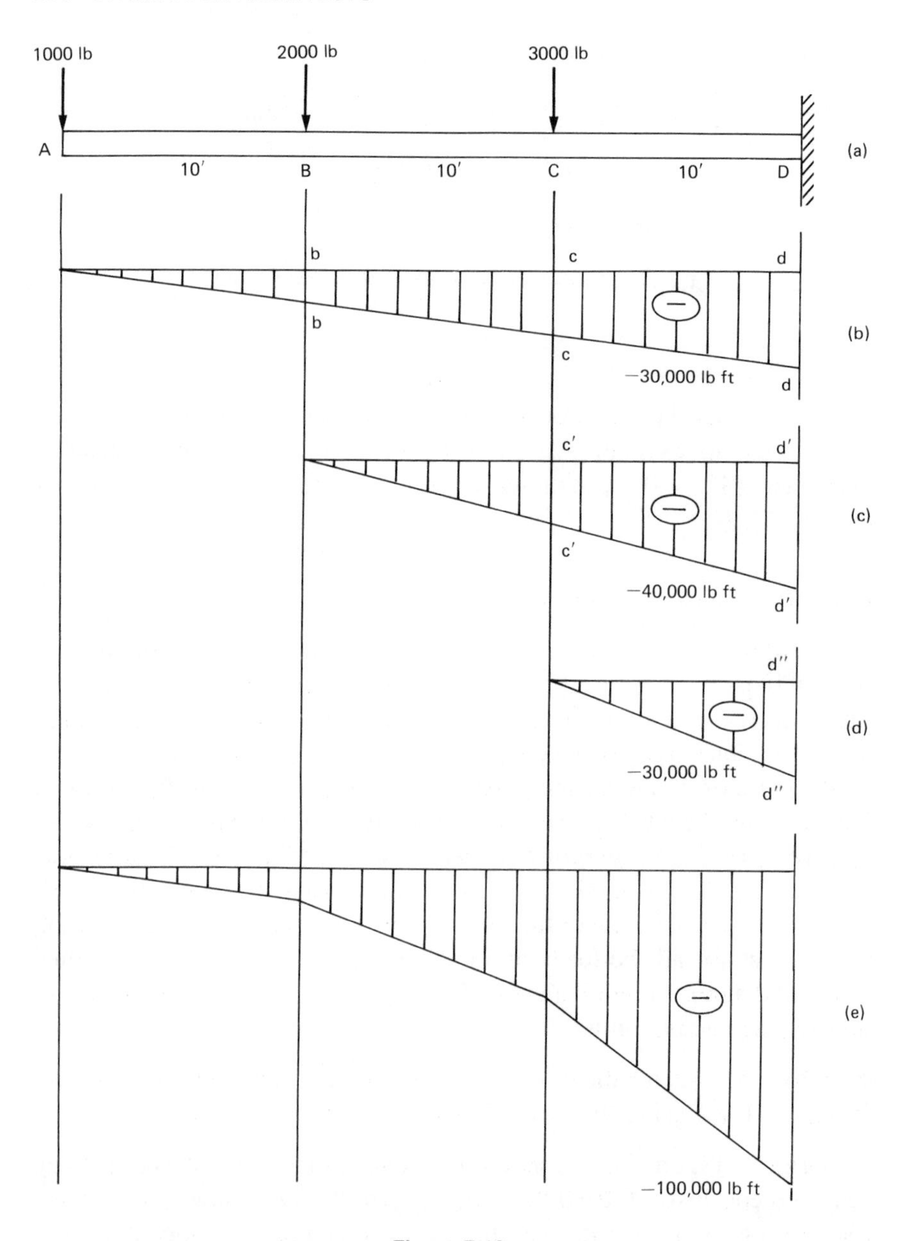
1000 lb
2000 lb
3000 lb
A
B
C
D
10′
10′
10′
(a)
b
c
d
−30,000 lb ft
(b)
c′
d′
−40,000 lb ft
(c)
d″
−30,000 lb ft
(d)
−100,000 lb ft
(e)

**Figure 7.13**

Finally, the moment diagram for the beam in Fig. 7.13(a) can be obtained by adding the three individual moment diagrams (superposition). To add these three diagrams, draw vertical lines at *A*, *B*, *C*, and *D* and read the values of the moment on these vertical lines from each individual moment diagram and add them up algebraically.

At point *A*, the moment is zero; at point *B*, from Fig. 7.13(b), the value of the moment is equal to $-10{,}000$ lb ft (this can be obtained from the similarity of the triangles in this case or simply by measuring *bb* using the scale used to plot the diagrams); there is no contribution from Figs. 7.13(c) and (d) at this station. Similarly, the value of the moment at *C* is equal to the sum of the values of individual moments from Figs. 7.13(b) and (c), or $(-20{,}000) + (-20{,}000) = -40{,}000$ lb ft, with no contribution from Fig. 7.13(d). Or, similarly, one can measure *cc* from Fig. 7.13(b) and *c'c'* from Fig. 7.13(c), which are both equal to 2000 lb ft, and find their algebraic sum. Finally, the value of moment at *D* is equal to $(-30{,}000) + (-40{,}000) + (-30{,}000) = -100{,}000$ lb ft, which can also be obtained similarly by adding lengths *dd*, *d'd'*, and *d"d"*.

In between these points, the moments are added according to the form of the individual moment distribution. In this case, since the individual moment distributions are all linear, the sum will also be linear (the sum of several straight-line distributions is another straight-line distribution). Figure 7.13(e) shows the moment diagram achieved by the method of superposition for the beam shown in Figure 7.13(a). For the purpose of determining the deflection of the beam, the moment of the area of this moment diagram has to be calculated. Since this area is the sum of the moment areas in Figs. 7.13(b), (c), and (d), its moment is therefore equal to the sum of the moments of these areas. Now that each diagram is a simple area with known centroid, calculations become simpler.

EXAMPLE 6: Draw the moment diagram of the beam shown in Fig. 7.14(a) using the method of superposition.

SOLUTION: Figure 7.14(b) shows the moment diagram of the beam with a concentrated load of only 500 lb on it. The distributed load is the sum of rectangular and triangular distributed loads, or a trapezoidal load. Considering each one of these loads on the beam separately results in the bending moment diagrams shown in Figs. 7.14(c) and (d). The moment diagram of Fig. 7.14(c) is parabolic in a way similar to those studied earlier. The moment diagram in Fig. 14(d), however, is cubic, which

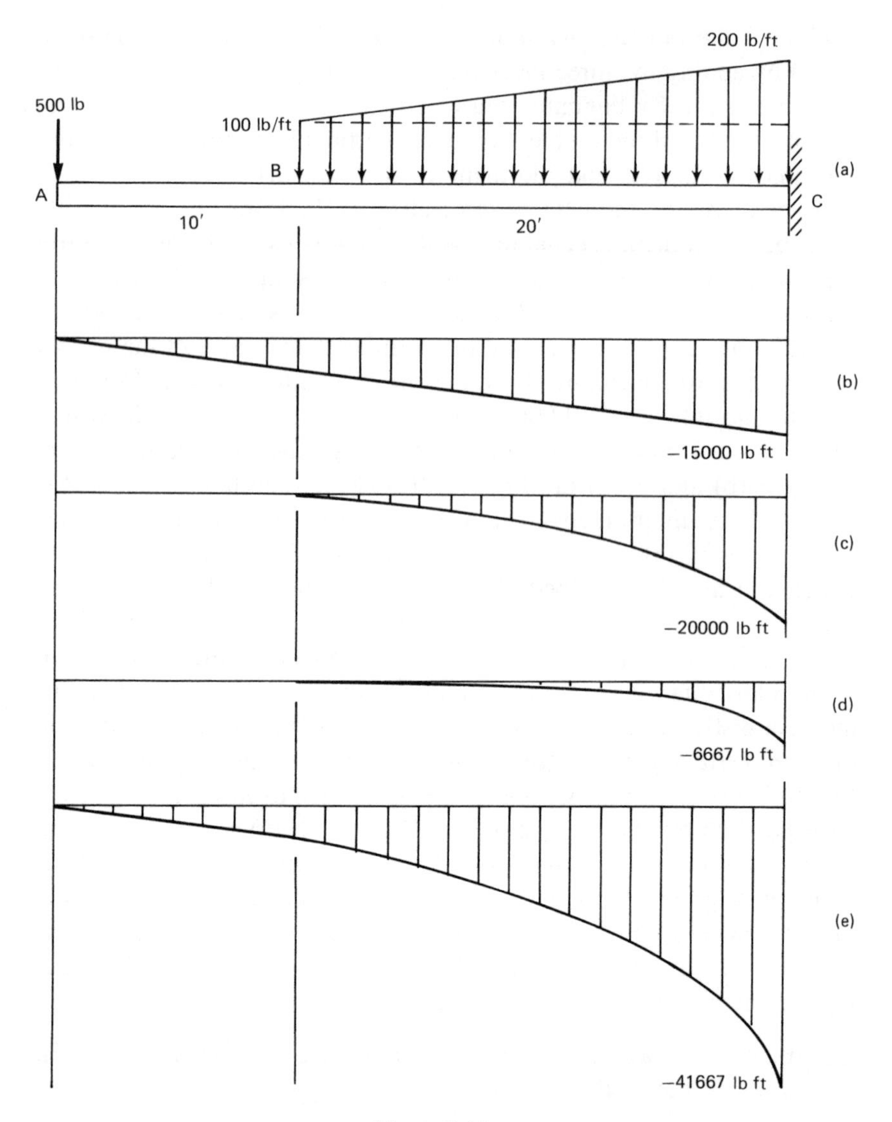


**Figure 7.14**

will be discussed in detail in the next example. The moment diagram in Fig. 7.11(e) is the sum of the three individual diagrams, 7.14(b), (c), and (d). Notice that in region *A–B* we have only a straight line, which is duplicated in Fig. 7.14(e). In region *B–C*, diagrams of linear, parabolic, and cubic distributions appear, resulting in a final cubic moment distribution, as shown in Fig. 7.14(e). To draw this curve accurately, read the values of moment from each individual moment diagram at various points between $B$ and $C$, add them up individually, plot them, and then pass a smooth curve between them.

The problem of plotting moment diagrams by the method of superposition can be made much simpler by merely sketching them approximately. Since the loads applied to beams are of a limited number of types, one can select the appropriate diagrams from those given in Fig. 7.15.

EXAMPLE 7: Plot the shear and moment diagram for the beam shown in Fig. 7.16(a).

SOLUTION: This type of distributed load is called a *triangular load* and it is expressed as the intensity of the load in the same way as the uniformly distributed load discussed earlier. In this case the maximum intensity of the load as shown measured from $B$ to $C$ is 200 lb/ft. The total amount of the distributed load is equal to the area of the load triangle, or $(200 \times 6)/2 = 600$ lb. Using the free-body diagram of the entire beam, the reactions of the beam can be calculated as: $R_A = 150$ lb and $R_C = 450$ lb.

To draw the shear and moment diagrams, the beam should be divided into two regions, one from $A$ to $B$ and the other from $B$ to $C$. In the first region, the shear has a constant value, $V_1 = +150$ lb, whereas the moment changes linearly since $M_2 = 150x$ lb ft.

The shear will start to decrease from point $B$. To derive the equation for shear, consider section $a$–$a$ between points $B$ and $C$. The free-body diagram of the portion of the beam to the left of $a$–$a$ is shown in Fig. 7.16(b). The total amount of distributed load is equal to the area of triangle $BEF$, but this triangle is proportional to the triangle $BCD$ in Fig. 7.16(a). Therefore,

$$\frac{BF}{BC} = \frac{EF}{DC}$$

After substituting for the known values,

$$\frac{x - 2}{6} = \frac{w(x)}{200}$$

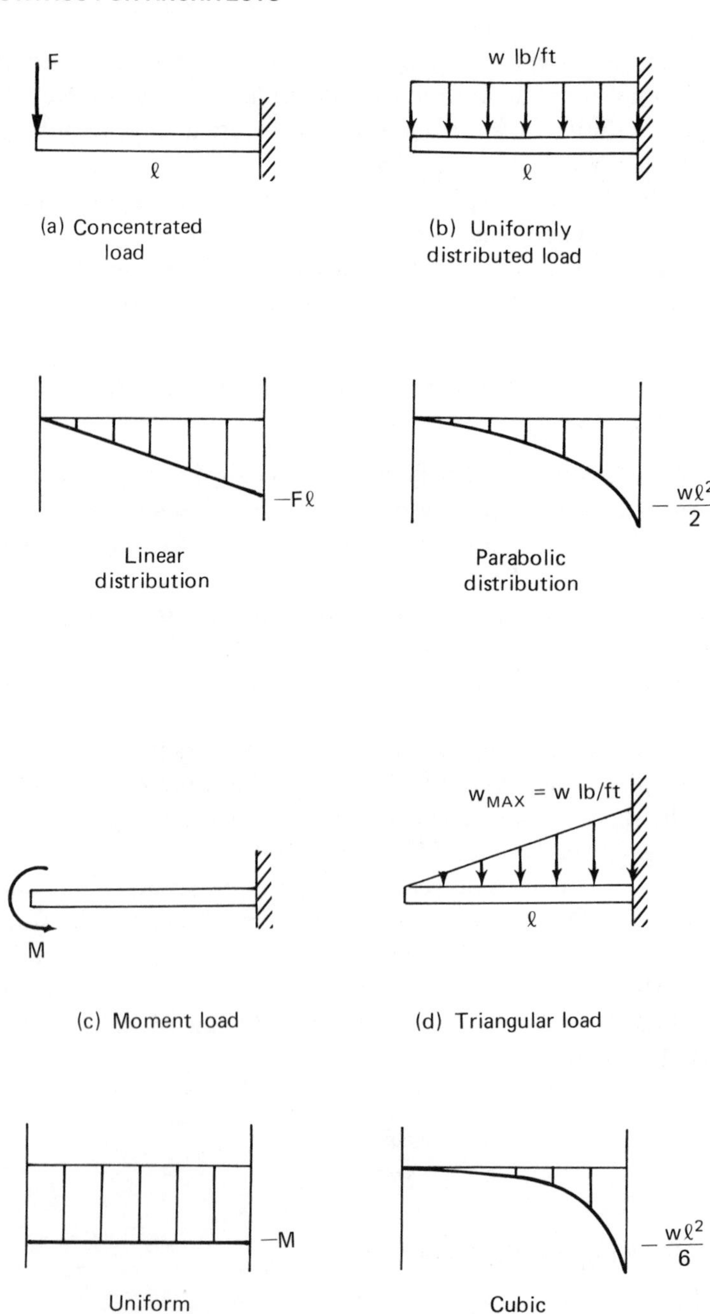
F
ℓ
(a) Concentrated load
w lb/ft
ℓ
(b) Uniformly distributed load
−Fℓ
Linear distribution
− wℓ²/2
Parabolic distribution
M
(c) Moment load
w_MAX = w lb/ft
ℓ
(d) Triangular load
−M
Uniform distribution
− wℓ²/6
Cubic distribution

**Figure 7.15**

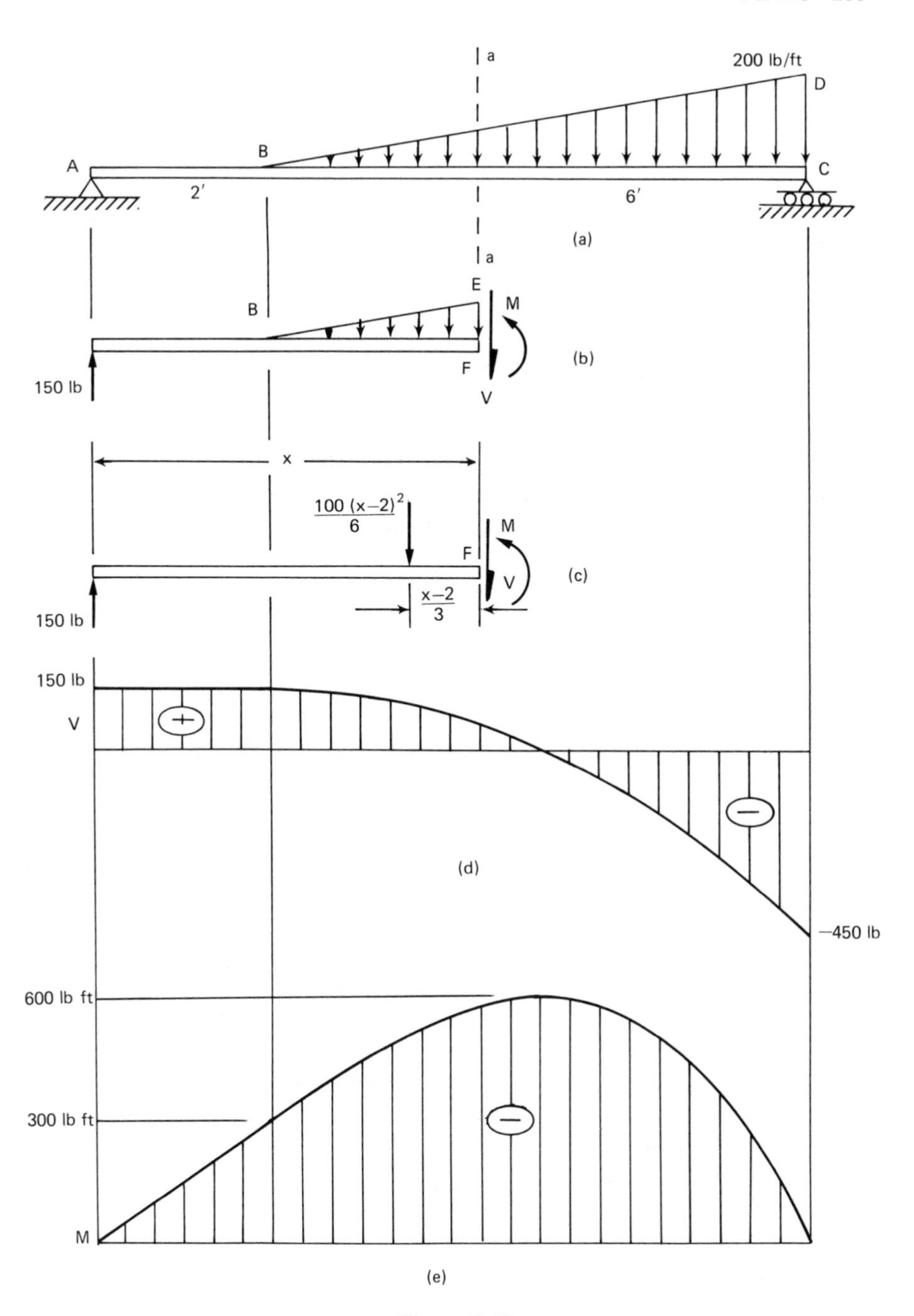
a
200 lb/ft
D
B
A
C
2′
6′
(a)
a
E
B
M
(b)
F
150 lb
V
x
$\frac{100(x-2)^2}{6}$
M
F
(c)
V
$\frac{x-2}{3}$
150 lb
150 lb
V
+
−
(d)
−450 lb
600 lb ft
300 lb ft
−
M
(e)

**Figure 7.16**

Solving for $w(x)$,

$$w(x) = \frac{200}{6}(x - 2)$$

The area of triangle *BEF*, which can now be calculated, is equal to the total concentrated load equivalent to the triangular distributed load (*BEF*), or

$$\frac{1}{2}w(x)(x - 2) = \frac{1}{2}\frac{200}{6}(x - 2)(x - 2) = \frac{100}{6}(x - 2)^2$$

This concentrated load acts at the centroid of the triangle shown in Fig. 7.16(c). Using equilibrium equations, the expressions for $V$ and $M$ at section $a$–$a$ can now be derived, as follows:

$$+\uparrow\Sigma F_y = 0$$

$$150 - \frac{100}{6}(x - 2)^2 - V_2 = 0$$

$$V_2 = 150 - \frac{100}{6}(x - 2)^2$$

$$+\circlearrowright\Sigma M_F = 0$$

$$150x - \frac{100(x - 2)^2}{6}\left(\frac{x - 2}{3}\right) - M_2 = 0$$

$$M_2 = 150x - \frac{100}{18}(x - 2)^3$$

All four expressions for the two regions are now listed as follows:

$$\begin{cases} V_1 = 150 \text{ lb} \\ M_1 = 150x \text{ lb ft} \end{cases} \qquad 0' \leq x \leq 2'$$

$$\begin{cases} V_2 = 150 - \dfrac{100}{6}(x - 2)^2 \\ M_2 = 150x - \dfrac{100}{18}(x - 2)^3 \end{cases} \qquad 2' \leq x \leq 8'$$

The above equations can now be plotted. The shear is constant in the first region and has parabolic form in the second (because the highest power of $x$ in $V_2$ is equal to 2). The distribution of moment in the first region is the equation of straight line, whereas the distribution of moment in the second region is cubic (because the highest power of $x$ in $M_2$ is equal to 3). Figures 7.16(d) and (e) show the shear and moment diagrams for this beam.

EXAMPLE 8: Plot the shear and moment diagram for the beam shown in Fig. 7.17(a).

SOLUTION: This beam is called a "hinged beam." In fact, there are two beams, $AC$ and $CE$, hinged at point $C$ (the construction of such beams is usually resorted to for economic reasons). Point $C$ acts like a hinge, i.e., $M_C = 0$. This new condition provides one more equation.

To calculate reactions at $A$, the free-body diagrams of $AC$ and $CE$ are separately drawn, as shown in Fig. 7.17(b) and (c). Notice that in Fig. 7.17(c), no moment appears at $C$. Using the equations of equilibrium for the free-body diagram of $CDE$ first, the value of $V_C$ can be calculated:

$$+\circlearrowright \Sigma M_C = 0$$

$$2000 \times 7.5 - R_D \times 5 = 0$$

$$R_D = 3000 \text{ lb}$$

$$\Sigma F_y = 0$$

$$R_D - V_C - 2000 = 0$$

$$V_C = 1000 \text{ lb}$$

Using value of $V_C = 1000$ lb, the equations of equilibrium can now be applied to the free-body diagram of $ABC$, as shown in Fig. 7.17(b):

$$\Sigma F_y = 0$$

$$V_C - R_A = 0$$

$$R_A = 1000 \text{ lb}$$

$$+\circlearrowright \Sigma M_C = 0$$

$$M_A + 1000 - R_A \times 10 = 0$$

$$M_A = 9000 \text{ lb ft}$$

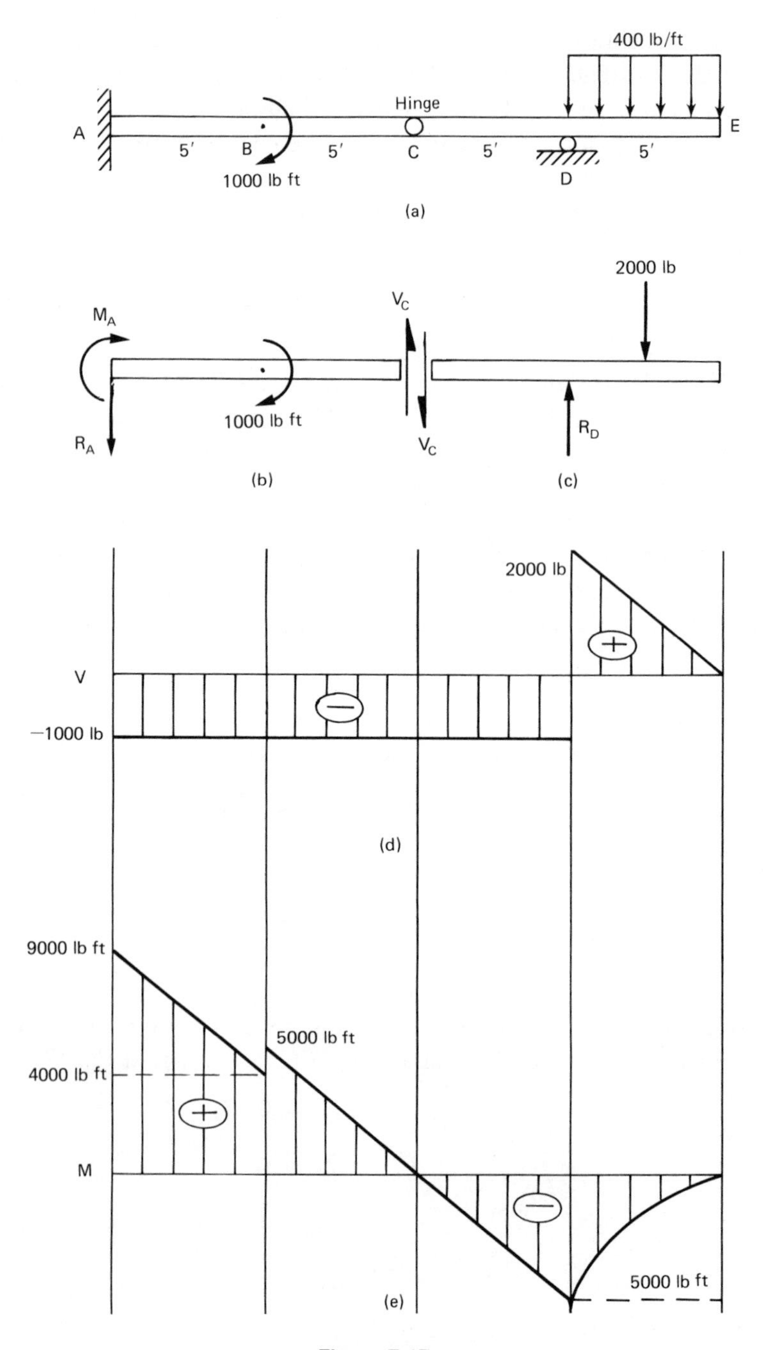
400 lb/ft
Hinge
A
E
5′
B
5′
C
5′
5′
D
1000 lb ft
(a)
2000 lb
$M_A$
$V_C$
1000 lb ft
$R_A$
$V_C$
$R_D$
(b)
(c)
2000 lb
V
−1000 lb
(d)
9000 lb ft
5000 lb ft
4000 lb ft
M
5000 lb ft
(e)

**Figure 7.17**

After calculation of reactions, the beam can be divided into four regions: $AB$, $BC$, $CD$, and $DE$. As in the previous problems, the shear and moment can be calculated for each region using the relationships between $w$, $V$, and $M$, and the diagrams plotted as shown in Figs. 7.17(d) and (e).

We start with the value of shear at $A$, as follows:

$$V_A = R_A = -1000 \text{ lb (the reaction at } A)$$

and since there is no distributed load between $A$ and $D$, the value of shear is constant from $A$ to $D$. At point $D$, the shear will be increased by 3000 lb (the reaction at $D$). Thus,

$$V_D = -1000 + 3000 = 2000 \text{ lb}$$

$$V_E - V_D = -[\text{area}_{ED}]_{\text{load}}$$

$$V_E - 2000 = -400 \times 5 = -2000 \text{ lb}$$

$$V_E = 0$$

which also checks with the fact that $E$ is a free end and that at free end shear is equal to zero.

Between points $D$ and $E$ shear varies linearly since its derivative (the distributed load) is constant. Similarly, the values of moment at different points can be calculated, as shown below:

$$M_A = 9000 \text{ lb ft (reactive moment at } A)$$

$$M_B - M_A = [\text{area}_{BA}]_{\text{shear}} = -1000 \times 5 = -5000$$

$$M_B = 4000 \text{ lb ft}$$

From $A$ to $B$, moment diminishes linearly since its derivative (the shear) is constant and has a negative value.

At point $B$, a positive moment of 1000 lb ft will increase the value of the moment to $M_B = 4000 + 1000 = 5000$ lb ft. Since there is no other externally applied moment from $B$ to $D$, and since the value of shear is constant between $B$ and $D$, the moment changes linearly from $B$ to $D$.

$$M_D - M_B = [\text{area}_{DB}]_{\text{shear}}$$

$$M_D - 5000 = -1000 \times 10 = -10{,}000 \text{ lb ft}$$

$$M_D = -5000 \text{ lb ft}$$

$$M_E - M_D = [\text{area}_{ED}]_{\text{shear}}$$

$$M_E - (-5000) = (1/2)(2000)(5) = 5000 \text{ lb ft}$$

$$M_E = 0$$

which again checks with the fact that $E$ is a free end and the value of moment at a free end is equal to zero. Since the shear has a linear distribution, moment should have parabolic distribution and since shear from $D$ to $E$ is decreasing, it indicates that moment has diminishing slope (its shape is concave up).

The purpose of solving this problem was:

1. To introduce a method of handling "hinged beam."
2. To introduce an externally applied moment to the beam and examine its influence in moment and shear diagrams.
3. To show that a nonzero moment can exist at a hinge support. Notice that in all other problems, we stated that the value of moment at a hinge support is equal to zero. This is true provided the hinge support is an end support. As is evident from the moment diagram of Fig. 7.17(e), a hinge support ($D$) that is not an end support can have nonzero moment.

## 7.7 Beams with Axial Loads

In Sec. 7.2, *beams* were defined as long bars that sustain loads. These loads are usually normal to the axis of the beam. If the loads are aligned with the axis of the bar, the bar is called a *column*, and a bar that is under the combination of both normal and axial loads is called a *beam–column*. For various reasons besides the normally applied loads, axial loads are also present, and such beams are really beam-columns. Since their beam behavior is more dominant, however, they are still called beams rather than beam-columns. As far as the design and analysis of beam-columns are concerned, a third diagram, called a *normal force* or *axial force diagram*, is needed. Axial force diagrams are usually plotted along with the shear and bending diagrams. The following problems require axial load diagrams to be plotted.

EXAMPLE 9: Draw the axial force diagram for the beam shown in Fig. 7.18(a).

SOLUTION: The reactions of supports $A$ and $D$ can be determined by drawing the free-body diagram of the entire beam, as shown in Fig. 7.18(b), and using the equilibrium equations.

$$\Sigma F_x = 0$$

$$A_x - 1000 = 0$$

$$A_x = 1000 \text{ lb}$$

Since we are interested in plotting the axial load diagram only, reactions $A_y$ and $D$ are not needed and are therefore not calculated.

As in shear and moment diagrams, the beam is divided into regions; in this case, only three of them: $AB$, $BC$, and $CD$. In each region, the beam is sectioned, and the values of axial force $N$ ($N$ stands for normal force, because axial load is normal to the cross-sectional area of the beam). If the beam is sectioned somewhere between $A$ and $B$, the free-body diagram of the portion of the beam to the left of the cutting plane will appear as in Fig. 7.18(c). Because of the axial force, $A_x = 1000$ lb, there will be a normal force, $N = 1000$ lb, at the section. Shear and moment also exist at the section but are not used in this problem. Forces $A_x$ and $N$ are compressing the portion of the beam shown in Fig. 7.18(c). If the portion of the beam is under compressive load, the sign of $N$ is usually taken as negative. If the beam is sectioned in regions $BC$ and $CD$, the value of normal load will be equal to zero because $A_x = 1000$ lb and cancels out the horizontally applied load of 1000 lb at $B$. The diagram for the axial load is shown in Fig. 7.18(d).

EXAMPLE 10: Draw the axial (or normal) force diagram for the beam shown in Fig. 7.19(a).

SOLUTION: The reaction at $A$ is calculated using the free-body diagram of the entire beam and the equilibrium equations [we need only $A_x$ to

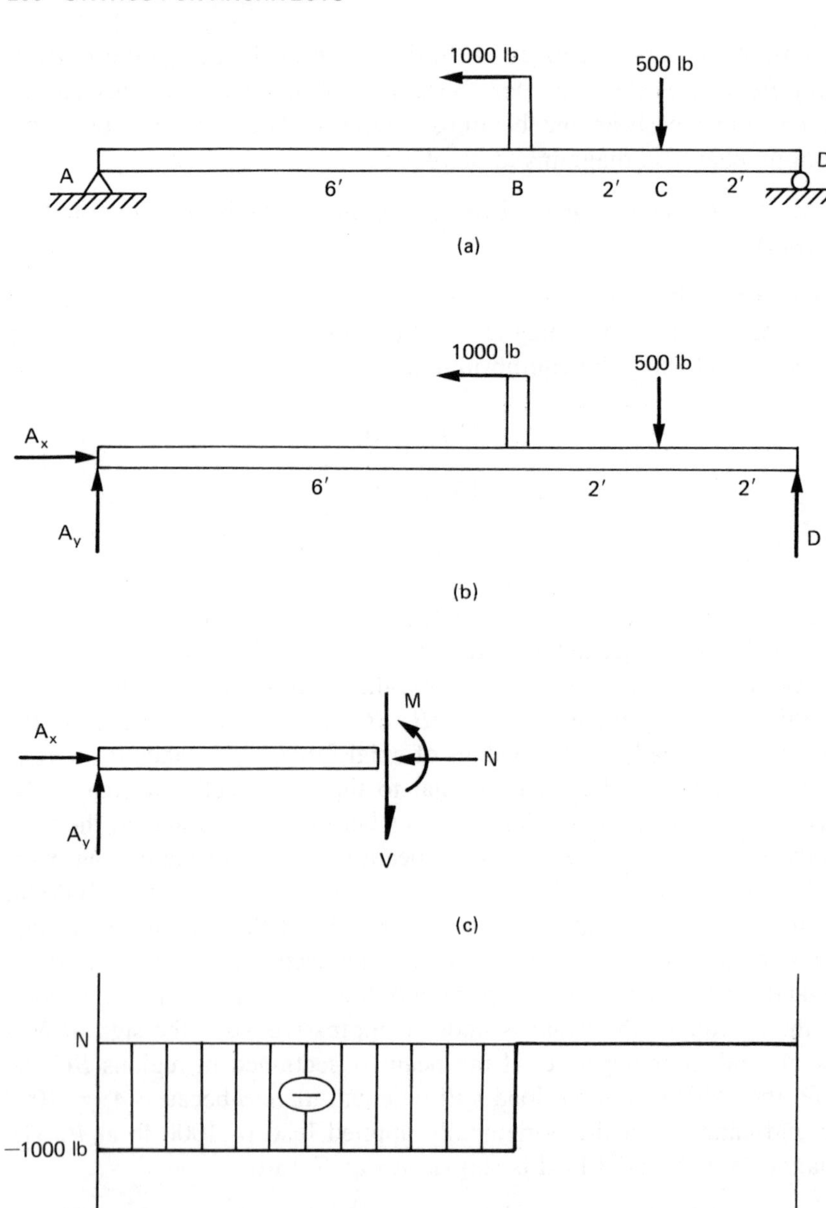
1000 lb
500 lb
A
D
6′
B
2′
C
2′
(a)
1000 lb
500 lb
$A_x$
6′
2′
2′
$A_y$
D
(b)
M
$A_x$
N
$A_y$
V
(c)
N
−1000 lb
(d)

**Figure 7.18**

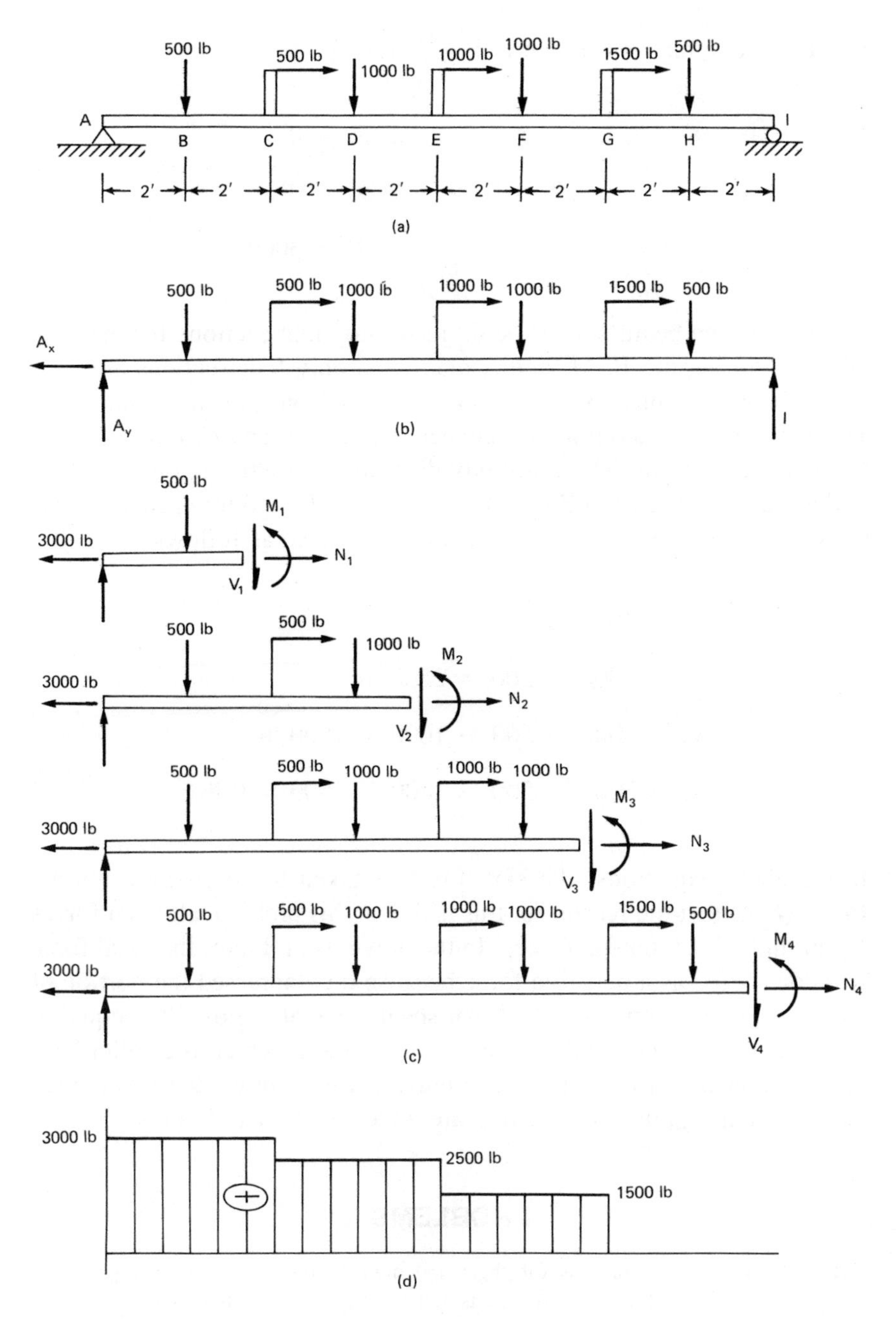

500 lb
500 lb
1000 lb
1000 lb
1000 lb
1500 lb
500 lb
A
B
C
D
E
F
G
H
I
2′
(a)
$A_x$
$A_y$
(b)
I
3000 lb
$M_1$
$N_1$
$V_1$
$M_2$
$N_2$
$V_2$
$M_3$
$N_3$
$V_3$
$M_4$
$N_4$
$V_4$
(c)
3000 lb
2500 lb
1500 lb
+
(d)


**Figure 7.19**

plot the $N$ diagram, as shown in Fig. 7.19(b)]:

$$\Sigma F_x = 0$$

$$-A_x + 500 + 1000 + 1500 = 0$$

$$A_x = 3000 \text{ lb}$$

Although the beam should be divided into eight sections for the purpose of plotting the $V$, $M$, and $N$ diagrams, only four sections are necessary in this problem since we are interested only in the $N$ diagram. This is because the axial loads between $A$ and $C$, $C$ and $E$, $E$ and $G$, and $G$ and $I$ are uniform. The free-body diagrams of each sectioned portion of the beam are shown in Fig. 7.19(c). Using $\Sigma F_x = 0$ for each diagram, the values for each normal force can be calculated as follows:

$$N_1 = 3000 \text{ lb}$$

$$N_2 = 3000 - 500 = 2500 \text{ lb}$$

$$N_3 = 3000 - 500 - 1000 = 1500 \text{ lb}$$

$$N_4 = 3000 - 500 - 1000 - 1500 = 0 \text{ lb}$$

In the above equations, the sign for $N$ is taken to be positive normal force. (Note that the reason for this is that in this problem all axial forces $N_1$ through $N_4$ are tensile forces. In the previous problem, the axial force was considered as a negative force because it compressed the portion of the beam in question.) The fact that shear force at region $GI$ is equal to zero checks with the condition of the support at $I$, which is a roller having no resistance for axial load. Figure 7.19(d) shows the axial force diagram. All regions from $A$ to $G$ are under tensile axial loads.

## PROBLEMS

**7.1–7.10.** Write the equations for shear and moment for the beams in Figs. P7.1 through P7.10. Plot these diagrams and identify points of maximum shear and moment as well as all the changing points on the diagrams.

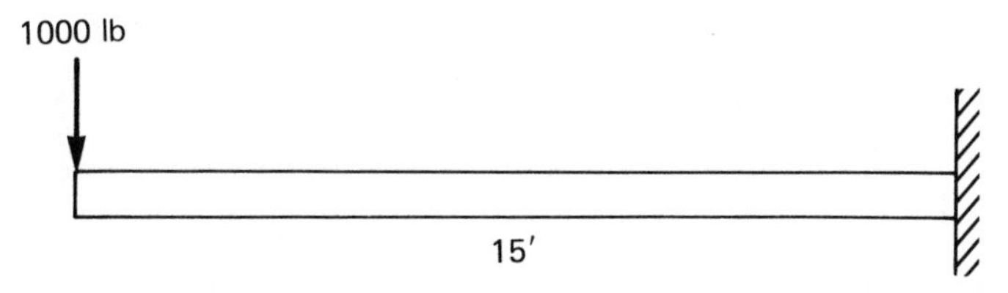


**Figure P7.1**

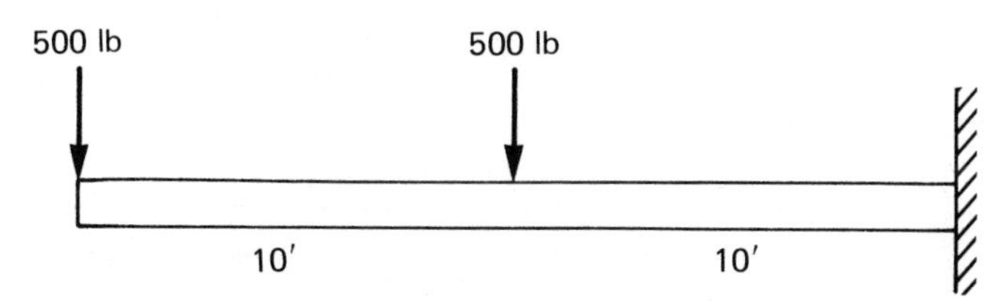


**Figure P7.2**

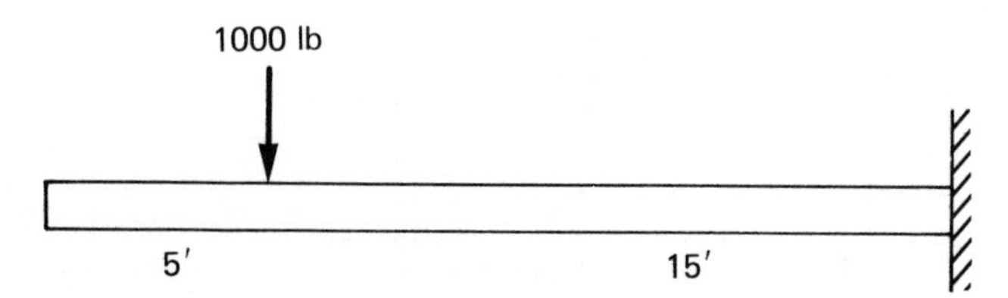


**Figure P7.3**

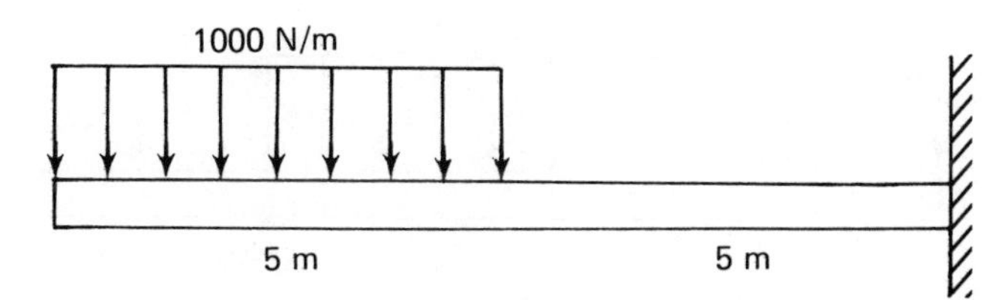


**Figure P7.4**

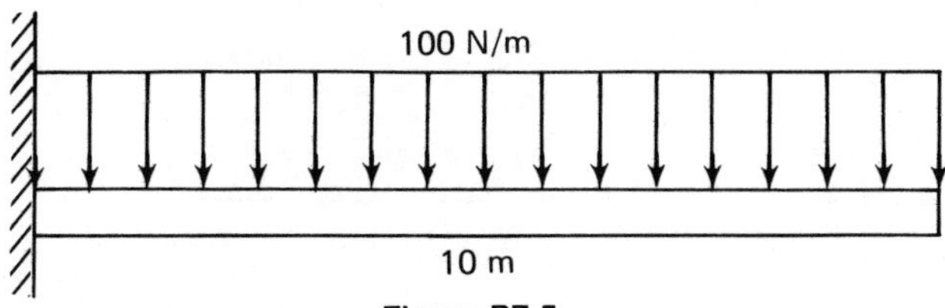


**Figure P7.5**

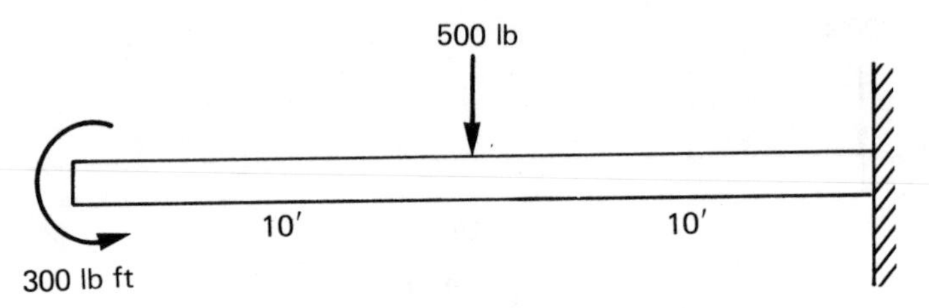


Figure P7.6

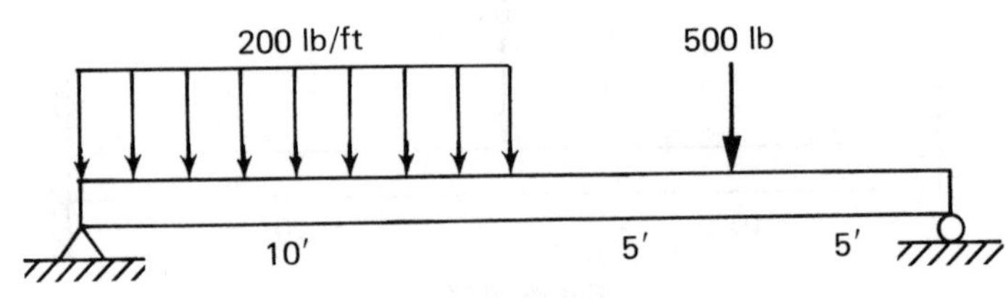


Figure P7.7

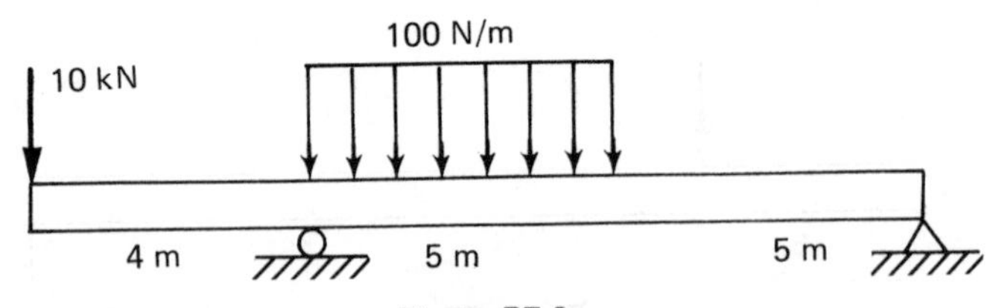


Figure P7.8

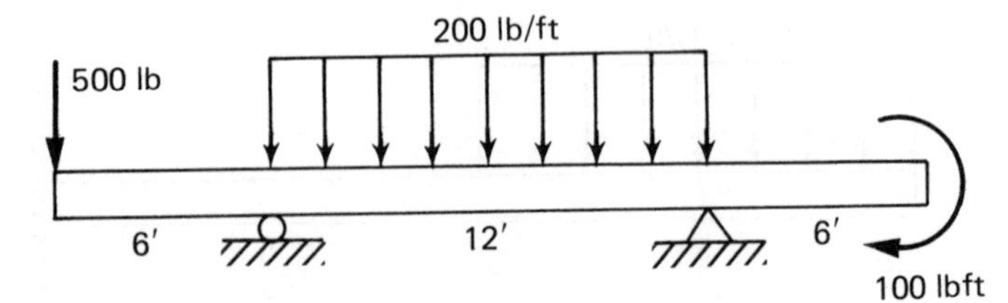


Figure P7.9

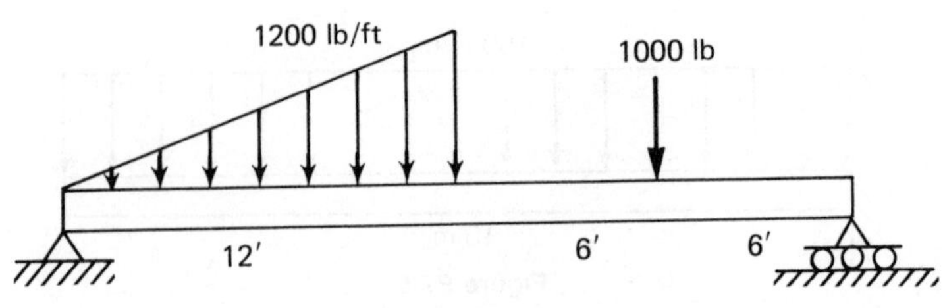


Figure P7.10

**7.11–7.15.** In the beams shown in Figs. P7.11 through P7.15, draw shear and bending moment diagrams using the relationships between load, shear and moment.

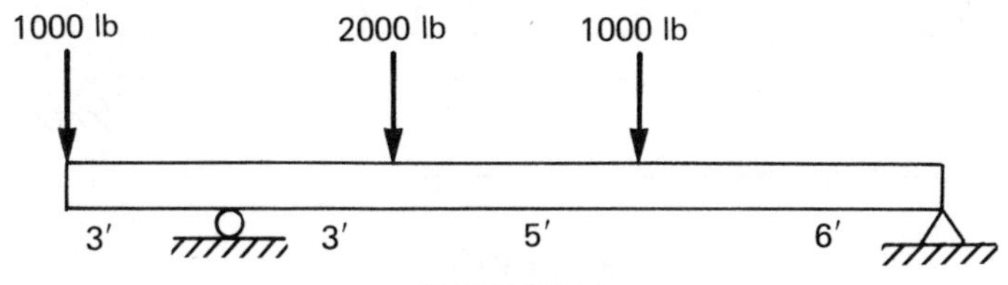


**Figure P7.11**

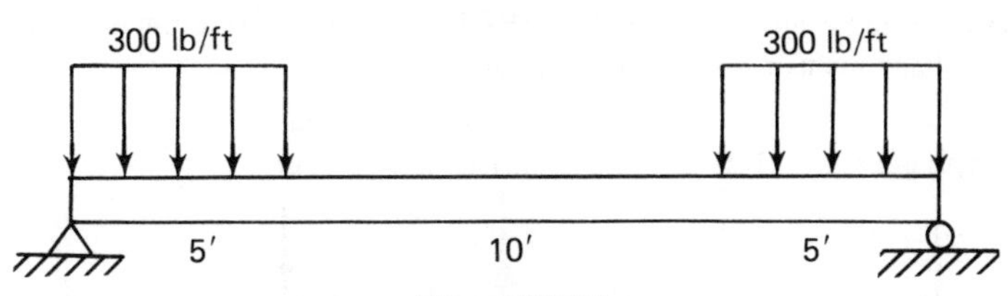


**Figure P7.12**

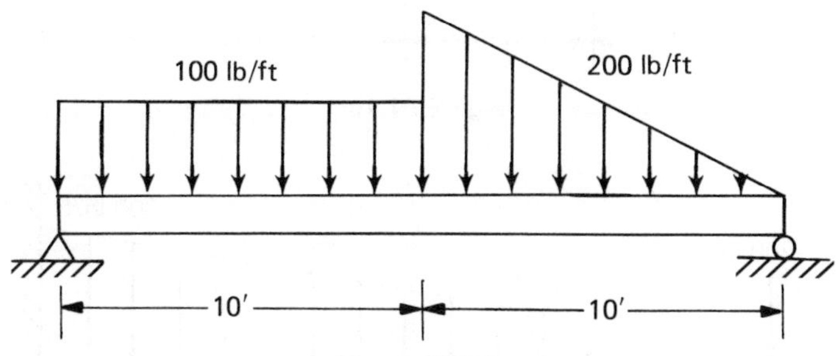


**Figure P7.13**

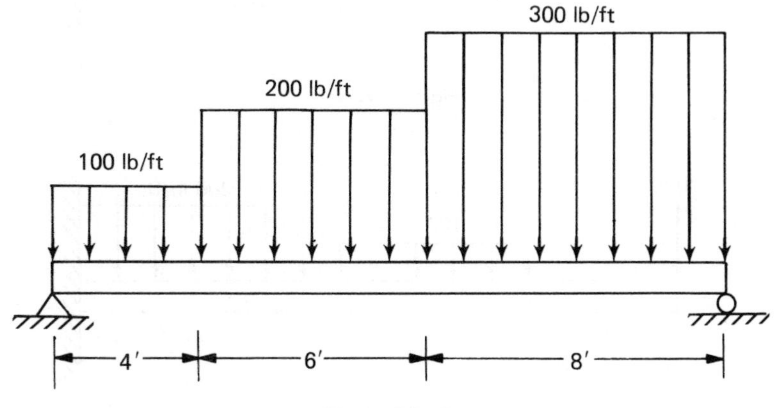


**Figure P7.14**

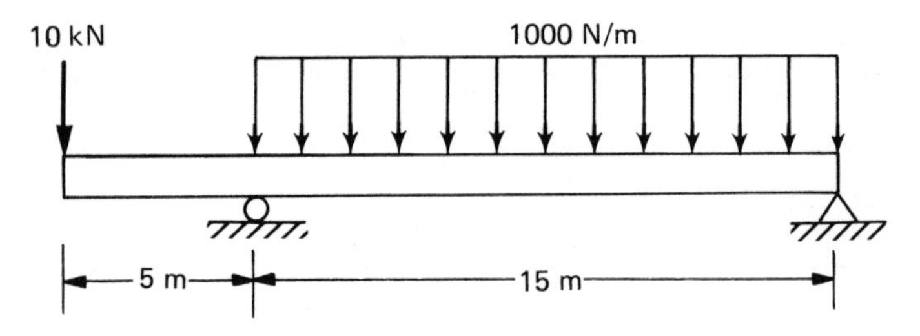


Figure P7.15

**7.16–7.20.** For the beams shown in Figs. P7.16 through P7.20, draw the bending moment diagrams using the method of superposition.

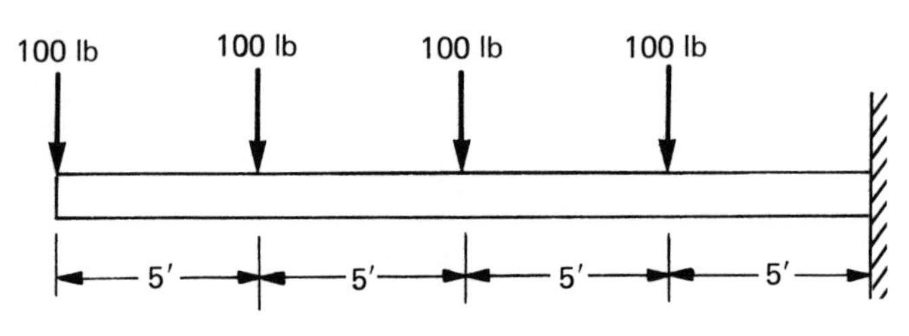


Figure P7.16

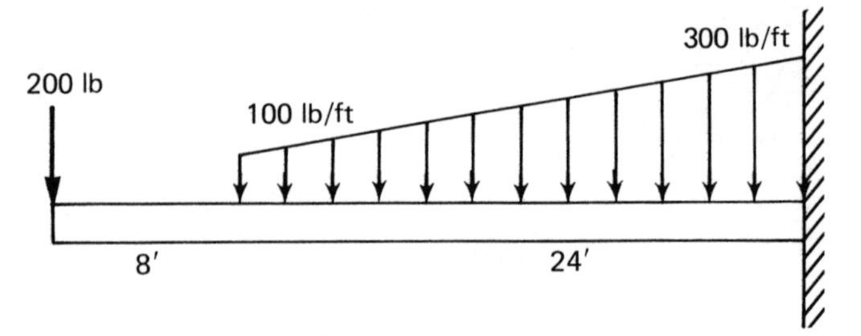


Figure P7.17

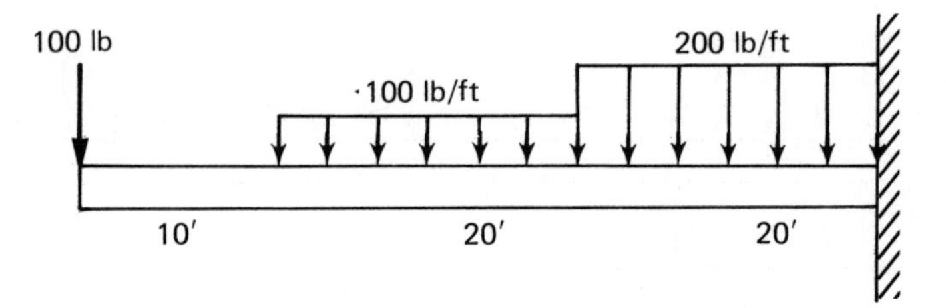


Figure P7.18

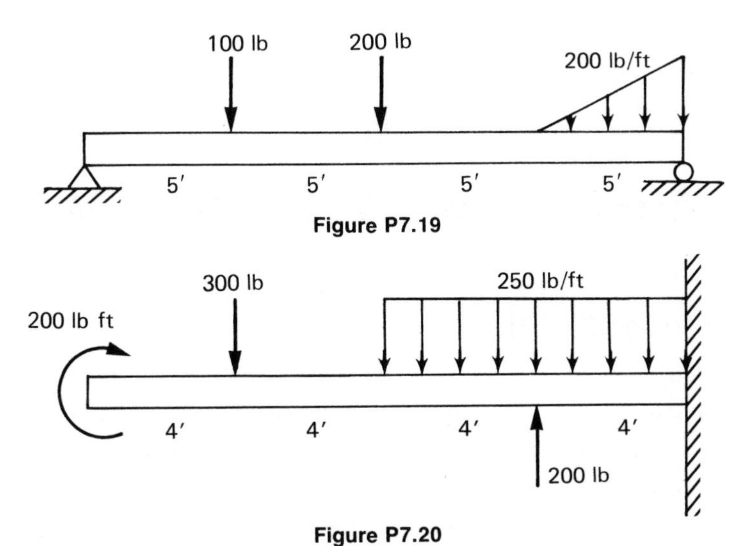


Figure P7.19

Figure P7.20

**7.21–7.23.** For the beams shown in Figs. P7.21, P7.22, and P7.23, draw the axial load diagrams.

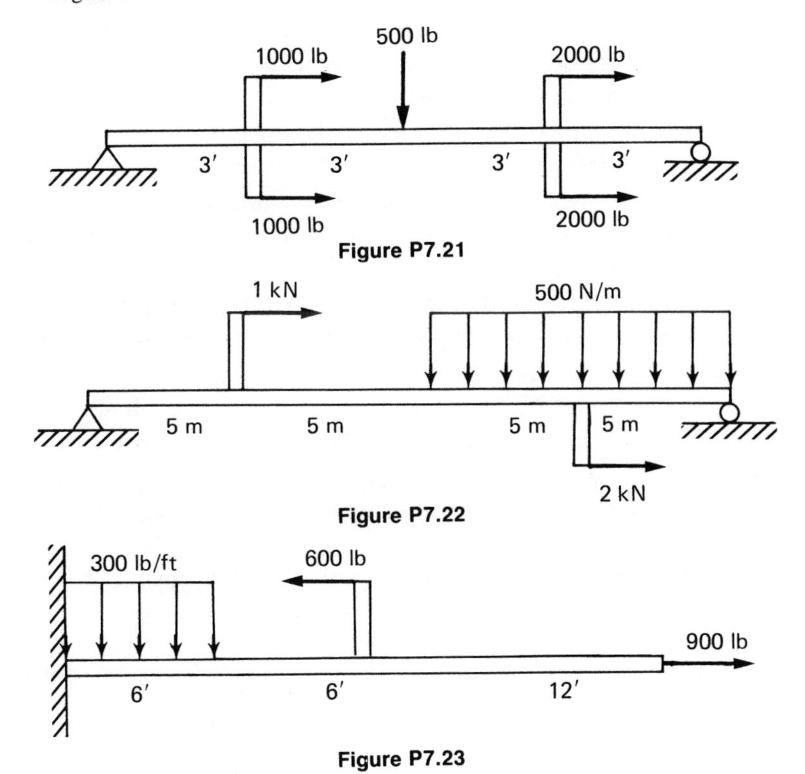


Figure P7.21

Figure P7.22

Figure P7.23

# 8

# AREA MOMENT OF INERTIA

## 8.1 INTRODUCTION

Often in many problems in mechanics (strength of materials or structures) certain quantities that are purely geometric appear over and over. To simplify the solution of such problems, quantities that have common forms of this kind are calculated once and for all for repeated use.

In the formulations for bending a beam, loading a column, applying torsional load to a rod, etc., these quantities appear as properties of the cross-sectional areas of the beam, column, rod, etc. Although of a numerical nature in that they are related to the values of the dimensions of these cross-sectional areas, yet they cannot be measured with any measuring device. These quantities are called *area moment of inertia*, *second moment of area*, or simply *moment of inertia.*

Calculation of the moment of inertia of the area of any arbitrary shape requires use of the integration method that has been avoided in this book. Fortunately, since most areas involved in architectural applications are limited to standard cross-sectional shapes, the use of direct integration is not necessary.

This chapter presents a simple method for calculating moments of inertia of simple as well as composite areas used in architecture. These are extremely important properties; in fact, one of the very first steps in the analysis and design of beams, columns, etc., is to evaluate the moment of inertia of the areas involved. Therefore, one cannot overemphasize the fact that students must be very comfortable and fluent in calculating moments of inertia of various applicable areas.

## 8.2 DEFINITION OF THE MOMENT OF INERTIA OF AN AREA

The moment of inertia of an area is always calculated with reference to a point, a line, or a plane. It is indicated by the capital letter $I$ and a subscript that represents, say, the particular axis to which the calculation of $I$ is referred, such as $I_x$, $I_{a-a}$, etc. In particular, if it is measured about the centroidal axes (coordinate axes passing through the centroid of the area), it is shown with a bar on top of the letter $I$, as $\bar{I}_x$, $\bar{I}_{a-a}$, etc.

Now consider the area ($A$) shown in Fig. 8.1, along with a rectangular coordinate system ($Ox$, $Oy$) used as reference axes, and a small area element $\Delta A$ ($\Delta A$ does not mean $\Delta$ multiplied by $A$; $\Delta A$, collectively, simply means a small area element called $\Delta A$; whenever a small amount or small variation of a quantity $A$ is to be considered, it is usually indicated as $\Delta A$). This small, generic area element, $\Delta A$, with respect to the coordinate system shown, has coordinates $x$ and $y$. By multiplying the value of this small area by the square of its $y$ coordinate, we get the term,

$$\Delta I_x = y^2 \Delta A$$

which is called the *moment of inertia* of the area $\Delta A$ with respect to the $x$ axis. Similarly, we can multiply $\Delta A$ by the square of its $x$ coordinate to obtain the moment of inertia of area $\Delta A$ about the $y$ axis, as follows:

$$\Delta I_y = x^2 \Delta A$$

Now suppose that we take another small area element next to element $\Delta A$ and construct its moments of inertia about the $x$ and $y$ axes by mul-

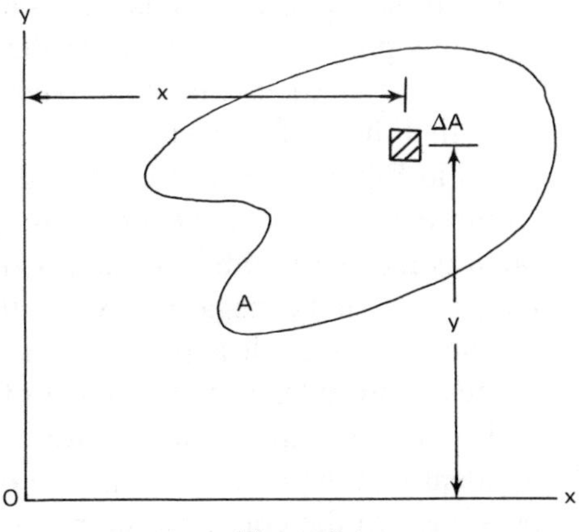


**Figure 8.1**

tiplying the value of the new small area by the square of its $x$ or $y$ coordinates; similarly, we keep choosing more and more elements to cover the total area $A$ and calculate the moments of intertia of each new element about the $x$ and $y$ axes, finally adding all moments of inertias obtained. The new values are the moments of inertia of the total area $A$ about the $x$ and $y$ axes. Note that in the process we choose enough small elements to cover the entire area $A$ without any gap or overlap. In adding such terms, we use the Greek letter $\Sigma$ to imply "sum." We can therefore write

$$\begin{aligned} I_x &= \Sigma\, y^2 \Delta A \\ I_y &= \Sigma\, x^2 \Delta A \end{aligned} \tag{8.1}$$

These two formulas, which constitute the definition of moment of inertia of an area, are the basis for calculating the moment of inertia of *any given area about any axis*. Before applying them to the cross-sectional areas of architectural structures, let's consider the first equation in Eq. 8.1, which can be written as follows:

$$I_x = \Sigma\, y(y\Delta A)$$

The final term, $y\Delta A$, is known as the moment of the area, $\Delta A$, about the $x$ axis. It is obtained by multiplying the value of the area (analogous to force) by a perpendicular distance $y$. Since such a product is by definition analogous to moment, that's why $y\Delta A$ is called a *moment of area* (about the $x$ axis in this case). Similarly, $y(y\Delta A)$ is the moment of $(y\Delta A)$ about the $x$ axis, and since in this process we have taken the moment of area $\Delta A$ twice, the term $y(y\Delta A)$ or $y^2\Delta A$, is called the *second moment of the area*, whereas $y\Delta A$ is called the first moment of the area, $\Delta A$.

In the example shown in Fig. 8.1, area $A$ is located in the first quadrant (both $x$ and $y$ coordinates of all its elements are positive). If the entire or part of an area is located in other quadrants, the $x$ and/or $y$ of some or all elements of $A$ will be negative. Since the coordinate is squared, however, the negative sign disappears. Consequently, an area with negative coordinates contributes to the moment of inertia as much as an equivalent area located the same distance away with positive coordinates. It is also evident that the moment of inertia of any area about any axis is always positive. On the contrary, the moments of area used in determination of centroids can have positive, negative, or even zero value.

The dimension of moment of inertia is clearly a dimension of length that has been raised to the fourth power, since area has the dimension of length squared, and to become a moment of inertia, must be multiplied by $x^2$ or $y^2$, which also represents a dimension of length squared.

In the U.S. customary unit system, the unit for moment of inertia is $(\text{inch})^4$, which is called a *quartic inch*; in the S.I. system, the unit is $(\text{m})^4$ or $(\text{mm})^4$, which are known as a *quartic meter* and a *quartic millimeter*, respectively.

## 8.3 DETERMINATION OF MOMENT OF INERTIA USING THE APPROXIMATE METHOD

As shown earlier, use of Eq. 8.1 requires one to divide the given area into small area elements, $\Delta A$, find the moment of inertia of each of these, and then add the results. It is not always possible to cover an irregularly shaped area completely. In the area shown in Fig. 8.1, for example, an approximation is involved because it is not possible to enclose an area with curved boundary with square elements. The smaller the elements $\Delta A$, however, the more accurate the results, and for some areas without curved boundaries, the approximate method will give values very close to the exact values.

EXAMPLE 1: Determine the value of $I_x$ for the $5'' \times 10''$ rectangular area shown in Fig. 8.2.

SOLUTION: In this case, a $5'' \times 1''$ strip can be taken for the $\Delta A$ element; the entire $5'' \times 10''$ rectangle can be covered using 10 such strips, as shown in the figure. The area of each element is thus equal to 5 in.$^2$ The distance of each from the $x$ axis is measured from the centroid of each rectangular strip with reference to the $x$ axis. This value for strip element No. 1 is 0.5″; for strip element No. 2, 1.5″; etc. Using these values and the first part of Eq. 8.1, we can write

$$\begin{aligned} I_x = \Sigma\, y^2 \Delta A &= (0.5)^2(5) + (1.5)^2(5) + (2.5)^2(5) + (3.5)^2(5) \\ &\quad + (4.5)^2(5) + (5.5)^2(5) + (6.5)^2(5) + (7.5)^2(5) \\ &\quad + (8.5)^2(5) + (9.5)^2(5) = 1662.5 \text{ in.}^4 \end{aligned}$$

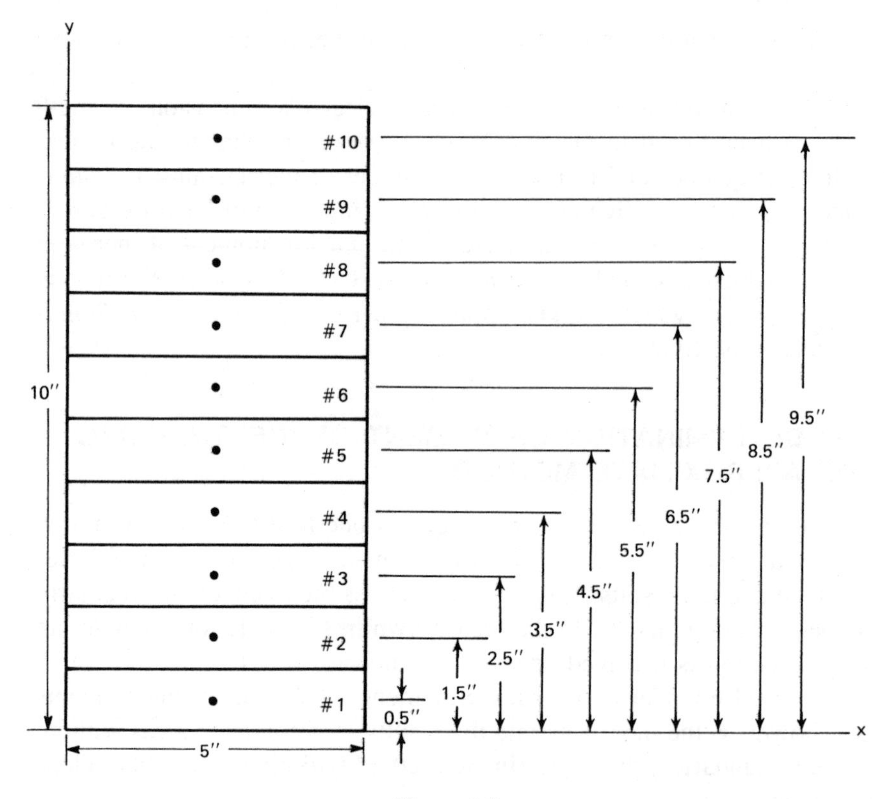


**Figure 8.2**

Using the integration method or Table 8.1, the exact value for $I_x$ is obtained as 1667 in.[4] In the approximate method used here, the error is about 0.3 percent, which is fairly good. By taking narrower strips, of course, the accuracy can be improved. Similarly, $I_y$ can be calculated by taking vertical strips. The calculation of $I_y$ is left for students as a homework problem.

## 8.4 PARALLEL AXIS THEOREM

As pointed out earlier, the area moment of inertia is a specific property of an area and is also interpreted as the second moment of that area about a specified axis. This means that the value of the moment of inertia of an area depends on the location of that specified axis (the distance of the area in question from the axis, or moment arm). This may create a problem since, although the area moment of inertia is a "specific property" of the area in question, its value is not unique and will have different values about different reference axes. The problem is not as serious as it sounds, however, because a well established relationship exists between the values of moment of inertia of an area taken about different axes. This relationship is derived by means of the *parallel axis theorem*.

This theorem, which will be discussed in detail in this section, helps us to calculate the value of the moment of inertia of an area about any desired axis from the value given for a particular axis. To unify the calculations, all tabulations of the values of the moment of inertia of common areas take the centroidal axes as the reference axes. Remember that the centroidal axes of an area are coordinate axes that pass through its centroid. Depending on our needs, most probably we will use axes other than the centroidal axes. By knowing the moment of inertia of an area about its centroidal axis, however, one can calculate its value about any axis parallel to the centroidal axis, as will now be demonstrated.

Consider area $A$ shown in Fig. 8.3. Point $C$ represents the centroid of this area. The two axes, $Cx_0$ and $Cy_0$, are the centroidal axes (they pass

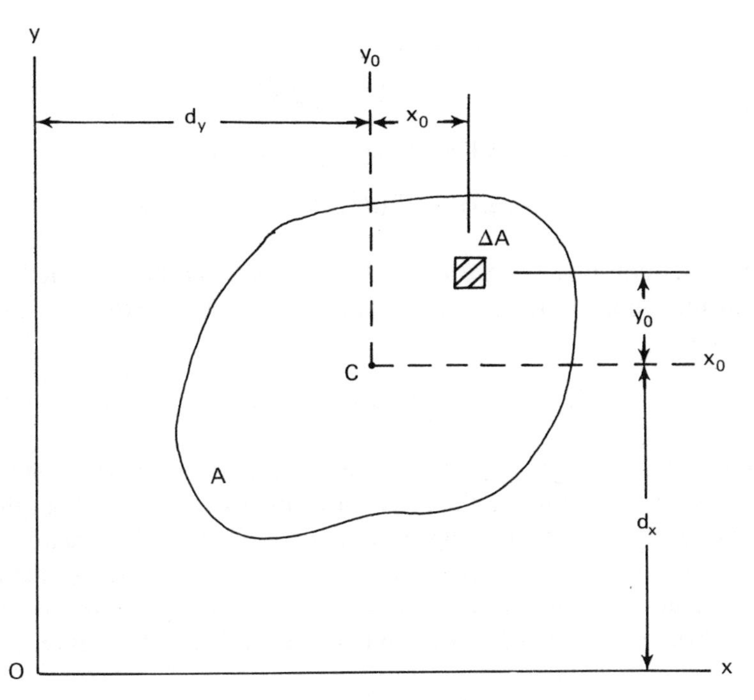


Figure 8.3

through the centroid of the area). Now consider two other axes, $Ox$ and $Oy$, which are parallel to $Cx_0$ and $Cy_0$, respectively. These axes are at distances, say, $d_x$ and $d_y$ away from the centroidal axes.

As defined earlier, the moment of inertia of a small area $\Delta A$ about $Ox$ is defined as

$$\Delta I_x = (y_o + d_x)^2 \, \Delta A$$

where $y_o + d_x$ is the distance of the small area $\Delta A$ from the $x$ axis (moment arm). Using this formula and the summation sign, $\Sigma$, we can find the moment of inertia of the whole area $A$ about the $x$ axis as follows:

$$I_x = \Sigma \, (y_o + d_x)^2 \, \Delta A$$

Raising the term inside the parentheses to the second power, it can be written as follows:

$$(y_o + d_x)^2 = y_o^2 + d_x^2 + 2y_o d_x$$

or

$$\begin{aligned} I_x &= \Sigma \, (y_o^2 + d_x^2 + 2y_o d_x) \, \Delta A \\ &= \Sigma \, (y_o^2 \Delta A + d_x^2 \Delta A + 2y_o d_x \Delta A) \\ &= \Sigma \, (y_o^2 \Delta A) + \Sigma \, (d_x^2 \Delta A) + \Sigma \, (2y_o d_x \Delta A) \end{aligned}$$

Since $d_x$ is constant (as are $d_x^2$ and $2d_x$), they can be extracted from within the summation sign, and the above can be reexpressed as the following:

$$I_x = \Sigma \, (y_o^2 \Delta A) + d_x^2 \, \Sigma \, (\Delta A) + 2d_x \, \Sigma \, (y_o \Delta A) \qquad (8.2)$$

Now, let's consider each term of the above expression. The first term, $\Sigma \, (y_o^2 \Delta A)$, is by definition the moment of inertia of area $A$ about the $x_o$ axis (sum of terms comprising an area multiplied by the square of its distance from the $x_o$ axis). Since axis $x_o$ is a centroidal axis for the area $A$, the value of this moment of inertia about the centroidal axis is called the *centroidal moment of inertia* and is denoted by $\bar{I}_x$. Therefore,

$$\Sigma \, (y_o^2 \Delta A) = \bar{I}_x \qquad (8.2a)$$

The second term, $d_x^2 \Sigma (\Delta A)$, is equal to $d_x^2 A$, since $\Sigma (\Delta A)$ means the sum of all the small elements that fully cover the area $A$ and thus is obviously equal to the total area $A$. Therefore,

$$d_x^2 \Sigma (\Delta A) = d_x^2 A \qquad (8.2b)$$

The third term, $2d_x \Sigma (y_o \Delta A)$, is equal to zero. Remember from Chap. 5 that in order to determine the centroid of an area, we multiplied the value of a small area element $\Delta A$ by its distance from an axis and then summed all such terms for all areas $\Delta A$ to cover the total area $A$. The result gave $\bar{x}$ and $\bar{y}$ of the centroid. Therefore, if we choose our reference axis to pass through the centroid, the values of $\bar{x}$ and $\bar{y}$ will be equal to zero and thus

$$2d_x \Sigma (y_o \Delta A) = 0 \qquad (8.2c)$$

Substituting 8.2(a), 8.2(b) and 8.2(c) into 8.2 will give

$$I_x = \bar{I}_x + d_x^2 A \qquad (8.3)$$

This last equation is the relationship between the two values of the moment of inertia of area $A$ calculated about two different axes—one passing through the centroid of the area $A$ (centroidal axis) and the other parallel and at distance $d_x$ away from the centroidal axis. Similarly, the moment of inertia about the $y$ axis can be obtained, as follows:

$$I_y = \bar{I}_y + d_y^2 A \qquad (8.4)$$

Both Eqs. 8.3 and 8.4 simply state that if the moment of inertia of area $A$ about its centroidal axis, $\bar{I}$, is given, to obtain the moment of inertia, $I$, of the same area about any axis parallel to the centroidal axis, simply add a term $d^2A$ to $\bar{I}$, where $d$ is the distance separating the two axes. It can easily be seen that the moment of inertia of an area has the smallest value when it is referred to the centroidal axis. *This smallest value for I is known as the centroidal moment of inertia of an area.*

EXAMPLE 2: Determine the moment of inertia of the rectangle shown in Fig. 8.4(a) about the $x$ and $y$ axes.

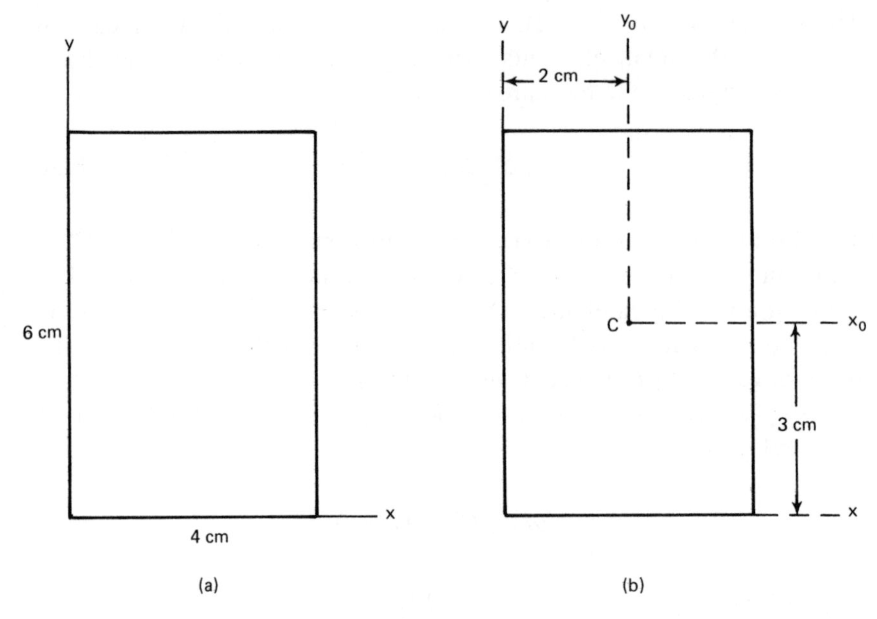


**Figure 8.4**

SOLUTION: The centroidal moment of inertia of a rectangular area can be calculated by using formulas from Table 8.1,* as follows:

$$\bar{I}_x = \tfrac{1}{12}bh^3 = \tfrac{1}{12}(4)(6)^3 = 72 \text{ cm}^4$$

$$\bar{I}_y = \tfrac{1}{12}hb^3 = \tfrac{1}{12}(6)(4)^3 = 32 \text{ cm}^4$$

Substituting these values into Eqs. 8.3 and 8.4 and knowing that, as shown in Fig. 8.4(b), $d_x = 3$ cm, $d_y = 2$ cm, and $A = 4 \times 6 = 24$ cm$^2$, the values of the moment of inertia of the given rectangle about the $x$ and $y$ axes can be calculated as follows:

$$I_x = \bar{I}_x + d_x^2 A = 72 + [(3)^2 \times (24)] = 72 + 216 = 288 \text{ cm}^4$$

$$I_y = \bar{I}_y + d_y^2 A = 32 + [(2)^2 \times (24)] = 32 + 96 = 128 \text{ cm}^4$$

*The values obtained from the equations in this table can also be obtained using integration, a field of mathematics avoided in this book.

**TABLE 8.1. Moment of Inertia of Common Areas**

| | | $\bar{I}$ | $I$ |
|---|---|---|---|
| 1. Rectangle | 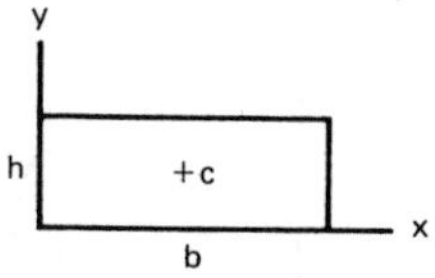 | $\bar{I}_x = \frac{1}{12} bh^3$<br>$\bar{I}_y = \frac{1}{12} hb^3$ | $I_x = \frac{1}{3} bh^3$<br>$I_y = \frac{1}{3} hb^3$ |
| 2. Square | 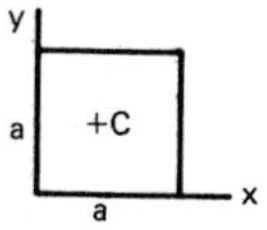 | $\bar{I}_x = \bar{I}_y = \frac{1}{12} a^4$ | $I_x = I_y = \frac{1}{3} a^4$ |
| 3. Circle | 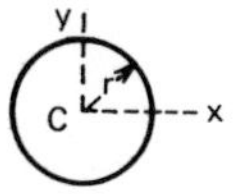 | $\bar{I}_x = \bar{I}_y = \frac{1}{4} \pi r^4$ | — |
| 4. Semicircle | 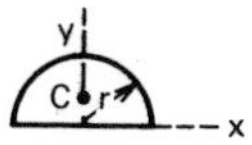 | — | $I_x = I_y = \frac{1}{8} \pi r^4$ |
| 5. Quadrant | 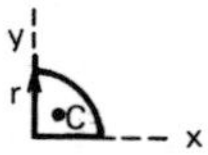 | — | $I_x = I_y = \frac{1}{16} \pi r^4$ |
| 6. Right triangle | 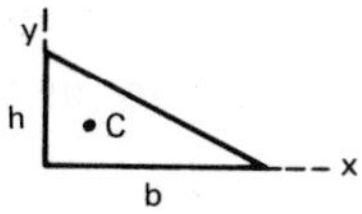 | $\bar{I}_x = \frac{1}{36} bh^3$<br>$\bar{I}_y = \frac{1}{36} hb^3$ | $I_x = \frac{1}{12} bh^3$<br>$I_y = \frac{1}{12} hb^3$ |
| 7. General triangle | 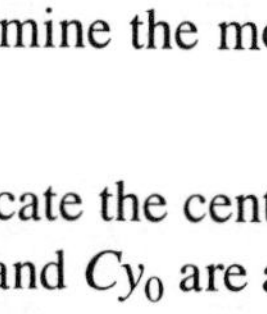 | $\bar{I}_x = \frac{1}{36} bh^3$ | |

EXAMPLE 3: Determine the moment of inertia of the triangle shown in Fig. 8.5(a).

SOLUTION: First locate the centroid of the triangle using Table 5.1. The centroidal axes $Cx_0$ and $Cy_0$ are axes parallel to the $x$ and $y$ axes as shown in Fig. 8.5(b) and their distance from the $x$ and $y$ axes can be calculated as follows:

$$d_x = 2 + \tfrac{1}{3}(9) = 5 \text{ in.}$$

$$d_y = 4 + \tfrac{2}{3}(3) = 6 \text{ in.}$$

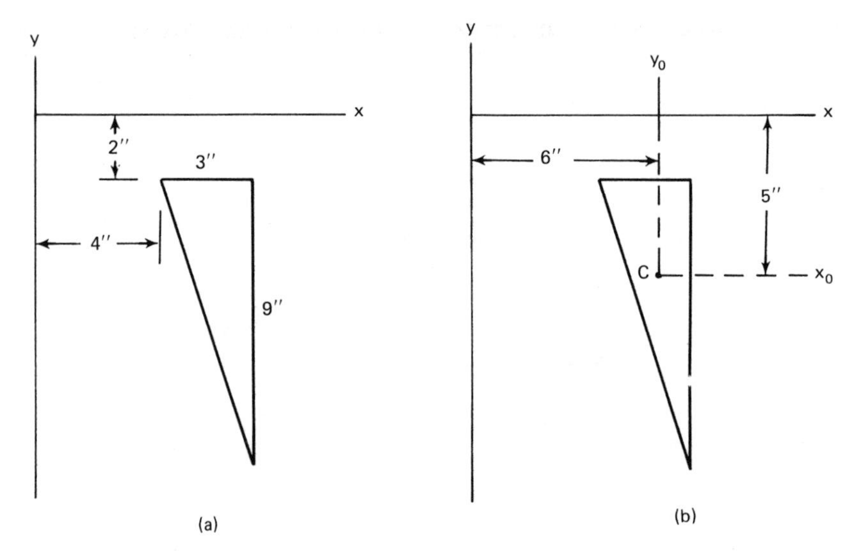


Figure 8.5

The area of the triangle is

$$A = \tfrac{1}{2}(3)(9) = 13.5 \text{ in.}^2$$

Using Table 8.1, $\bar{I}_x$ and $\bar{I}_y$ for the triangle are as follows:

$$\bar{I}_x = \tfrac{1}{36}\, bh^3 = \tfrac{1}{36}(3)(9)^3 = 60.8 \text{ in.}^4$$

$$\bar{I}_y = \tfrac{1}{36}\, hb^3 = \tfrac{1}{36}(9)(3)^3 = 6.8 \text{ in.}^4$$

Substituting the above values into Eqs. 8.3 and 8.4 will give the values of the moment of inertia of the given triangle about the $x$ and $y$ axes, as follows:

$$I_x = \bar{I}_x + d_x^2 A = 60.8 + (5)^2(13.5) = 398.3 \text{ in.}^4$$

$$I_y = \bar{I}_y + d_y^2 A = 6.8 + (6)^2(13.5) = 492.8 \text{ in.}^4$$

## 8.5 OTHER PROPERTIES OF AREA

Other properties of an area that can be derived from the moment of inertia of the area also have frequent applications. These properties will now be examined.

## Polar Moment of Inertia of an Area

Consider the area shown in Fig. 8.6. Similar to the definition introduced for moments of inertia of an area about the $x$ and $y$ axes in Eq. 8.1, one can define one more moment of inertia of area $A$ about the $z$ axis (or about pole $O$, the $z$ axis being perpendicular to the $x$–$y$ plane at point $O$), as follows:

$$J_o = \Sigma\,(r^2 \Delta A) \tag{8.5}$$

where $J_o$ is known as the *polar moment of inertia* and $r$ is the radial distance of $\Delta A$ from point (pole) $O$. In general, it is very difficult to calculate the polar moment of inertia of an area using Eq. 8.5 (especially in terms of a rectangular coordinate system). The equation, however, can be simplified by using the following argument.

Consider the right triangle formed in Fig. 8.6 by the hypotenuse $r$, which is the distance from pole $O$ to the center of the small area $\Delta A$. In this right triangle, one can write

$$r^2 = x^2 + y^2$$

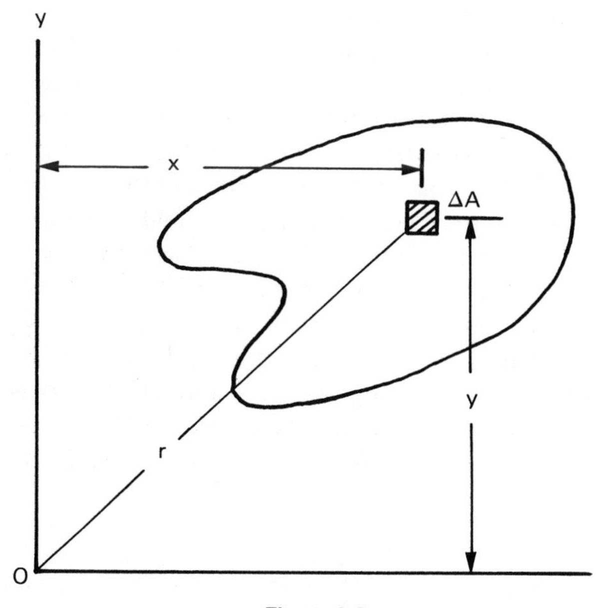


Figure 8.6

Substituting this value of $r^2$ in Eq. 8.5 yields the following:

$$J_o = \Sigma\,(r^2 \Delta A) = \Sigma\,(x^2 + y^2)\,\Delta A = \Sigma\,(x^2 \Delta A) + \Sigma\,(y^2 \Delta A)$$

But $\Sigma\,(x^2 \Delta A) = I_y$ and $\Sigma\,(y^2 \Delta A) = I_x$ and therefore

$$J_o = I_x + I_y \tag{8.6}$$

This equation indicates that the polar moment of inertia of an area can be obtained by adding the values of the moments of inertia of the area about the $x$ and $y$ axes.

EXAMPLE 4: Determine the polar moment of inertia of the rectangle shown in Fig. 8.4(a).

SOLUTION: The values of $I_x$ and $I_y$ for this problem were calculated in Example 2 to be $I_x = 288$ cm$^4$ and $I_y = 128$ cm$^4$. Equation 8.6 thus gives the polar moment of inertia as

$$J_o = I_x + I_y = 288 + 128 = 416 \text{ cm}^4.$$

## Section Modulus of an Area

In bending a beam, it can be shown that the stresses on the top or bottom of the beam are tensile or compressive, depending on how the beam is loaded. Going from tensile stress on one side to compressive stress on the other, one has to traverse a plane of zero stress that passes through the centroid of the cross-sectional area of the beam. This plane or axis is known as the *neutral axis*. It can be shown that the fibers of the beam at maximum distance from the neutral axis are under algebraically maximum stress. In order to calculate this maximum stress, one needs to know the value of the moment of inertia of the cross-sectional area of the beam about the neutral axis as well as the distance $c$, which is the distance of the fibers farthest away from the neutral axis. Defining a new quantity related to $I$ further simplifies calculations. This new quantity, which is a purely numerical value related only to the geometry of the cross-sectional area, is called the *section modulus* and is given by the equation,

$$Z = \frac{I}{c} \tag{8.7}$$

EXAMPLE 5: Calculate the section modulus about the $x_o$ axis for the rectangle shown in Fig. 8.7.

SOLUTION: The centroid of the rectangle is located at $C$, with $x = 2.5$ in. and $y = 5$ in. The centroidal axes are parallel to the $x$ and $y$ axes, with their origin at $C$. The top and bottom sides of the rectangle are at the greatest distance from the neutral axis, $x_o$. Therefore, as previously defined, $c$ in this case is equal to

$$c = \tfrac{10}{2} = 5 \text{ in.}$$

and

$$\bar{I}_{x_o} = \bar{I}_x = \tfrac{1}{12}\, bh^3 = \tfrac{1}{12}(5)(10)^3 = 416.7 \text{ in.}^4$$

Using Eq. 8.7, the section modulus can be determined as follows:

$$Z_x = \frac{\bar{I}_x}{c} = \frac{416.7}{5} = 83.3 \text{ in.}^3$$

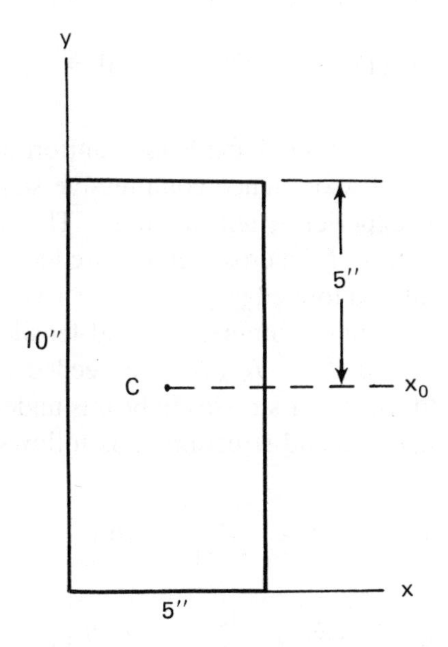


Figure 8.7

Notice that the unit for $Z$ is the unit of length cubed, or in.$^3$ in U.S. customary units, and mm$^3$ or cm$^3$ in the S.I. system.

EXAMPLE 6: The cross-sectional area of a beam is shown in Fig. 8.8(a). Calculate the section modulus about its neutral axis.

SOLUTION: The cross-sectional area is composed of two rectangles 1″ × 8″ and 1″ × 6″. The neutral axis (N.A.), which is the same as the centroidal axis, can be easily determined as described in Chap. 5:

$$\bar{y} = \frac{A_1 y_1 + A_2 y_2}{A_1 + A_2} = \frac{[(1 \times 8) \times 5] + [(1 \times 6) \times 0.5]}{(1 \times 8) + (1 \times 6)} = 3.07''$$

Therefore, the N.A. of the area is located 3.07″ above the $x$ axis (the bottom edge of the area). The top edge of the area is located 5.93″ above the N.A. (9″ − 3.07″). The centroid of the upper rectangle is 1.93″ above the N.A. and of the lower rectangle, 2.57″ below the N.A., as shown in Fig. 8.8(b). The centroidal moment of inertia of the entire cross-sectional area can then be calculated as the following sum:

$$I_{\text{N.A.}} = (I_1)_{\text{N.A.}} + (I_2)_{\text{N.A.}} = [\tfrac{1}{12}(1)(8)^3 + (1.93)^2(1 \times 8)]$$

$$+ [\tfrac{1}{12}(6)(1)^3 + (2.57)^2(1 \times 6)] = 112.6 \text{ in.}^4$$

If the beam is simply supported and loaded uniformly on top, say, the area above the N.A. will experience compressive stress while the area below the N.A. will experience tensile stress. The most distant compressive and tensile fibers of the cross section are located at the top edge ($c_1 = 5.93''$) and at the bottom edge ($c_2 = 3.07''$) as measured from the N.A. To calculate maximum compressive and tensile stresses, section moduli $Z_1 = I_{\text{N.A.}}/c_1$ and $Z_2 = I_{\text{N.A.}}/c_2$ are needed (these stresses are covered in detail in the study of stresses in beams undertaken in a course on the strength of materials and structures), as follows:

$$Z_1 = \frac{I_{\text{N.A.}}}{c_1} = \frac{112.6}{5.93} = 19.0 \text{ in.}^3$$

$$Z_2 = \frac{I_{\text{N.A.}}}{c_2} = \frac{112.6}{3.07} = 36.7 \text{ in.}^3$$

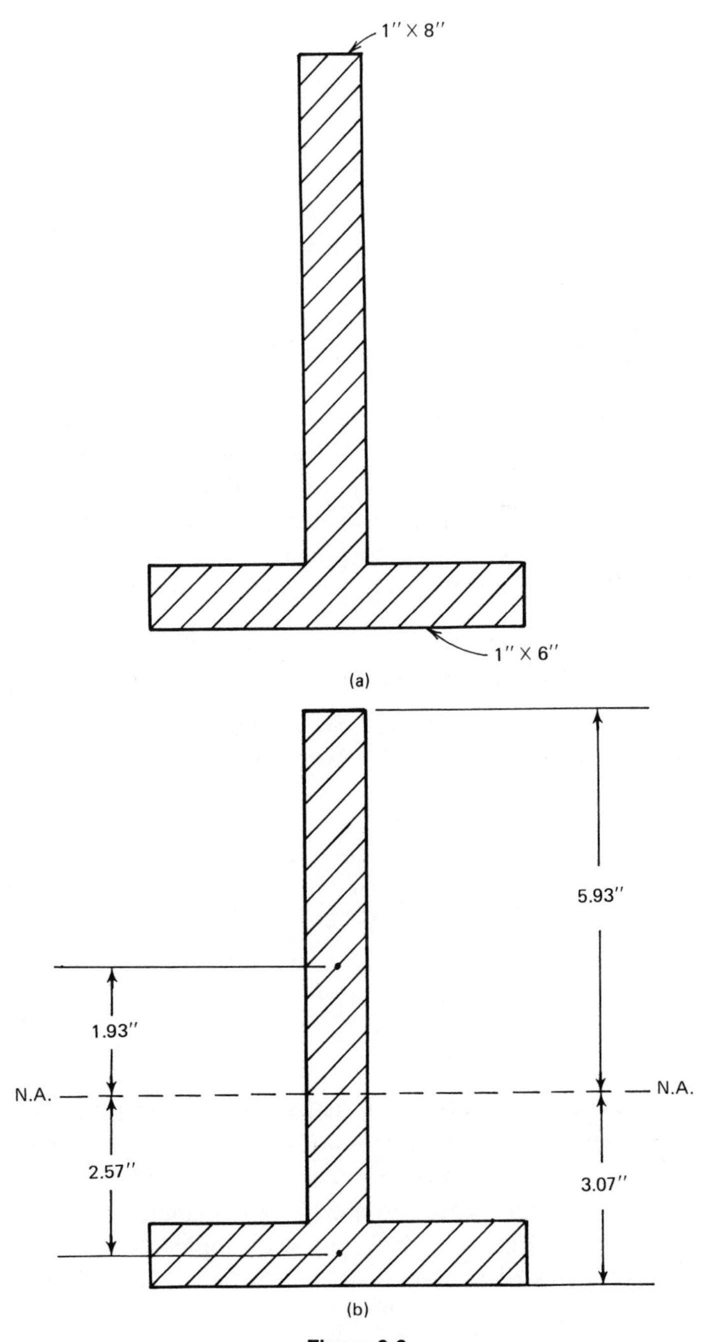


**Figure 8.8**

## Radius of Gyration

The radius of gyration, $r$, is derived from the moment of inertia and is defined as follows:

$$r = \sqrt{\frac{I}{A}} \tag{8.8}$$

Thus, the value of the radius of gyration is such that, if its square is multiplied by the value of the area, it will be equal to the value of the moment of inertia of that area, or $I = r^2A$.

Since we have defined three moments of inertia, two rectangular and one polar, there will be three different values for the radius of gyration, one for each moment of inertia. The radius of gyration has many applications in mechanics, especially in dynamics; for our purpose, it will be used in calculations involving the analysis of columns (which is also a subject of a course on the strength of materials and structures).

EXAMPLE 7: Determine the radius of gyration with respect to the $x$ and $y$ axes of the rectangle in Fig. 8.4(a).

SOLUTION: In Example 2, $I_x$ and $I_y$ were calculated for Fig. 8.4(a) as: $I_x = 288 \text{ cm}^4$ and $I_y = 128 \text{ cm}.^4$ Substituting these values, along with $A = 24 \text{ cm}^2$, yields

$$r_x = \sqrt{\frac{I_x}{A}} = \sqrt{\frac{288}{24}} = 3.46 \text{ cm}$$

$$r_y = \sqrt{\frac{I_y}{A}} = \sqrt{\frac{128}{24}} = 2.31 \text{ cm}$$

Table 8.1 shows those cross-sectional areas most commonly used and the formulas for $\bar{I}$ and $I$. As discussed in Chap. 5, the commercially available structural shapes have rounds and fillets. The data for these modified structural shapes are provided by the manufacturers and tabulated in handbooks; Table 8.2 shows a few examples. Appendix A gives more extended tables on structural shapes. Notice the coordinate system used to obtain the values for each cross section.

**TABLE 8.2. Properties of Structural Shapes**

| Section | Weight Per Ft (lb) | $A$ (in.)$^2$ | $\bar{x}$ (in.) | $\bar{y}$ (in.) | $I_x$ (in.)$^4$ | $I_y$ (in.)$^4$ |
|---|---|---|---|---|---|---|
| W36×300 | 300 | 88.3 | 0 | 0 | 20300 | 1300 |
| W24×104 | 104 | 30.6 | 0 | 0 | 3100 | 259 |
| W12×14 | 14 | 4.16 | 0 | 0 | 88.6 | 2.36 |
| C15×50.0 | 50.00 | 14.7 | 0.799 | 0 | 404 | 11.0 |
| C8×18.75 | 18.75 | 5.51 | 0.565 | 0 | 44 | 1.98 |
| C3×4.1 | 4.1 | 1.21 | 0.437 | 0 | 1.66 | 0.197 |
| WT18×150 | 150.0 | 44.1 | 0 | 4.13 | 1230 | 648 |
| WT12×34 | 34.0 | 10.0 | 0 | 3.06 | 137 | 35.2 |
| WT4×20 | 20.0 | 5.87 | 0 | 0.735 | 5.73 | 24.5 |
| L9×4×5/8 | 26.3 | 7.73 | 1.04 | 3.36 | 64.9 | 8.32 |
| L5×3×1/4 | 6.6 | 1.94 | 0.657 | 1.66 | 5.11 | 1.44 |
| L3×2×3/8 | 5.9 | 1.73 | 0.539 | 1.04 | 1.53 | 0.543 |

## 8.6 MOMENT OF INERTIA OF COMPOSITE AREAS

A similar method to that used in Chap. 5 to determine the centroid of composite areas will be used here to determine the moment of inertia of composite areas. Such a method was used in the previous example for a T-shaped cross-sectional area and in the following examples will be demonstrated for an L-shaped area as well as built-up sections composed of an I and C-channel:

EXAMPLE 8: Determine the centroidal moment of inertia of the L-shaped channel shown in Fig. 8.9(a).

SOLUTION: The first step in determining the moment of inertia of composite figures is to subdivide the area into simple areas with known centroids and moments of inertia. This L-shaped channel can be divided into two rectangles, with centroids $C_1$ and $C_2$ as shown in Figs. 8.9(b) and (c). The next step is to locate the centroidal axes of the entire cross section and then perform routine calculations, as follows:

$$\bar{x} = \frac{A_1x_1 + A_2x_2}{A_1 + A_2} = \frac{(4 \times 1)(2) + (5 \times 1)(0.5)}{(4 \times 1) + (5 \times 1)} = 1.17''$$

$$\bar{y} = \frac{A_1y_1 + A_2y_2}{A_1 + A_2} = \frac{(4 \times 1)(0.5) + (5 \times 1)(3.5)}{(4 \times 1) + (5 \times 1)} = 2.17''$$

Since both areas are rectangles, $I_x$ and $I_y$ are calculated from the formulas,

$$I_x = \frac{bh^3}{12}$$

$$I_y = \frac{hb^3}{12}$$

and since both areas are off-center (centroids $C_1$ and $C_2$ do not coincide with $C$), the parallel axis theorem for moment of inertia must be used with reference to the $x_o$, $y_o$ axes shown in Fig. 8.9(d).

The moment of inertia of the L-shaped area shown in Fig. 8.9(a) is equal to the sum of the moments of inertia of the simple rectangular areas shown in Figs. 8.9(b) and (c), all with respect to the same coordinate axes:

$$\bar{I}_x = (\bar{I}_1)_x + (\bar{I}_2)_x$$

$$\bar{I}_y = (\bar{I}_1)_y + (\bar{I}_2)_y$$

The bar on top of $\bar{I}_x$ and $\bar{I}_y$ indicates that the moments of inertia are calculated about the centroidal axes passing through the centroid of the entire L-shaped area ($Cx_o$, $Cy_o$). ($I_1$) and ($I_2$) therefore have to be calculated with respect to these axes.

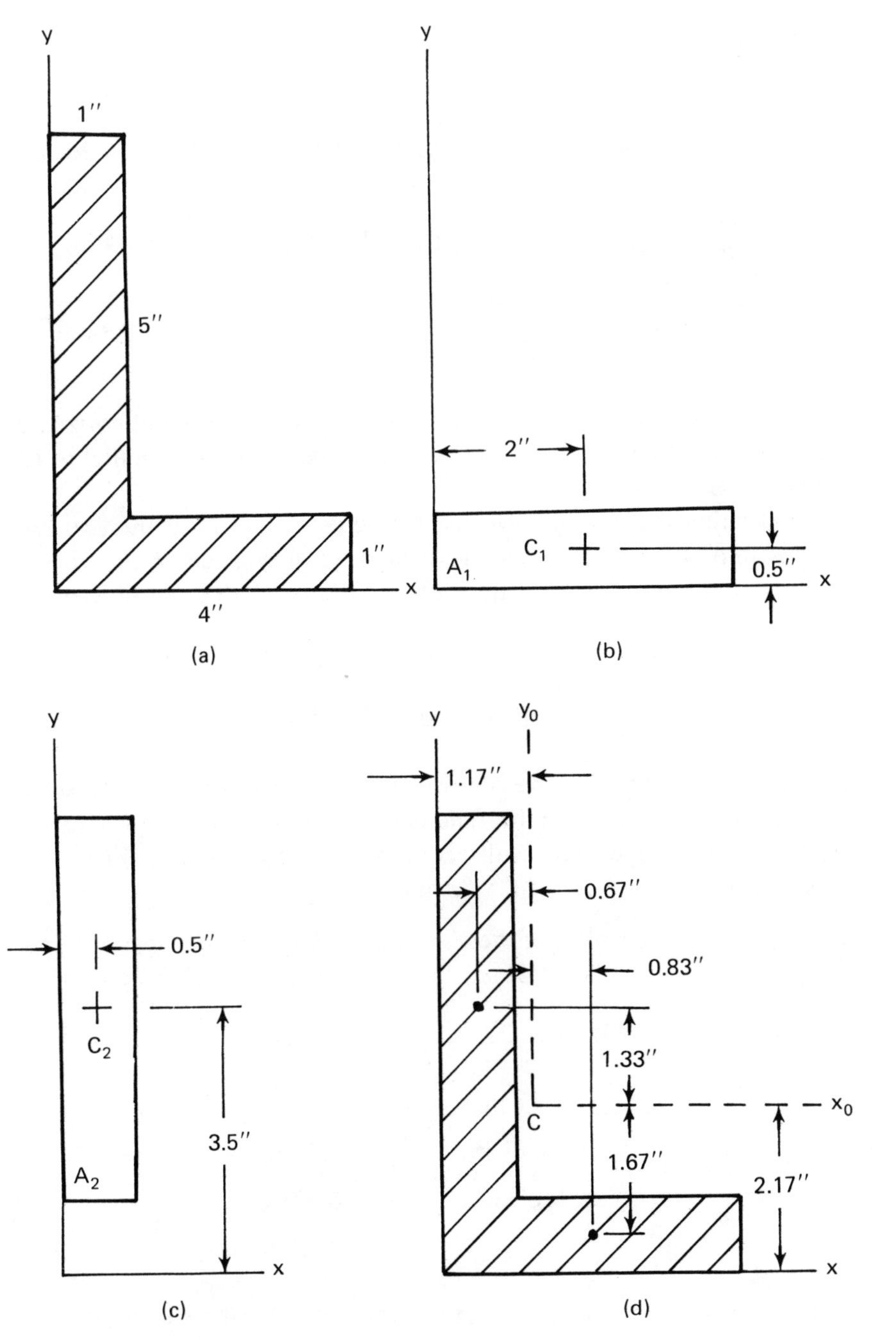
y
1″
5″
1″
4″
x
(a)
y
2″
C₁
A₁
0.5″
x
(b)
y
0.5″
C₂
3.5″
A₂
x
(c)
y
y₀
1.17″
0.67″
0.83″
1.33″
C
x₀
1.67″
2.17″
x
(d)

Figure 8.9

$$(\bar{I}_1)_x = \frac{bh^3}{12} + d_x^2 A = \frac{(4)\,(1)^3}{12} + (2.17 - 0.5)^2\,(4 \times 1) = 11.49 \text{ in.}^4$$

$$(\bar{I}_2)_x = \frac{bh^3}{12} + d_x^2 A = \frac{(1)\,(5)^3}{12} + (3.5 - 2.17)^2\,(5 \times 1) = 19.26 \text{ in.}^4$$

$$(\bar{I}_1)_y = \frac{hb^3}{12} + d_y^2 A = \frac{(1)\,(4)^3}{12} + (2 - 1.17)^2\,(4 \times 1) = 8.09 \text{ in.}^4$$

$$(\bar{I}_2)_y = \frac{hb^3}{12} + d_y^2 A = \frac{(5)\,(1)^3}{12} + (1.17 - 0.5)^2\,(5 \times 1) = 2.66 \text{ in.}^4$$

Note that $d_x$ and $d_y$ for the two rectangles are the vertical and horizontal distances of $C$ from $C_1$ and $C_2$ measured in the directions parallel to the $x$ and $y$ axes.

Substituting these values for the moments of inertia of the two rectangles into the equations for the composite area given in Fig. 8.9(a) will result in the following values of the centroidal moment of inertia:

$$\bar{I}_x = 11.49 + 19.26 = 30.75 \text{ in.}^4$$

$$\bar{I}_y = 8.09 + 2.66 = 10.75 \text{ in.}^4$$

EXAMPLE 9: Determine the centroidal moment of inertia for the structural shape given in Fig. 8.10. The neutral axis is located at $\bar{y} = 6.71''$ from the lower edge. The built-up beam is composed of a $W10 \times 22$ (wide flange) and a $C8 \times 11.5$ (channel).

SOLUTION: Using tables for these structural shapes in the Appendix, the following values can be obtained for the wide flange and the channel.

| $W10 \times 22$ | $C8 \times 11.5$ |
|---|---|
| $A = 6.49$ in.$^2$ | $A = 3.38$ in.$^2$ |
| $d = 10.17$ in. | $\bar{I}_y = 1.32$ in.$^4$ |
| $\bar{I}_x = 118$ in.$^4$ | $\bar{x} = 0.571$ in. |
| | $t_w = 0.22$ in. |

The centroidal moment of inertia can be calculated by using the parallel axis theorem and the above values (see Fig. 8.10):

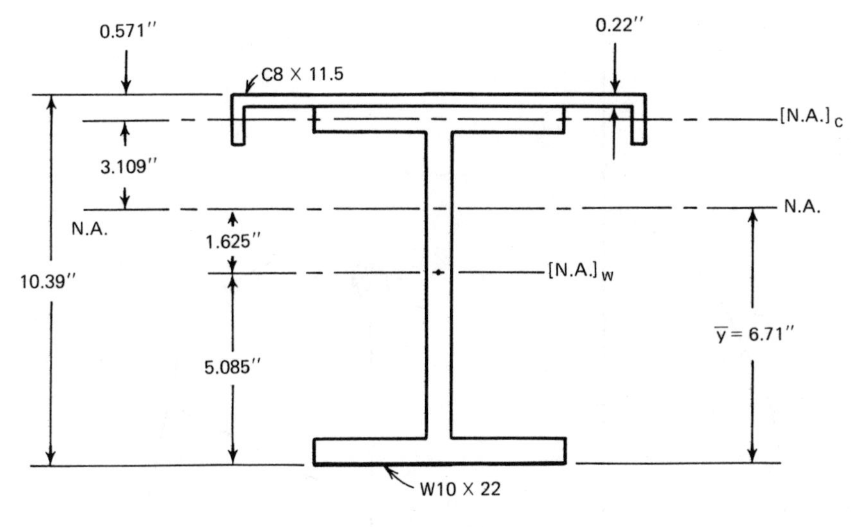


**Figure 8.10**

$$I_{\text{N.A.}} = 118 + [(6.71 - 5.085)^2 \times 6.49] + 1.32 + [(3.109)^2 \times 3.38] = 169 \text{ in.}^4$$

Notice that since the *C*-channel used in this problem rotated 90° with respect to the diagram used in the table, $\bar{I}_y$ is used for $\bar{I}_x$ and, similarly, $\bar{x}$ for $\bar{y}$.

## 8.7 PRODUCT OF INERTIA

Consider the area *A* shown in Fig. 8.11. At any arbitrary point such as point *O*, one can draw two axes such as *Ox* and *Oy*. If these axes are rotated counterclockwise through angle $\alpha$, two more coordinate axes are produced, labeled *On* and *Ot* in the figure. According to the definitions given in Sec. 8.2, the moments of inertia of the area *A* about *Ox* and *Oy* are given by

$$I_x = \Sigma y^2 \Delta A$$

$$I_y = \Sigma x^2 \Delta A$$

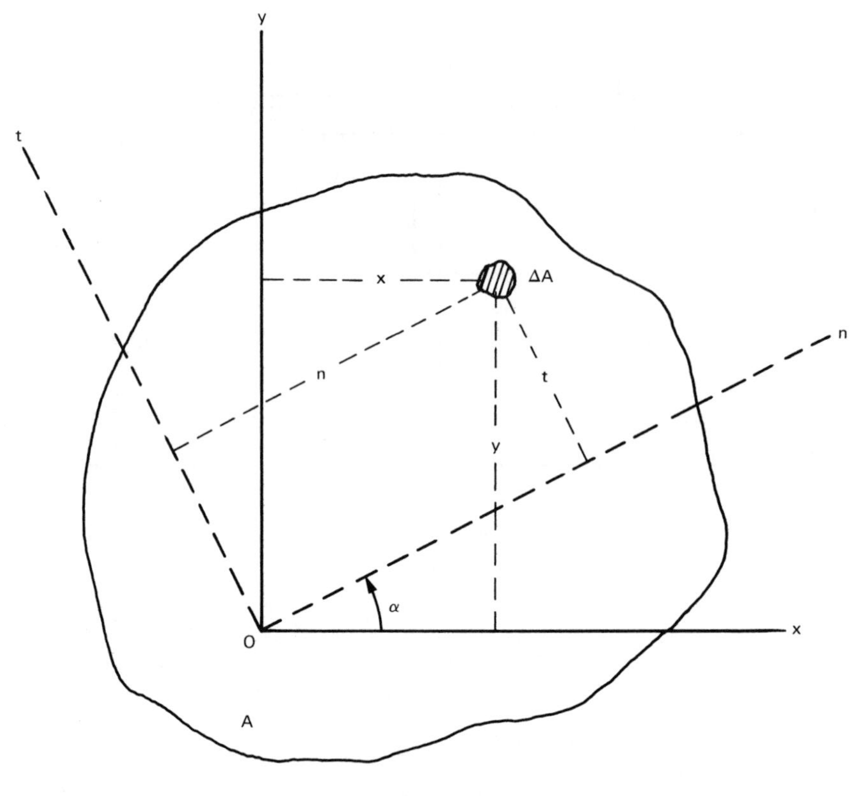


**Figure 8.11**

Similarly, the moments of inertia of the area $A$ can be calculated with reference to axes $On$ and $Ot$ as follows:

$$
\begin{aligned}
I_n &= \Sigma t^2 \Delta A \\
I_t &= \Sigma n^2 \Delta A
\end{aligned}
\qquad (8.9)
$$

where $n$ and $t$, similar to $x$ and $y$, represent the perpendicular distances of the small generic area element $\Delta A$ from the $n$ and $t$ axis. The $n$–$t$ coordinate system is produced by rotating the $x$–$y$ coordinate system by the angle $\alpha$. In analytic geometry, well-known coordinate transformation formulas transform the $x$ and $y$ coordinates of a point from one system of coordinates to another if the rotation angle $\alpha$ is known. These formulas are

$$n = x \cos \alpha + y \sin \alpha$$
$$t = y \cos \alpha - x \sin \alpha \qquad (8.10)$$

Substituting Eq. 8.10 into the first part of Eq. 8.9 will result in the following:

$$\begin{aligned} I_n &= \Sigma t^2 \Delta A = \Sigma (y \cos \alpha - x \sin \alpha)^2 \, \Delta A \\ &= \Sigma (y^2 \cos^2 \alpha + x^2 \sin^2 \alpha - 2xy \sin \alpha \cos \alpha) \, \Delta A \\ &= \Sigma y^2 \cos^2 \alpha \, \Delta A + \Sigma x^2 \sin^2 \alpha \, \Delta A - \Sigma 2xy \sin \alpha \cos \alpha \, \Delta A \\ &= \cos^2 \alpha \Sigma y^2 \, \Delta A + \sin^2 \alpha \, \Sigma x^2 \, \Delta A - 2 \sin \alpha \cos \alpha \Sigma xy \Delta A \end{aligned}$$

Now, since $\Sigma y^2 \Delta A = I_x$, $\Sigma x^2 \Delta A = I_y$, and $\Sigma xy \Delta A$ may be replaced by $I_{xy}$, we have

$$I_n = I_x \cos^2 \alpha + I_y \sin^2 \alpha - 2 I_{xy} \sin \alpha \cos \alpha \qquad (8.11)$$

The term, $I_{xy} = \Sigma xy \Delta A$, is called the *product of inertia*. It can also be directly defined in a way similar to that used for definitions of $I_x$ and $I_y$. Consider the area element $\Delta A$. By taking its first moment about the $x$ axis, which gives $y \Delta A$, and then taking the moment of the result but this time about the $y$ axis, which gives $xy \Delta A$, and, finally, by doing this for all the small area elements needed to cover the entire area $A$, we have

$$I_{xy} = \Sigma xy \Delta A$$

In Eq. 8.11, the product and squares of the sine and cosine of angle $\alpha$ are involved. Trigonometric identities exist that relate these quantities to the sine and cosine of angle $2\alpha$. (These relationships were not reviewed in Chap. 1 because, except for the present instance, they were not applicable to the concerns of this book.) Substituting these identities in Eq. 8.11, we have

$$I_n = \frac{I_x + I_y}{2} + \frac{I_x - I_y}{2} \cos 2\alpha - I_{xy} \sin 2\alpha \qquad (8.12)$$

Using a similar procedure, the counterpart of Eq. 8.12 can be derived for $I_t$. Moreover, the product of inertia with respect to the $n$–$t$ coordinate system can be defined as $I_{nt} = \Sigma nt\Delta A$. Manipulation of this similar to that for $I_n$ will result in the formula shown below, together with the formulas for $I_n$ and $I_t$:

$$I_n = \frac{I_x + I_y}{2} + \frac{I_x - I_y}{2} \cos 2\alpha - I_{xy} \sin 2\alpha \quad (8.13a)$$

$$I_t = \frac{I_x + I_y}{2} - \frac{I_x - I_y}{2} \cos 2\alpha + I_{xy} \sin 2\alpha \quad (8.13b)$$

$$I_{nt} = \frac{I_x - I_y}{2} \sin 2\alpha + I_{xy} \cos 2\alpha \quad (8.13c)$$

These transformation formulas indicate that once the moments and product of inertia ($I_x$, $I_y$, and $I_{xy}$) of an area with respect to the $x$–$y$ coordinate system are known and the coordinate system is rotated by $\alpha$, the corresponding $I_n$, $I_t$, and $I_{nt}$ referred to in the new coordinate system $n$–$t$ can be calculated using these formulas.

It should be mentioned that the parallel-axes theorem can also be used for the product of inertia, as follows: The product of inertia of an area with respect to any coordinate axes, say $x$–$y$, is equal to the product of inertia of that area about centroidal axes parallel to the $x$–$y$ axes ($x_o$, $y_o$)—refer to Fig. 8.3—plus the product of the size of the area and the two distances $d_x$ and $d_y$, where $d_x$ and $d_y$ are the perpendicular distances of the $x$–$y$ axes and the two centroidal axes ($x_o$, $y_o$) parallel to the $x$–$y$ axes. Thus,

$$I_{xy} = I_{x_o y_o} + Ad_x d_y \quad (8.14)$$

Care has to be exercised in using the parallel-axes theorem for the product of inertia since—unlike Eq. 8.3, where the signs of $d_x$ and $d_y$ are of no importance because both are squared terms—Eq. 8.14 is sensitive to the signs of $d_x$ and $d_y$, and any mistake in sign will give wrong values for $I_{xy}$. If the centroidal axis $Cx_o$ is above the $x$ axis, then $d_x$ will have positive sign; if the centroidal axis, $Cy_o$ is to the right of the $y$ axis, then $d_y$ will have positive sign. Also note that $Cx_o$, $Cy_o$ must be the centroidal axes of the area for the parallel-axes theorem to be valid for the product of inertia.

As stated earlier, the moment of inertia of an area about any axis always has a positive value, but the product of inertia can have positive,

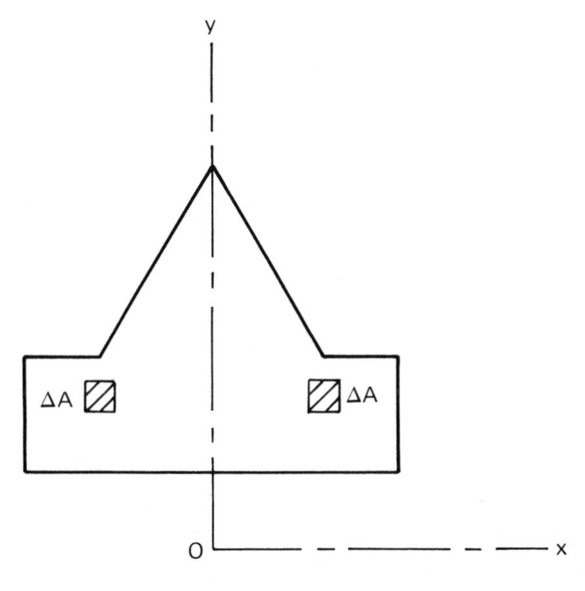


**Figure 8.12**

negative, or even zero value. Notice that only one axis is involved in the calculation of $I_x$ or $I_y$, but both axes are involved in the calculation of $I_{xy}$.

In the special case when one of the axes is also an axis of symmetry of the area, then $I_{xy} = 0$. This can easily be demonstrated by considering the area shown in Fig. 8.12, which has one axis of symmetry ($Oy$). For every area element, $\Delta A$, having coordinates $x$ and $y$, there exists another symmetrical area element, $\Delta A$, having the same $x$ and $y$ coordinates but with the $x$ coordinate of opposite sign. Therefore, the products of $xy\Delta A$ for each set of symmetrically located elements $\Delta A$ cancel each other out when the product of inertia is calculated using $I_{xy} = \Sigma xy\Delta A$. Thus, it can be generally stated that if an area has one or two axes of symmetry, the product of inertia about a coordinate axis $x$–$y$ with at least one of the axes coinciding with one of the axes of symmetry of the area is equal to zero.

EXAMPLE 9: Determine the product of inertia of the rectangle shown in Figs. 8.13(a), (b), and (c) with respect to the $x$–$y$ axes shown on each figure.

SOLUTION: The centroidal axes $Cx_o$, $Cy_o$ for the area are shown in the figures. Since they are also the symmetric axes of the area, $I_{x_o y_o} = 0$. In

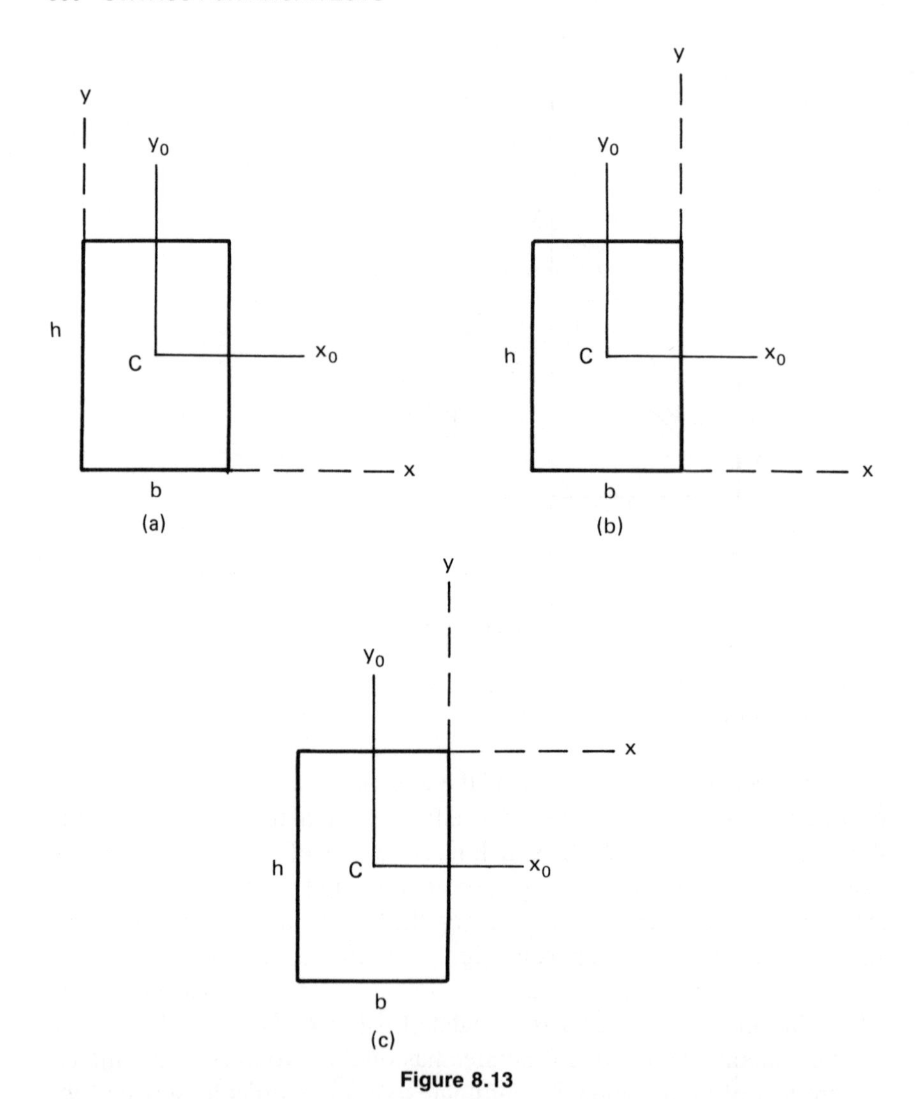


**Figure 8.13**

Fig. 8.13(a), the perpendicular distances between the centroidal axes and the $x$–$y$ axes are $d_x = +h/2$ and $d_y = +b/2$. Using the parallel axes theorem, we have

$$I_{xy} = I_{x_o y_o} + A d_x d_y$$

$$= 0 + (b \times h) \left(\frac{h}{2}\right) \left(\frac{b}{2}\right)$$

or

$$I_{xy} = \frac{b^2h^2}{4}$$

In Fig. 8.13(b), $d_x = +(h/2)$ and $d_y = -(b/2)$, and therefore the product of inertia will have negative value, as follows:

$$I_{xy} = 0 + (b \times h)\left(\frac{h}{2}\right)\left(\frac{-b}{2}\right) = -\frac{b^2h^2}{4}$$

In Fig. 8.13(c), since both distances are negative, that is, $d_x = -(h/2)$ and $d_y = -(b/2)$, the product of inertia with respect to the $x$–$y$ coordinate system will be positive, as follows:

$$I_{xy} = 0 + (b \times h)\left(-\frac{h}{2}\right)\left(-\frac{b}{2}\right) = \frac{b^2h^2}{4}$$

EXAMPLE 10: Determine the centroidal product of inertia of the cross-sectional area shown in Fig. 8.14(a).

SOLUTION: The composite area can be divided into two rectangles as shown in Fig. 8.14(b), with $A_1 = 0.02 \times 0.08\ m^2$ and $A_2 = 0.02 \times 0.10\ m^2$. The centroid, $C$, of the entire area can be found as follows:

$$\bar{x} = \frac{A_1x_1 + A_2x_2}{A_1 + A_2} = \frac{(0.02 \times 0.08)(0.04) + (0.02 \times 0.10)(0.07)}{0.0036}$$

$$= 0.0567\ m$$

$$\bar{y} = \frac{A_1y_1 + A_2y_2}{A_1 + A_2} = \frac{(0.02 \times 0.08)(0.11) + (0.02 \times 0.10)(0.05)}{0.0036}$$

$$= 0.0767\ m$$

Notice that each area, $A_1$ and $A_2$, has symmetric axes and that therefore their centroidal product of inertias about these axes are zero. Using the parallel-axes theorem, the centroidal product of inertia for the entire cross section with reference to the $Cx_o$ and $Cy_o$ axes can be calculated, as follows:

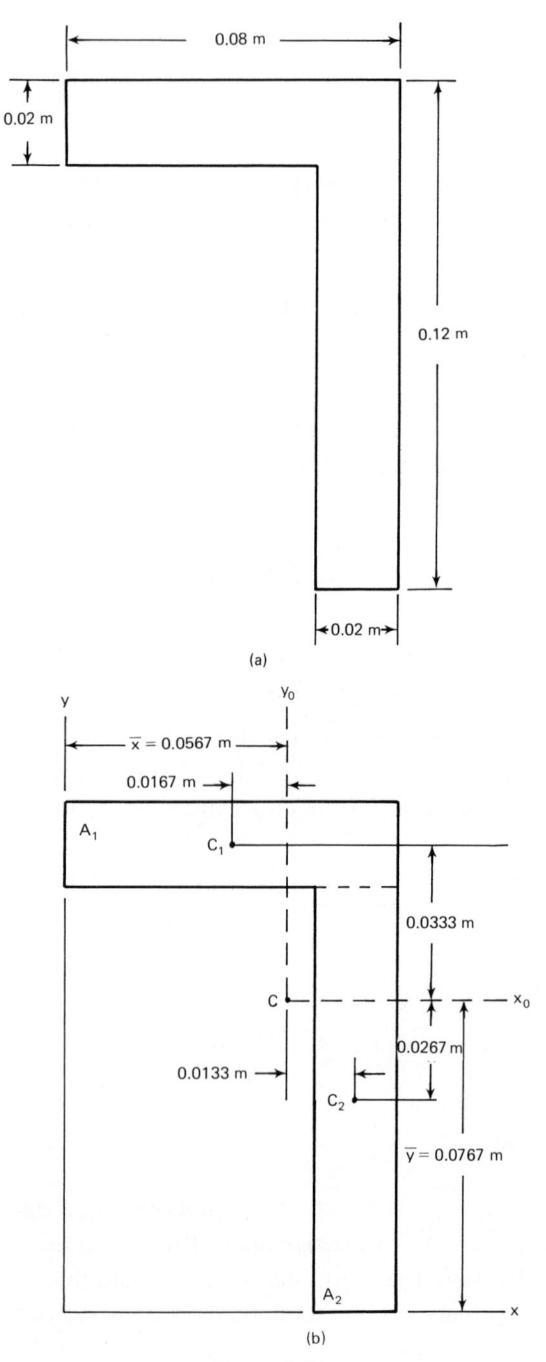
0.08 m
0.02 m
0.12 m
0.02 m
(a)
y
$y_0$
$\overline{x}$ = 0.0567 m
0.0167 m
$A_1$
$C_1$
0.0333 m
C
$x_0$
0.0267 m
0.0133 m
$C_2$
$\overline{y}$ = 0.0767 m
$A_2$
x
(b)

**Figure 8.14**

$$\begin{aligned} I_{x_o y_o} &= (I_{x_o y_o})_1 + (I_{x_o y_o})_2 \\ &= (\bar{I}_{x_o y_o} + Ad_x d_y)_1 + (\bar{I}_{x_o y_o} + Ad_x d_y)_2 \\ &= [0 + (0.02 \times 0.08)\,(0.0333)\,(-0.0167)] \\ &\quad + [0 + (0.02 \times 0.10)\,(-0.0267)\,(0.0133)] \\ &= -1.6 \times 10^{-6} m^4 \end{aligned}$$

## 8.8 PRINCIPAL AXES AND PRINCIPAL MOMENTS OF INERTIA

Let's consider Fig. 8.11 again with a set of coordinate axes $Ox$ and $Oy$ centered at an arbitrary point such as $O$. We saw that if $I_x$, $I_y$, and $I_{xy}$ of the area $A$ are known, the moments and product of inertia of the same area about any set of axes such as $On$, $Ot$ can be calculated using Eqs. 8.13(a), (b), and (c). Notice that the new coordinate axes, $On$, $Ot$, are produced by the counterclockwise rotation of $Ox$, $Oy$ axes through the angle $\alpha$. Obviously, the moments and product of inertia of the area $A$ about the new axes ($On$, $Ot$) will have different values from those obtained with reference to the $Ox$, $Oy$ axes. In fact, the moments of inertia of area $A$ will assume different values with respect to every different set of perpendicular axes at a given point. One of these sets of axes, at point $O$, however, will make the value of the moment of inertia either maximum or minimum. Such axes are called the *principal axes*, and the moments of inertia obtained with reference to these principal axes are called the *principal moments of inertia*.

Equation 8.12 gives the value of one of the moments of inertia of the area $A$ as a function of rotation angle $\alpha$. To find the particular value of angle $\alpha$ that causes this moment of inertia to assume maximum or minimum value, one can find the derivative of the moment of inertia, $I_n$,* set it equal to zero, and then solve it for $\alpha$, giving

$$\tan 2\alpha_m = -\frac{2I_{xy}}{I_x - I_y} \tag{8.15}$$

In this equation, the particular value of $\alpha$ that makes the moment of inertia become maximum or minimum is denoted by $\alpha_m$. This $\alpha_m$ cor-

---

*The complete derivation is not given here because, although simple, it requires taking derivatives involving trigonometric functions that are not needed nor covered by this book.

responds to the principal axes of the area $A$ at point $O$ (Fig. 8.11) that make the moments of inertia either maximum or minimum. Equation 8.15 gives two values for $\alpha_m$, one corresponding to the maximum and the other to the minimum moment of inertia.

To calculate the values of these maximum or minimum moments of inertia, one can substitute the values of $\alpha_m$ obtained from Eq. 8.15 into Eq. 8.13(a) and obtain the following:

$$(I)_{\substack{max \\ min}} = \frac{I_x + I_y}{2} \pm \sqrt{\left(\frac{I_x - I_y}{2}\right)^2 + (I_{xy})^2} \tag{8.16}$$

It should be noticed that if we use the equation for $I_t$ [Eq. 8.13(b)], a result identical to that shown above will be obtained. The two values given, $I_{max}$ and $I_{min}$, can also be written as follows:

$$I_{max} = I_x$$

$$I_{min} = I_y$$

It can be shown that the product of inertia vanishes when it is referred to the principal axes. This means that if one identifies the principal axis for an area at a point using Eq. 8.15 and then calculates the product of inertia of the area about that axis, the results would be zero. In fact, the truth of this is easy to see. Let's substitute $I_{nt} = 0$ in Eq. 8.13(c) and then divide both sides of the equation by cos $2\alpha$; the result will be Eq. 8.15. We can therefore state that *the product of inertia about principal axes is zero*.

It can also be shown that if an area has an axis of symmetry, the axis of symmetry and any axis perpendicular to it constitute a set of principal axes. Therefore, *any axis of symmetry is always a principal axis.* It should also be mentioned that any area has a unique set of principal axes at any point inside or outside the area. Furthermore, this is true regardless of whether the area does or does not have an axis or axes of symmetry. The principal axis, therefore, does not need to be an axis of symmetry.

In Fig. 8.11, if point $O$ coincides with point $C$, the centroid of the area, the principal axes at $C$ are known as the *principal centroidal axes.* As mentioned earlier, a given area will have an infinite number of sets

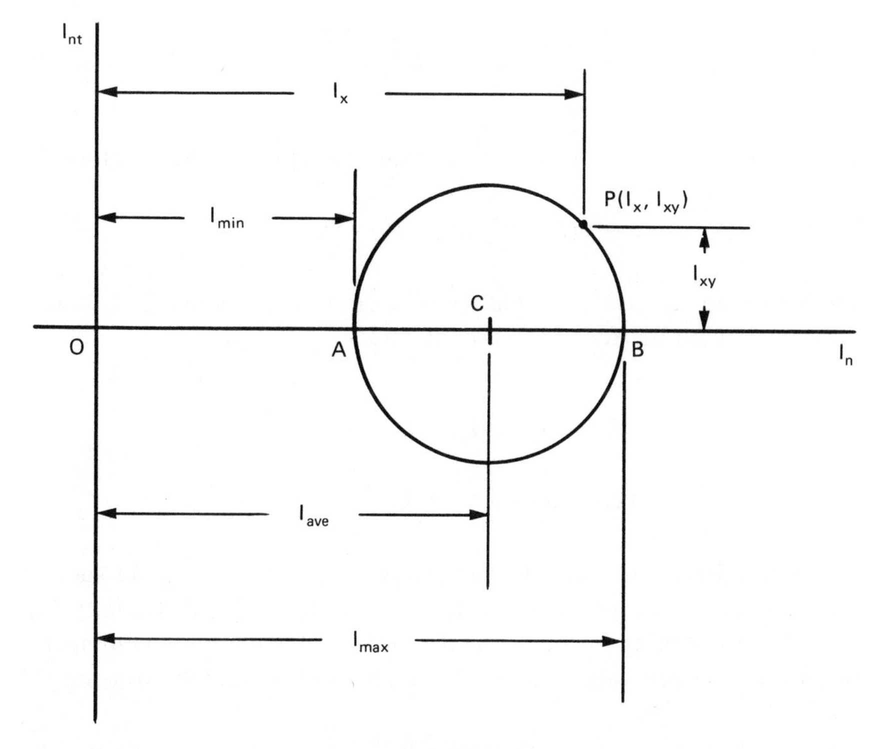


**Figure 8.15**

of principal axes, one set at any given point, but it has only one unique set of principal centroidal axes.

Now let's look at the whole subject with the help of graphical illustrations. Equations 8.13(a) and 8.13(c) are parametric equations of a circle (the parameter is the angle $\alpha$). This simply means that, by using the rectangular axes $I_n$ and $I_{nt}$ as a coordinate system, as shown in Fig. 8.15, a point $P$ with coordinates $I_x$ and $I_{xy}$ will fall on a circle that has the parametric equations given by Eqs. 8.13(a) and 8.13(c). All values of the parameter angle $\alpha$ correspond to points on the same circle. After eliminating $\alpha$ from Eqs. 8.13(a) and 8.13(c), one will obtain the following:

$$\left(I_n - \frac{I_x + I_y}{2}\right)^2 + I_{nt}^2 = \left(\frac{I_x - I_y}{2}\right)^2 + I_{xy}^2 \qquad (8.17)$$

For simplicity, let's set $I_{\text{ave}}$ and $R^2$ as follows:

$$I_{ave} = \frac{I_x + I_y}{2} \quad \text{and} \quad R^2 = \left(\frac{I_x - I_y}{2}\right)^2 + I_{xy}^2 \tag{8.18}$$

Using Eq. 8.18, Eq. 8.17 can be written in a simple form as follows:

$$(I_n - I_{ave})^2 + I_{nt}^2 = R^2 \tag{8.19}$$

The above equation is the equation of a circle with radius $R$, its center at point $C$, with the following coordinates:

$$\text{Abscissa} = I_n = I_{ave} = \frac{I_x + I_y}{2}$$

$$\text{Ordinate} = I_{nt} = 0$$

These coordinates are shown graphically in Fig. 8.15. A point such as $P$ on this circle has coordinates, $I_n = I_x$ and $I_{nt} = I_{xy}$, that satisfy Eq. 8.19. It is interesting to notice that the maximum and minimum moments of inertia as given earlier in Eq. 8.16 can now be visualized using the circle in Fig. 8.15.

As mentioned earlier, any point on this circle has an $x$ coordinate equal to $I_n$ and a $y$ coordinate equal to $I_{nt}$. In particular, points $A$ and $B$ both have a $y$ coordinate ($I_{nt}$) equal to zero. The $x$ coordinate of point $A$ corresponds to the minimum value of $I_n$, and the $x$ coordinate of point $B$ corresponds to the maximum value of $I_n$. Since the $y$ coordinate ($I_{nt}$) for both points $A$ and $B$ equals zero, substituting $I_{nt} = 0$ into Eq. 8.13(c) results in the following:

$$\frac{I_x - I_y}{2} \sin 2\alpha_m + I_{xy} \cos 2\alpha_m = 0$$

or

$$\tan 2\alpha_m = -\frac{2I_{xy}}{I_x - I_y} \tag{8.20}$$

This equation is the same as Eq. 8.15. Remember that Eq. 8.15 was a condition for obtaining the principal directions that belong to maximum and minimum moments of inertia. Equation 8.20 gives two values for

$2\alpha_m$ that are 180° apart, or, alternatively, we can say that it gives two angles $\alpha_m$ that are 90° apart. One angle corresponds to point $A$ and the other to point $B$, which give the minimum and maximum moments of inertia, respectively. Using Fig. 8.15, the values of maximum and minimum moments of inertia can be easily written as follows:

$$I_{\text{min}} = I_{\text{ave}} - R \qquad \text{and} \qquad I_{\text{max}} = I_{\text{ave}} + R$$

where $I_{\text{ave}}$ is the $x$ coordinate of the center and $R$ is the radius of the circle. Substituting their values from Eq. 8.18 results in the following:

$$I_{\text{min}} = \frac{I_x + I_y}{2} - \sqrt{\left(\frac{I_x - I_y}{2}\right)^2 + I_{xy}^2}$$

and

$$I_{\text{max}} = \frac{I_x + I_y}{2} + \sqrt{\left(\frac{I_x - I_y}{2}\right)^2 + I_{xy}^2}$$

These are the same as the expression obtained in Eq. 8.16.

## Use of Mohr's Circle

Now consider area $A$ in Fig. 8.11 again. One can calculate moments and the product of inertia—$I_x$, $I_y$, and $I_{xy}$—of the area $A$ about these axes as follows:

$$I_x = \Sigma y^2 \Delta A \qquad I_y = \Sigma x^2 \Delta A \qquad I_{xy} = \Sigma xy \Delta A$$

Now, if we rotate the $Ox$, $Oy$ axes about point $O$ through angle $\alpha$, two new axes—$On$, $Ot$—will be produced. As shown before, the moments and product of inertia of area $A$ can be calculated with respect to these new axes at $O$ by using the formulas given in Eqs. 8.13(a), (b), and (c). If we rotate the coordinate axes continuously, we will obtain different values for $I_n$, $I_t$, and $I_{nt}$. All these values are related to one another through the circle shown in Fig. 8.15, which is known as *Mohr's circle* (after Otto Mohr, a German engineer).

This section demonstrates that if the moments and product of inertia of an area $A$ are known with respect to any set of rectangular coordinate axes ($Ox$, $Oy$) at a point $O$, Mohr's circle can be used to determine the following graphically:

1. The principle axes of area $A$ referred to point $O$
2. The principle moments of inertia of area $A$ referred to point $O$
3. The moments and product of inertia of area $A$ with respect to any other set of rectangular axes at point $O$.

To demonstrate this, let's consider the hatched area shown in Fig. 8.16(a), with the rectangular coordinates, $Ox$, $Oy$. $I_x$, $I_y$, and $I_{xy}$ for this area are given. We want to obtain the principal axis at $O$, the principal moments of inertia of the area at $O$, and also the new values of the moments and the product of inertia of the area at $O$ with respect to new coordinate axes that are produced by rotating the old axes ($Ox$, $Oy$) through an angle $\alpha_m$. To obtain these values, one could use Eq. 8.15 to determine the principal axes, Eq. 8.16 to determine the principal moments of inertia, and Eqs. 8.13(a), (b), and (c) to calculate the moments and product of inertia with respect to any other set of axes, but Mohr's circle will determine all these graphically, as will now be demonstrated.

As its name implies, the Mohr circle is a circle, and to draw a circle, one must locate its center and know its radius. Since these parameters have not been given, one has to obtain them from the information that has been—that is, from $I_x$, $I_y$, and $I_{xy}$. How one does so will be demonstrated step by step by the following.

Take a set of coordinate axes with $I_n$ (for $I_x$ or $I_y$) as the horizontal axis and $I_{nt}$ (for $I_{xy}$) as the vertical axis, as shown in Fig. 8.16(b). Next, take coordinates $I_x$ and $I_{xy}$ to plot a point $P$ and coordinates $I_y$ and $-I_{xy}$ to plot a point $Q$. It is assumed here that $I_{xy}$ is positive, which means that point $P$ will be located above the horizontal axis and $Q$ below it. If $I_{xy}$ were negative, $Q$ would be located above the horizontal axis and $P$ below it. The procedure, however, will be identical whether $I_{xy}$ is positive or negative. A straight line joining $P$ and $Q$ intersects the horizontal axes at $C$, which is the center of the Mohr circle required. This circle can now be constructed using $C$ as its center and $CP$ or $CQ$ as its radius, as shown in Fig. 8.16(c).

Notice that if the coordinates (the $x$ coordinates in this case) of two points are known, the coordinates of the midpoint between these two points are equal to one-half of the sum of the coordinates of these two points. Since the $x$ coordinates of points $P$ and $Q$ are $I_x$ and $I_y$, respec-

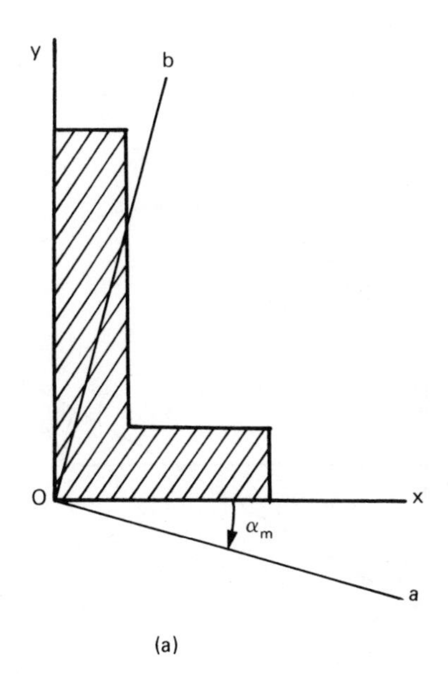


(a)

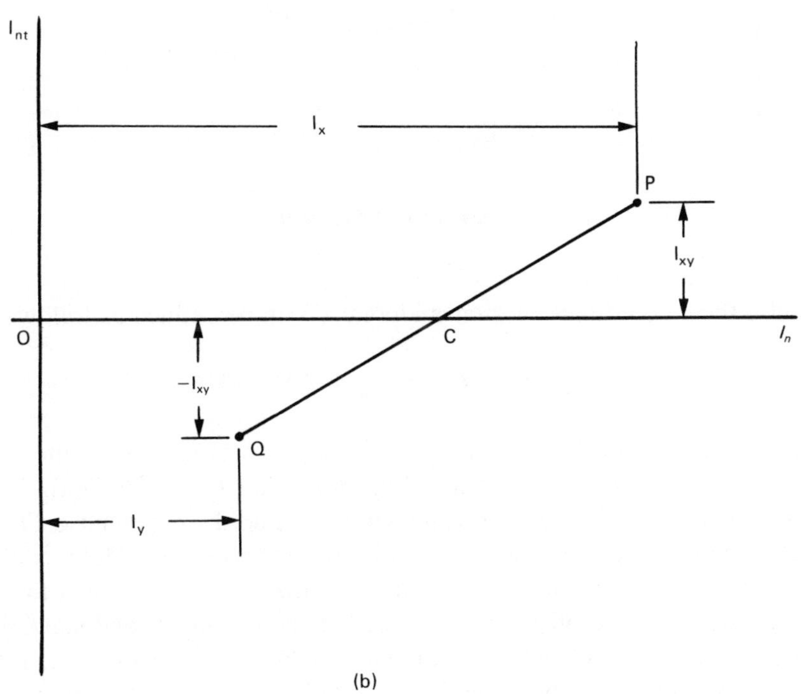


(b)

**Figure 8.16**

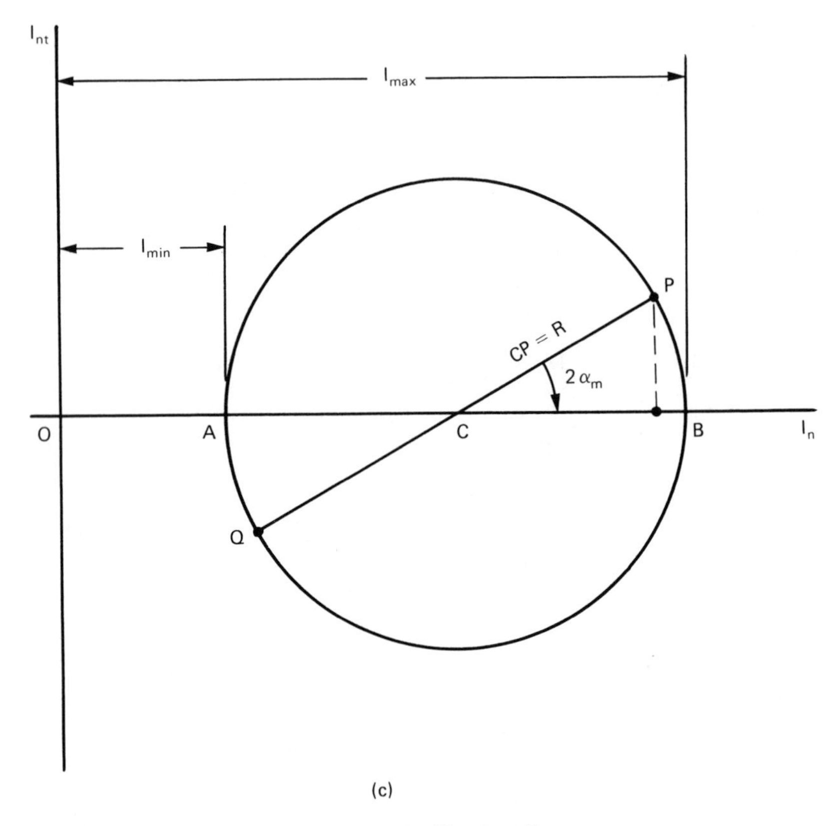


**Figure 8.16.** *(Continued)*

tively, the $x$ coordinate of their midpoint $C$ can be defined as follows:

$$x_C = OC = (I_x + I_y)/2$$

Also notice that in the right triangle in Fig. 8.16(c) formed by the vertices at $C$, $P$, and at point obtained by projecting $P$ onto the horizontal axis, one can use the Pythagorean theorem to find the hypotenuse $CP$ of this triangle, which is also equal to $R$, the radius of the Mohr circle. Since the vertical distance of $P$ from the horizontal axis is equal to $I_{xy}$, which is one of the other two sides of the right triangle, and since the third side is equal ($C$ being the midpoint of $PQ$) to one-half of the projection of $PQ$ on the horizontal axis, or $(I_x - I_y)/2$, the Pythagorean theorem tells us that the radius of the Mohr circle can be evaluated as

$$R = \sqrt{\left(\frac{I_x - I_y}{2}\right)^2 + I_{xy}^2}$$

With this determination of the center $C$ and the radius $R$, the Mohr circle can now be constructed as shown in Fig. 8.16(c). Notice that these values of $OC$ and $R$ are the same as those given by Eq. 8.18.

Now, much as in Fig. 8.15, points $A$ and $B$ correspond to the minimum and maximum values of the moments of inertia and are indicated accordingly in Fig. 8.16(c). Remember that this is the Mohr's circle for a *given* area at a *particular* point. Notice, furthermore, that in Fig. 8.16(c), the tangent of the angle $PCB$ is given as follows.

$$\tan PCB = \frac{I_{xy}}{(I_x - I_y)/2} = \frac{2I_{xy}}{I_x - I_y}$$

Comparing this with Eq. 8.20, we can say that angle $PCB$ corresponds to one of the angles $2\alpha_m$ (with the exception of the minus sign, since the right side of Eq. 8.20 is equal to the negative of the right side of this expression for tan $PCB$). Having this in mind, the clockwise angle $2\alpha_m$ in Fig. 8.16(c) corresponds to the principal axis. The same principal axis is shown in Fig. 8.16(a), there being produced by the clockwise rotation of $Ox$ through $\alpha_m$. Another way to look at this is that, by clockwise rotation of $PC$ through angle $2\alpha_m$ in Fig. 8.16(c), $PC$ coincides with $CB$. Since point $B$ corresponds to $I_{max}$, moreover, one can find one of the principal axes, $Oa$, of the area in Fig. 8.16(a) by clockwise rotation of $Ox$ through the angle $\alpha_m$. The other axis, $Ob$, is, of course, perpendicular to this axis.

It should be further noted that in the preceding discussion, $I_x > I_y$ and $I_{xy} > 0$. In this case, it is a clockwise rotation that lines up $PC$ with $CB$ and also a clockwise rotation that lines up $Ox$ with $Oa$ in Fig. 8.16(a). Therefore, it should be remembered that if a clockwise rotation of $2\alpha_m$ lines up $PC$ with $CB$ on Mohr's circle, it is likewise a clockwise rotation, but of $\alpha_m$, that lines up $Ox$ with the corresponding principal axis $Oa$.

Now consider the same area with another set of rectangular coordinate axes, shown as $Ox'$ and $Oy'$ in Fig. 8.17(a). Since Mohr's circle uniquely defines the relationship between moments and product of inertia of area $A$ at point $O$ with respect to any set of rectangular coordinate axes, the same Mohr's circle can be used for axes $Ox'$ and $Oy'$. Therefore, point

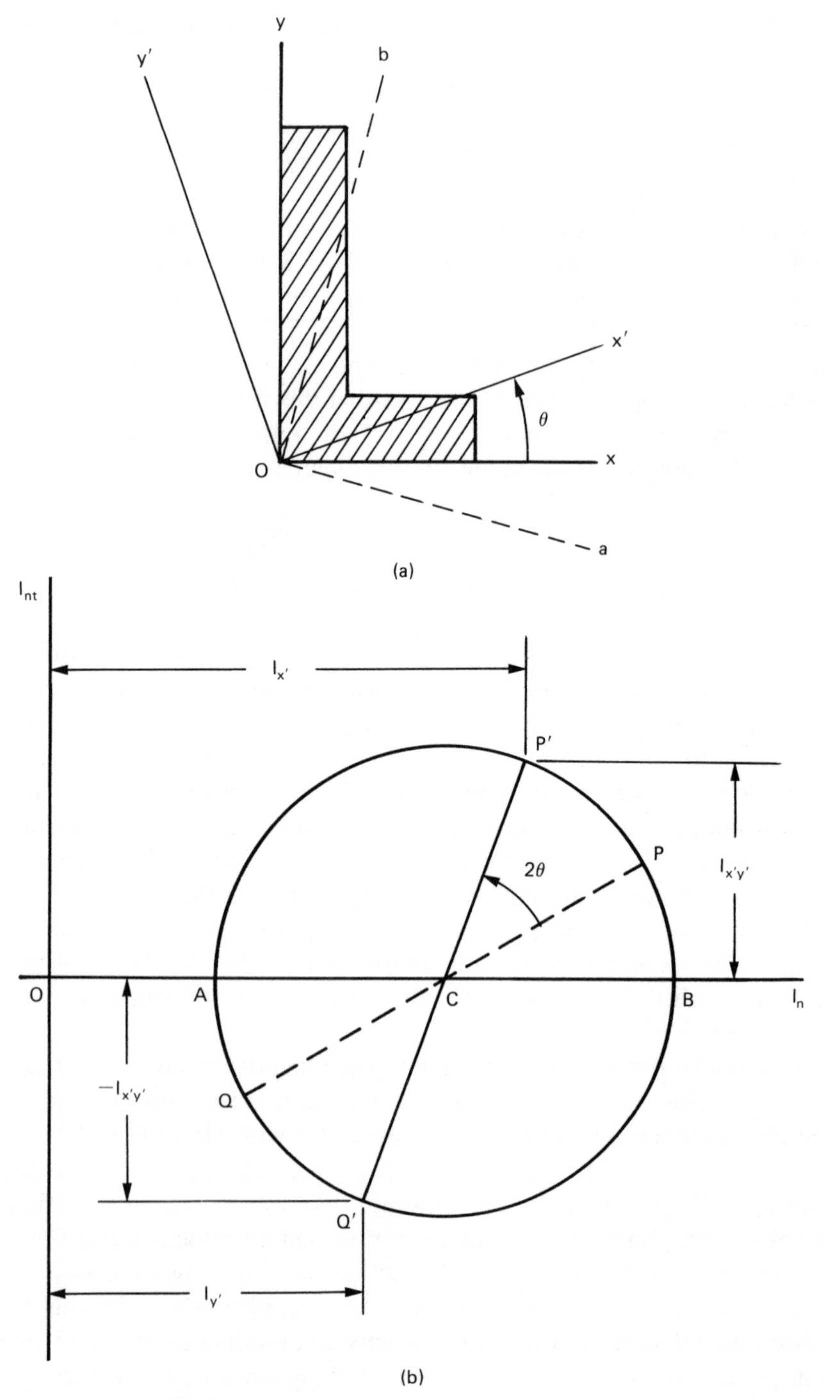


**Figure 8.17**

$P'$ (with coordinates $I_{x'}$ and $I_{x'y'}$) and point $Q'$ (with coordinates $I_y$ and $-I_{x'y'}$) are both located on the same Mohr's circle as that shown in Fig. 8.16(c), which is redrawn in Fig. 8.17(b). Angle $P'CB$ in Fig. 8.17(b) must be equal to twice angle $x'Oa$ in Fig. 8.17(a). This simply requires that angle $PCP'$ in Fig. 8.17(b) be twice angle $xOx'$ in Fig. 8.17(a). Therefore, we can conclude that if we rotate axis $Ox$ in Fig. 8.17(a) counterclockwise through angle $\theta$, the moments and product of inertia of area $A$ about the $Ox'$, $Oy'$ axes—$I_{x'}$, $I_{y'}$, and $I_{x'y'}$—can be obtained from Mohr's circle in Fig. 8.17(b) by rotating $PQ$ counterclockwise, but through angle $2\theta$. These values can be read off the Mohr's circle directly. Alternatively, $I_{x'}$, $I_{y'}$, and $I_{x'y'}$ can be expressed from Mohr's circle in terms of values of $OC$ and $R$.

In Chap. 2, we introduced graphical methods to determine the resultant of forces, etc. In such methods, all elements must be drawn to scale, and the accuracy of the results depends on how accurately the graphs are drawn. Although Mohr's circle is thought of as a graphical method one does not need to draw it to scale. One can merely sketch it, and the relationships obtained will still be precise.

EXAMPLE 11: In the Z-section given in Fig. 8.18, determine (1) the principal centroidal axes, and (2) the principal centroidal moments of inertia. The centroidal moments of inertia of the section are given as follows: $I_x = 4.9 \times 10^{-6}\ \text{m}^4$, $I_y = 1.4 \times 10^{-6}\ \text{m}^4$, and $I_{xy} = -1.6 \times 10^{-6}\ \text{m}^4$.

SOLUTION: The principal axes can be determined by evaluating the angle $\alpha$ corresponding to the principal axes by using Eq. 8.15:

$$\tan 2\alpha_m = -\frac{2I_{xy}}{I_x - I_y} = \frac{2 \times 1.6}{4.9 - 1.4} = 0.914$$

(in which the factor $10^{-6}$ is eliminated both in the numerator and denominator) or

$$2\alpha_m = 42.4°$$

$$\alpha_m = 21.2°$$

Notice that $\alpha_m = 111.2°$ is also a correct angle since it is obtained by adding 90° to 21.2° (it was mentioned earlier that the principal axes are 90° apart).

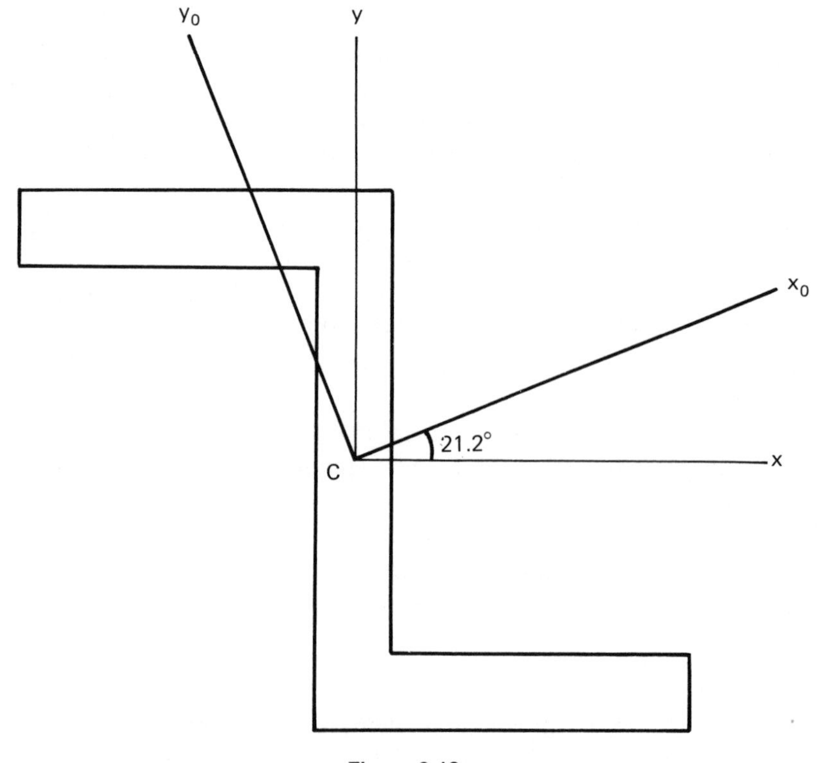


Figure 8.18

The principal centroidal moments of inertia can be calculated using Eq. 8.16, as follows:

$$(I)_{max} = 5.52 \times 10^{-6}\ m^4 \text{ (corresponding to } \alpha = 21.2°)$$

$$(I)_{min} = 0.78 \times 10^{-6}\ m^4 \text{ (corresponding to } \alpha = 111.2°)$$

EXAMPLE 12: For the area shown in Fig. 8.19(a), the centroidal moments of inertia are given as $I_x = 4.92 \times 10^{-6}$ m$^4$; $I_y = 1.72 \times 10^{-6}$ m$^4$; and $I_{xy} = -1.6 \times 10^{-6}$ m$^4$. Determine the principal centroidal axis as well as the principal moments of inertia using Mohr's circle.

SOLUTION: Draw horizontal and vertical coordinate axes, $I_n$ and $I_{nt}$, as shown in Fig. 8.19(b). Next, construct Mohr's circle using the given values of moments of inertia with the following points:

$$\text{Point } P \begin{cases} I_n = I_x = 4.92 \times 10^{-6}\ \text{m}^4 \\ I_{nt} = I_{xy} = -1.6 \times 10^{-6}\ \text{m}^4 \end{cases}$$

$$\text{Point } Q \begin{cases} I_n = I_y = 1.72 \times 10^{-6}\ \text{m}^4 \\ I_{nt} = -I_{xy} = 1.6 \times 10^{-6}\ \text{m}^4 \end{cases}$$

Since all four coordinates above have $10^{-6}$ as a common factor, for convenience in the construction of the Mohr circle this factor has been eliminated.

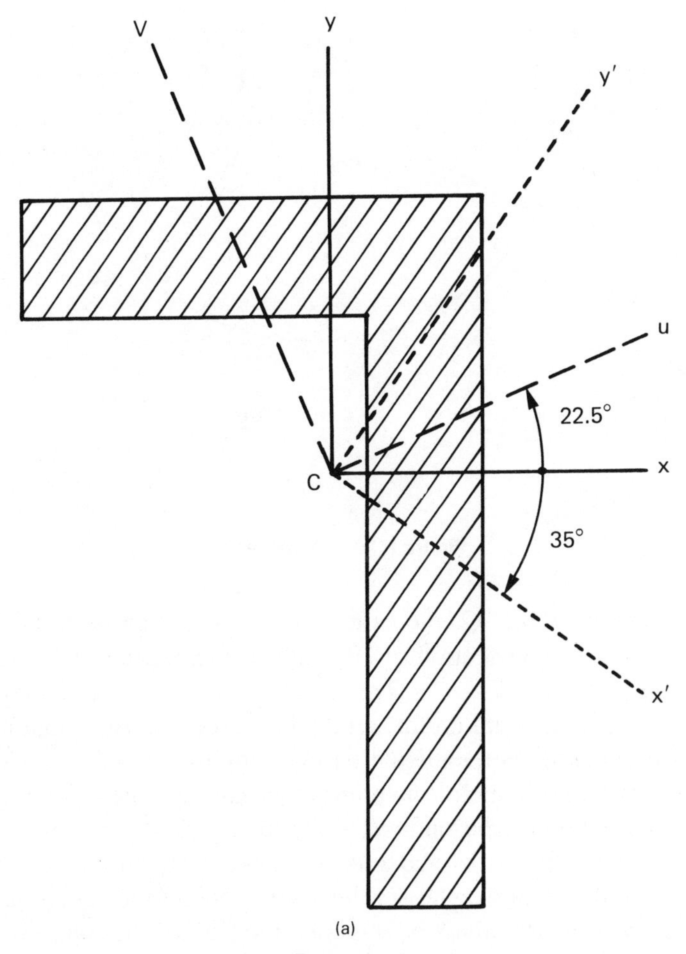


Figure 8.19

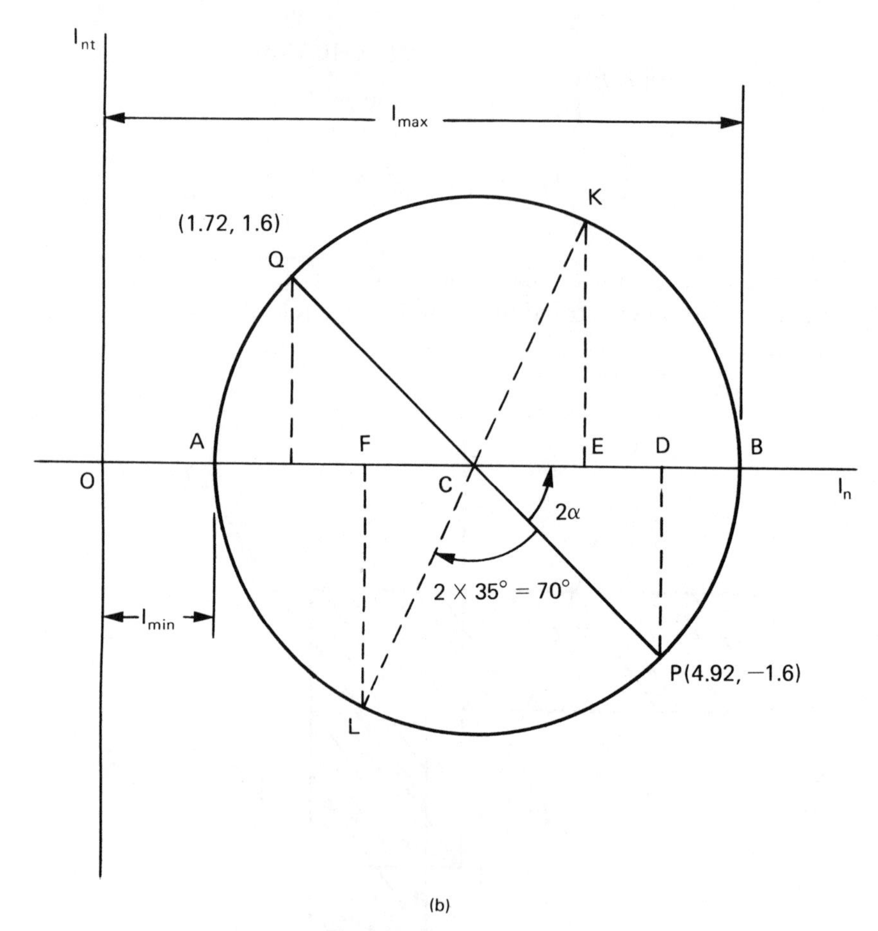


**Figure 8.19.** (*Continued*)

By connecting $P$ and $Q$, the center of the circle can be located, this being the point of intersection of $PQ$ with the horizontal axis. Using $C$ as the center and $R = CP = CQ$ as the radius, a complete circle can be circumscribed. Any point on this circle has two coordinates that may be obtained by drawing perpendiculars to the horizontal and vertical axes. The horizontal coordinate of the point is the same as the value of $I_n$ and the vertical coordinate the same as the value of $I_{nt}$. At the points where the circle crosses the horizontal axis, the value of the product of inertia is equal to zero; therefore, these values correspond to the maximum and minimum $I$, which are called the principal moments of inertia. Since they can be measured directly along the horizontal axis, we have the values

$$I_{\max} = OB = 5.6 \times 10^{-6}\ \text{m}^4$$

$$I_{\min} = OA = 1.06 \times 10^{-6}\ \text{m}^4$$

Alternatively, we can write analytic expressions for $I_{\max}$ and $I_{\min}$ using Mohr's circle as follows:

$$CD = (I_x - I_y)/2 = [(4.92 - 1.72) \times 10^{-6}]/2$$

$$= 1.6 \times 10^{-6}\ \text{m}$$

$$R = \sqrt{CD^2 + DP^2} = \sqrt{(1.6 \times 10^{-6})^2 + (-1.6 \times 10^{-6})^2}$$

$$= 2.26 \times 10^{-6}\ \text{m}$$

$$OC = I_{\text{ave}} = \frac{I_x + I_y}{2} = [(4.92 + 1.72) \times 10^{-6}]/2$$

$$= 3.32 \times 10^{-6}\ \text{m}^4$$

$$I_{\max} = OB = OC + R = (3.32 \times 10^{-6}) + (2.26 \times 10^{-6})$$

$$= 5.58 \times 10^{-6}\ \text{m}^4$$

$$I_{\min} = OA = OC - R = (3.32 \times 10^{-6}) - (2.26 \times 10^{-6})$$

$$= 1.06 \times 10^{-6}\ \text{m}^4$$

Using triangle $CDP$, we can write:

$$\tan 2\alpha = \frac{DP}{CD} = \frac{1.6 \times 10^{-6}}{1.6 \times 10^{-6}} = 1$$

Thus, if $\tan 2\alpha = 1$, $2\alpha = 45°$ and $\alpha = 22.5°$.

Therefore, the principal axes are 22.5° *ccw* (counterclockwise) from the $x$ axis and 90° + 22.5°, or 112.5° *ccw* from the $x$ axis. These axes are shown in Fig. 8.19(a) as $u$–$v$ axes.

EXAMPLE 13: Determine the centroidal moments and product of inertia of the area shown in Fig. 8.19(a) with respect to centroidal axes $Cx'$, $Cy'$; the latter are produced by rotating $Cx$, $Cy$ axes by 35° clockwise.

SOLUTION: In Fig. 8.19(a), the axes $Cx'$, $Cy'$ are produced by clockwise rotation of the $Cx$, $Cy$ axes through an angle of 35°. In the Mohr's circle shown in Fig. 8.19(b), $PQ$ corresponds to the $Cx$, $Cy$ system, and, as we learned earlier, a clockwise rotation of 35° in Fig. 8.19(a) means that $PQ$ must also rotate clockwise, but by 2 × 35°, or 70°. This new line is shown as $KL$ in Fig. 8.19(b). The values of $I$ and $I_{nt}$ corresponding to points $K$ and $L$ can be obtained by drawing perpendicular lines from $K$ and $L$ to the horizontal axes, as shown in Fig. 8.19(b), or, alternatively, by using calculations as before. Since from the previous problem we learned that angle $PCD$ = 45° and knowing that angle $LCF$ = angle $KCE$ = 180° − 70° − 45° = 65°, we can express $I_{x'}$, $I_{y'}$ and $I_{x'y'}$ as follows:

$$I_{x'} = OF = OC - CF = OC - CL \cos 65°$$

$$= (3.32 \times 10^{-6}) - (2.26 \times 10^{-6}) \cos 65°$$

$$= 2.36 \times 10^{-6}\ \text{m}^4$$

$$I_{y'} = OE = OC + CE = OC + CK \cos 65°$$

$$= (3.32 \times 10^{-6}) + (2.26 \times 10^{-6}) \cos 65°$$

$$= 4.28 \times 10^{-6}\ \text{m}^4$$

$$I_{x'y'} = KE = KC \sin 65° = (2.26 \times 10^{-6}) \sin 65°$$

$$= 2.05 \times 10^{-6}\ \text{m}^4$$

## PROBLEMS

**8.1.–8.6.** In Figs. P8.1 through P8.6, calculate the moments of inertia of the areas shown both about the $x$ and $y$ axes.

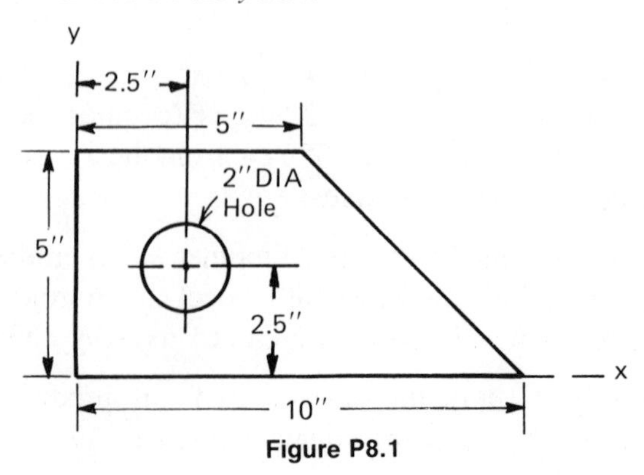


Figure P8.1

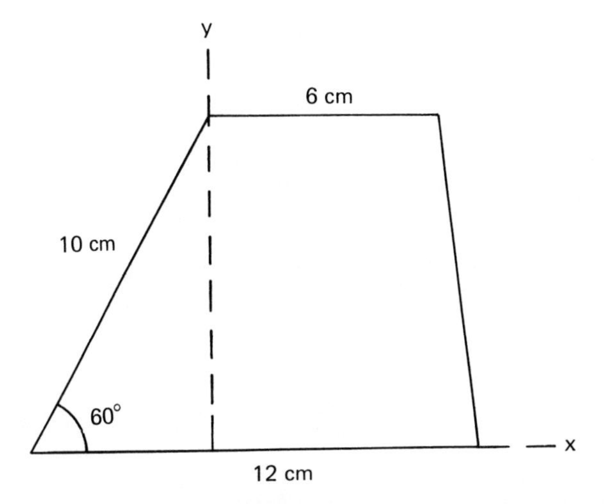


**Figure P8.2**

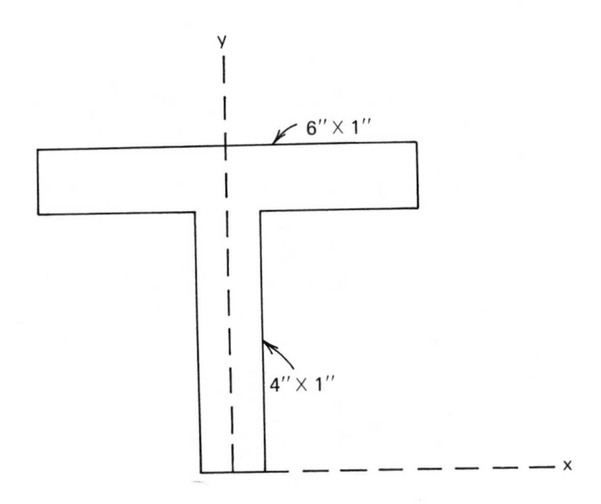


**Figure P8.3**

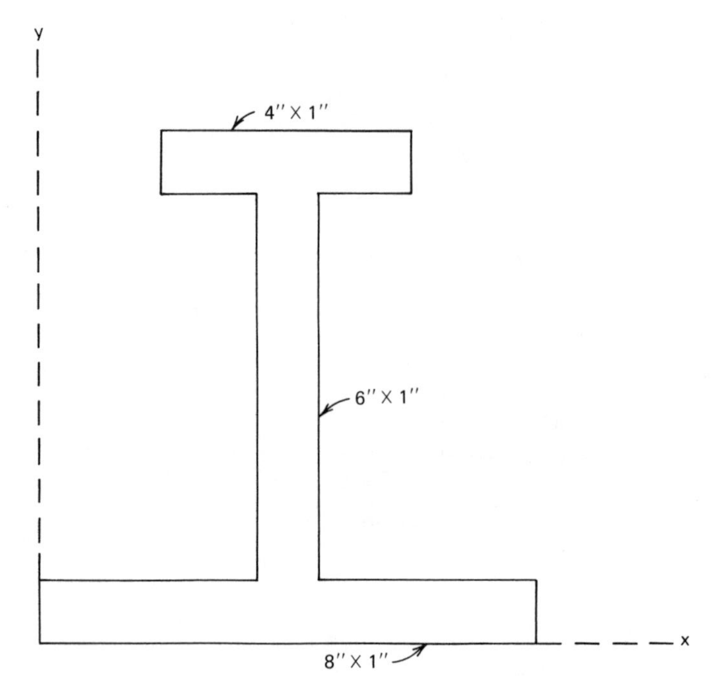


**Figure P8.4**

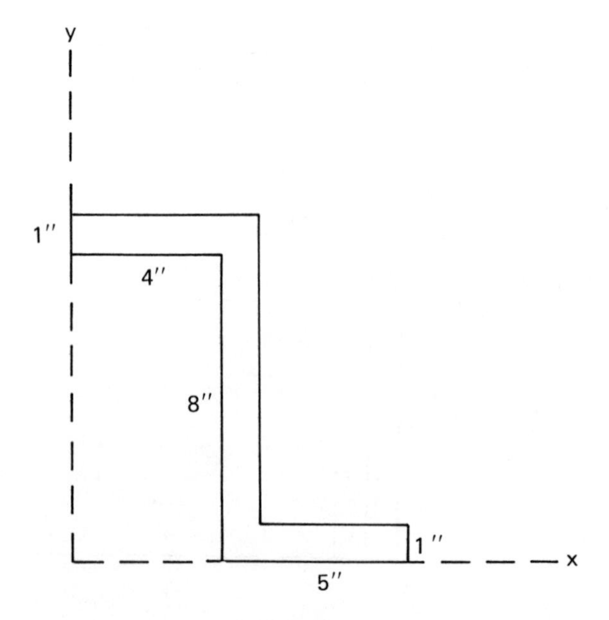


**Figure P8.5**

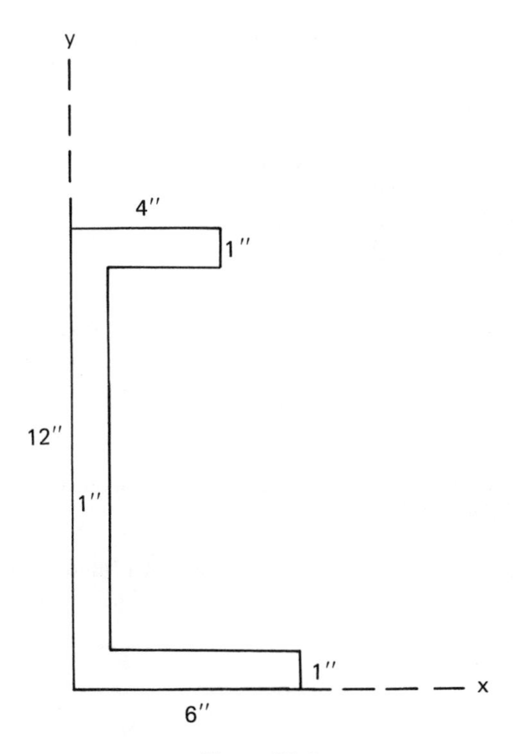


Figure P8.6

**8.7.–8.10.** In Figs. P8.3 through P8.6, determine the centroidal moment of inertia.

**8.11.** Determine the radius of gyration and section modulus for the areas in Figs. P8.4 through P8.6.

**8.12.–8.13.** Determine the centroidal moments of inertia ($\bar{I}_x$ and $\bar{I}_y$) for the structural shapes shown in Figs. P8.7 and P8.8.

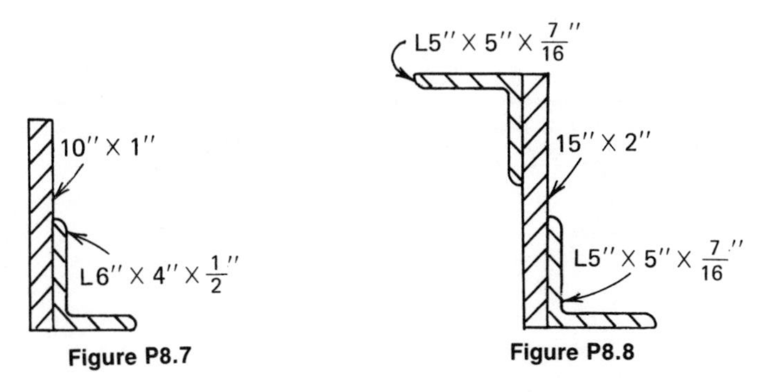


Figure P8.7

Figure P8.8

**8.14.** A beam is fabricated using three C9 × 13.4 channels, as shown in Fig. P8.9. Determine $\bar{I}_x$ and $\bar{I}_y$.

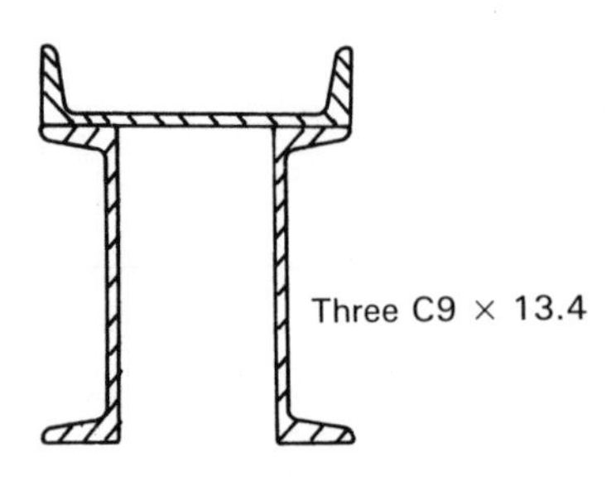


**Figure P8.9**

**8.15.–8.18.** Determine $I_x$, $I_y$, and $I_{xy}$ for the areas shown about the centroidal axes of the areas in Figs. P8.3 through P8.6.

**8.19.–8.22.** Using the values obtained in Problems 8.15 through 8.18, calculate the principal moments of inertia of the areas involved using equations as well as the Mohr circle.

# APPENDIX A

## PROPERTIES OF COMMON STEEL STRUCTURAL SHAPES

# WIDE FLANGE SHAPES

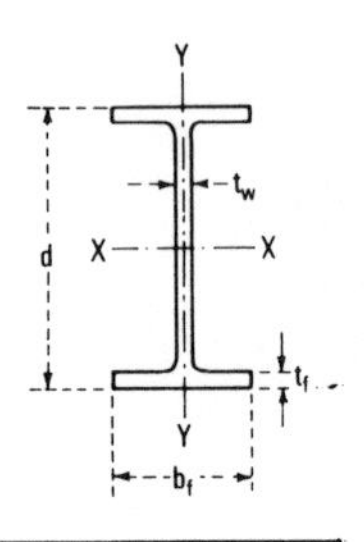


Theoretical Dimensions and Properties for **Designing**

| Section Number | Weight per Foot | Area of Section A | Depth of Section d | Flange Width $b_f$ | Flange Thickness $t_f$ | Web Thickness $t_w$ | $I_x$ | $Z_x$ | $r_x$ | $I_y$ | $Z_y$ | $r_y$ |
|---|---|---|---|---|---|---|---|---|---|---|---|---|
| | lb | in.² | in. | in. | in. | in. | in.⁴ | in.³ | in. | in.⁴ | in.³ | in. |
| **W36 x** | **300** | 88.3 | 36.74 | 16.655 | 1.680 | 0.945 | 20300 | 1110 | 15.2 | 1300 | 156 | 3.83 |
| | **280** | 82.4 | 36.52 | 16.595 | 1.570 | 0.885 | 18900 | 1030 | 15.1 | 1200 | 144 | 3.81 |
| | **260** | 76.5 | 36.26 | 16.550 | 1.440 | 0.840 | 17300 | 953 | 15.0 | 1090 | 132 | 3.78 |
| | **245** | 72.1 | 36.08 | 16.510 | 1.350 | 0.800 | 16100 | 895 | 15.0 | 1010 | 123 | 3.75 |
| | **230** | 67.6 | 35.90 | 16.470 | 1.260 | 0.760 | 15000 | 837 | 14.9 | 940 | 114 | 3.73 |
| **W36 x** | **210** | 61.8 | 36.69 | 12.180 | 1.360 | 0.830 | 13200 | 719 | 14.6 | 411 | 67.5 | 2.58 |
| | **194** | 57.0 | 36.49 | 12.115 | 1.260 | 0.765 | 12100 | 664 | 14.6 | 375 | 61.9 | 2.56 |
| | **182** | 53.6 | 36.33 | 12.075 | 1.180 | 0.725 | 11300 | 623 | 14.5 | 347 | 57.6 | 2.55 |
| | **170** | 50.0 | 36.17 | 12.030 | 1.100 | 0.680 | 10500 | 580 | 14.5 | 320 | 53.2 | 2.53 |
| | **160** | 47.0 | 36.01 | 12.000 | 1.020 | 0.650 | 9750 | 542 | 14.4 | 295 | 49.1 | 2.50 |
| | **150** | 44.2 | 35.85 | 11.975 | 0.940 | 0.625 | 9040 | 504 | 14.3 | 270 | 45.1 | 2.47 |
| | **135** | 39.7 | 35.55 | 11.950 | 0.790 | 0.600 | 7800 | 439 | 14.0 | 225 | 37.7 | 2.38 |
| **W33 x** | **241** | 70.9 | 34.18 | 15.860 | 1.400 | 0.830 | 14200 | 829 | 14.1 | 932 | 118 | 3.63 |
| | **221** | 65.0 | 33.93 | 15.805 | 1.275 | 0.775 | 12800 | 757 | 14.1 | 840 | 106 | 3.59 |
| | **201** | 59.1 | 33.68 | 15.745 | 1.150 | 0.715 | 11500 | 684 | 14.0 | 749 | 95.2 | 3.56 |
| **W33 x** | **152** | 44.7 | 33.49 | 11.565 | 1.055 | 0.635 | 8160 | 487 | 13.5 | 273 | 47.2 | 2.47 |
| | **141** | 41.6 | 33.30 | 11.535 | 0.960 | 0.605 | 7450 | 448 | 13.4 | 246 | 42.7 | 2.43 |
| | **130** | 38.3 | 33.09 | 11.510 | 0.855 | 0.580 | 6710 | 406 | 13.2 | 218 | 37.9 | 2.39 |
| | **118** | 34.7 | 32.86 | 11.480 | 0.740 | 0.550 | 5900 | 359 | 13.0 | 187 | 32.6 | 2.32 |
| **W30 x** | **211** | 62.0 | 30.94 | 15.105 | 1.315 | 0.775 | 10300 | 663 | 12.9 | 757 | 100 | 3.49 |
| | **191** | 56.1 | 30.68 | 15.040 | 1.185 | 0.710 | 9170 | 598 | 12.8 | 673 | 89.5 | 3.46 |
| | **173** | 50.8 | 30.44 | 14.985 | 1.065 | 0.655 | 8200 | 539 | 12.7 | 598 | 79.8 | 3.43 |
| **W30 x** | **132** | 38.9 | 30.31 | 10.545 | 1.000 | 0.615 | 5770 | 380 | 12.2 | 196 | 37.2 | 2.25 |
| | **124** | 36.5 | 30.17 | 10.515 | 0.930 | 0.585 | 5360 | 355 | 12.1 | 181 | 34.4 | 2.23 |
| | **116** | 34.2 | 30.01 | 10.495 | 0.850 | 0.565 | 4930 | 329 | 12.0 | 164 | 31.3 | 2.19 |
| | **108** | 31.7 | 29.83 | 10.475 | 0.760 | 0.545 | 4470 | 299 | 11.9 | 146 | 27.9 | 2.15 |
| | **99** | 29.1 | 29.65 | 10.450 | 0.670 | 0.520 | 3990 | 269 | 11.7 | 128 | 24.5 | 2.10 |

All shapes on these pages have parallel-faced flanges.

# WIDE FLANGE SHAPES

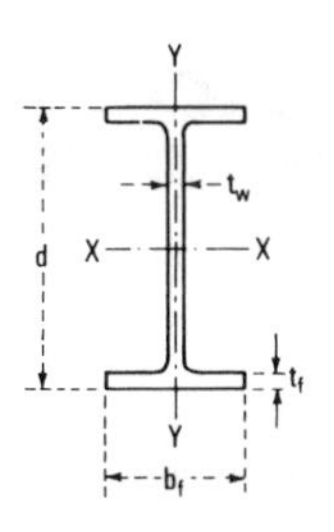


Theoretical Dimensions and Properties for **Designing**

| Section Number | Weight per Foot | Area of Section | Depth of Section | Flange | | Web Thickness | | | | | | |
|---|---|---|---|---|---|---|---|---|---|---|---|---|
| | | | | Width | Thickness | | $I_x$ | $Z_x$ | $r_x$ | $I_y$ | $Z_y$ | $r_y$ |
| | | A | d | $b_f$ | $t_f$ | $t_w$ | | | | | | |
| | lb | in.² | in. | in. | in. | in. | in.⁴ | in.³ | in. | in.⁴ | in.³ | in. |
| **W27 x** | **178** | 52.3 | 27.81 | 14.085 | 1.190 | 0.725 | 6990 | 502 | 11.6 | 555 | 78.8 | 3.26 |
| | **161** | 47.4 | 27.59 | 14.020 | 1.080 | 0.660 | 6280 | 455 | 11.5 | 497 | 70.9 | 3.24 |
| | **146** | 42.9 | 27.38 | 13.965 | 0.975 | 0.605 | 5630 | 411 | 11.4 | 443 | 63.5 | 3.21 |
| **W27 x** | **114** | 33.5 | 27.29 | 10.070 | 0.930 | 0.570 | 4090 | 299 | 11.0 | 159 | 31.5 | 2.18 |
| | **102** | 30.0 | 27.09 | 10.015 | 0.830 | 0.515 | 3620 | 267 | 11.0 | 139 | 27.8 | 2.15 |
| | **94** | 27.7 | 26.92 | 9.990 | 0.745 | 0.490 | 3270 | 243 | 10.9 | 124 | 24.8 | 2.12 |
| | **84** | 24.8 | 26.71 | 9.960 | 0.640 | 0.460 | 2850 | 213 | 10.7 | 106 | 21.2 | 2.07 |
| **W24 x** | **162** | 47.7 | 25.00 | 12.955 | 1.220 | 0.705 | 5170 | 414 | 10.4 | 443 | 68.4 | 3.05 |
| | **146** | 43.0 | 24.74 | 12.900 | 1.090 | 0.650 | 4580 | 371 | 10.3 | 391 | 60.5 | 3.01 |
| | **131** | 38.5 | 24.48 | 12.855 | 0.960 | 0.605 | 4020 | 329 | 10.2 | 340 | 53.0 | 2.97 |
| | **117** | 34.4 | 24.26 | 12.800 | 0.850 | 0.550 | 3540 | 291 | 10.1 | 297 | 46.5 | 2.94 |
| | **104** | 30.6 | 24.06 | 12.750 | 0.750 | 0.500 | 3100 | 258 | 10.1 | 259 | 40.7 | 2.91 |
| **W24 x** | **94** | 27.7 | 24.31 | 9.065 | 0.875 | 0.515 | 2700 | 222 | 9.87 | 109 | 24.0 | 1.98 |
| | **84** | 24.7 | 24.10 | 9.020 | 0.770 | 0.470 | 2370 | 196 | 9.79 | 94.4 | 20.9 | 1.95 |
| | **76** | 22.4 | 23.92 | 8.990 | 0.680 | 0.440 | 2100 | 176 | 9.69 | 82.5 | 18.4 | 1.92 |
| | **68** | 20.1 | 23.73 | 8.965 | 0.585 | 0.415 | 1830 | 154 | 9.55 | 70.4 | 15.7 | 1.87 |
| **W24 x** | **62** | 18.2 | 23.74 | 7.040 | 0.590 | 0.430 | 1550 | 131 | 9.23 | 34.5 | 9.80 | 1.38 |
| | **55** | 16.2 | 23.57 | 7.005 | 0.505 | 0.395 | 1350 | 114 | 9.11 | 29.1 | 8.30 | 1.34 |
| **W21 x** | **147** | 43.2 | 22.06 | 12.510 | 1.150 | 0.720 | 3630 | 329 | 9.17 | 376 | 60.1 | 2.95 |
| | **132** | 38.8 | 21.83 | 12.440 | 1.035 | 0.650 | 3220 | 295 | 9.12 | 333 | 53.5 | 2.93 |
| | **122** | 35.9 | 21.68 | 12.390 | 0.960 | 0.600 | 2960 | 273 | 9.09 | 305 | 49.2 | 2.92 |
| | **111** | 32.7 | 21.51 | 12.340 | 0.875 | 0.550 | 2670 | 249 | 9.05 | 274 | 44.5 | 2.90 |
| | **101** | 29.8 | 21.36 | 12.290 | 0.800 | 0.500 | 2420 | 227 | 9.02 | 248 | 40.3 | 2.89 |
| **W21 x** | **93** | 27.3 | 21.62 | 8.420 | 0.930 | 0.580 | 2070 | 192 | 8.70 | 92.9 | 22.1 | 1.84 |
| | **83** | 24.3 | 21.43 | 8.355 | 0.835 | 0.515 | 1830 | 171 | 8.67 | 81.4 | 19.5 | 1.83 |
| | **73** | 21.5 | 21.24 | 8.295 | 0.740 | 0.455 | 1600 | 151 | 8.64 | 70.6 | 17.0 | 1.81 |
| | **68** | 20.0 | 21.13 | 8.270 | 0.685 | 0.430 | 1480 | 140 | 8.60 | 64.7 | 15.7 | 1.80 |
| | **62** | 18.3 | 20.99 | 8.240 | 0.615 | 0.400 | 1330 | 127 | 8.54 | 57.5 | 13.9 | 1.77 |
| **W21 x** | **57** | 16.7 | 21.06 | 6.555 | 0.650 | 0.405 | 1170 | 111 | 8.36 | 30.6 | 9.35 | 1.35 |
| | **50** | 14.7 | 20.83 | 6.530 | 0.535 | 0.380 | 984 | 94.5 | 8.18 | 24.9 | 7.64 | 1.30 |
| | **44** | 13.0 | 20.66 | 6.500 | 0.450 | 0.350 | 843 | 81.6 | 8.06 | 20.7 | 6.36 | 1.26 |

All shapes on these pages have parallel-faced flanges.

# WIDE FLANGE SHAPES

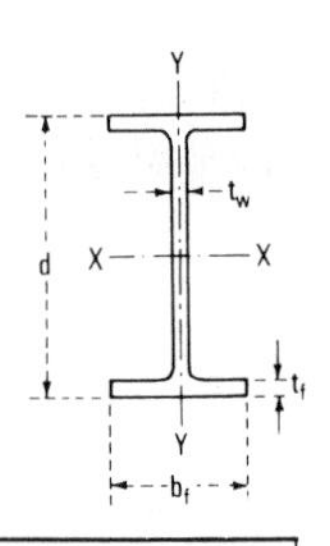


Theoretical Dimensions and Properties for **Designing**

| Section Number | Weight per Foot | Area of Section | Depth of Section | Flange | | Web Thickness | | | | | | |
|---|---|---|---|---|---|---|---|---|---|---|---|---|
| | | | | Width | Thickness | | $I_x$ | $Z_x$ | $r_x$ | $I_y$ | $Z_y$ | $r_y$ |
| | | A | d | $b_f$ | $t_f$ | $t_w$ | | | | | | |
| | lb | in.² | in. | in. | in. | in. | in.⁴ | in.³ | in. | in.⁴ | in.³ | in. |
| **W18 x** | **119** | 35.1 | 18.97 | 11.265 | 1.060 | 0.655 | 2190 | 231 | 7.90 | 253 | 44.9 | 2.69 |
| | **106** | 31.1 | 18.73 | 11.200 | 0.940 | 0.590 | 1910 | 204 | 7.84 | 220 | 39.4 | 2.66 |
| | **97** | 28.5 | 18.59 | 11.145 | 0.870 | 0.535 | 1750 | 188 | 7.82 | 201 | 36.1 | 2.65 |
| | **86** | 25.3 | 18.39 | 11.090 | 0.770 | 0.480 | 1530 | 166 | 7.77 | 175 | 31.6 | 2.63 |
| | **76** | 22.3 | 18.21 | 11.035 | 0.680 | 0.425 | 1330 | 146 | 7.73 | 152 | 27.6 | 2.61 |
| **W18 x** | **71** | 20.8 | 18.47 | 7.635 | 0.810 | 0.495 | 1170 | 127 | 7.50 | 60.3 | 15.8 | 1.70 |
| | **65** | 19.1 | 18.35 | 7.590 | 0.750 | 0.450 | 1070 | 117 | 7.49 | 54.8 | 14.4 | 1.69 |
| | **60** | 17.6 | 18.24 | 7.555 | 0.695 | 0.415 | 984 | 108 | 7.47 | 50.1 | 13.3 | 1.69 |
| | **55** | 16.2 | 18.11 | 7.530 | 0.630 | 0.390 | 890 | 98.3 | 7.41 | 44.9 | 11.9 | 1.67 |
| | **50** | 14.7 | 17.99 | 7.495 | 0.570 | 0.355 | 800 | 88.9 | 7.38 | 40.1 | 10.7 | 1.65 |
| **W18 x** | **46** | 13.5 | 18.06 | 6.060 | 0.605 | 0.360 | 712 | 78.8 | 7.25 | 22.5 | 7.43 | 1.29 |
| | **40** | 11.8 | 17.90 | 6.015 | 0.525 | 0.315 | 612 | 68.4 | 7.21 | 19.1 | 6.35 | 1.27 |
| | **35** | 10.3 | 17.70 | 6.000 | 0.425 | 0.300 | 510 | 57.6 | 7.04 | 15.3 | 5.12 | 1.22 |
| **W16 x** | **100** | 29.4 | 16.97 | 10.425 | 0.985 | 0.585 | 1490 | 175 | 7.10 | 186 | 35.7 | 2.52 |
| | **89** | 26.2 | 16.75 | 10.365 | 0.875 | 0.525 | 1300 | 155 | 7.05 | 163 | 31.4 | 2.49 |
| | **77** | 22.6 | 16.52 | 10.295 | 0.760 | 0.455 | 1110 | 134 | 7.00 | 138 | 26.9 | 2.47 |
| | **67** | 19.7 | 16.33 | 10.235 | 0.665 | 0.395 | 954 | 117 | 6.96 | 119 | 23.2 | 2.46 |
| **W16 x** | **57** | 16.8 | 16.43 | 7.120 | 0.715 | 0.430 | 758 | 92.2 | 6.72 | 43.1 | 12.1 | 1.60 |
| | **50** | 14.7 | 16.26 | 7.070 | 0.630 | 0.380 | 659 | 81.0 | 6.68 | 37.2 | 10.5 | 1.59 |
| | **45** | 13.3 | 16.13 | 7.035 | 0.565 | 0.345 | 586 | 72.7 | 6.65 | 32.8 | 9.34 | 1.57 |
| | **40** | 11.8 | 16.01 | 6.995 | 0.505 | 0.305 | 518 | 64.7 | 6.63 | 28.9 | 8.25 | 1.57 |
| | **36** | 10.6 | 15.86 | 6.985 | 0.430 | 0.295 | 448 | 56.5 | 6.51 | 24.5 | 7.00 | 1.52 |
| **W16 x** | **31** | 9.12 | 15.88 | 5.525 | 0.440 | 0.275 | 375 | 47.2 | 6.41 | 12.4 | 4.49 | 1.17 |
| | **26** | 7.68 | 15.69 | 5.500 | 0.345 | 0.250 | 301 | 38.4 | 6.26 | 9.59 | 3.49 | 1.12 |

All shapes on these pages have parallel faced flanges.

# WIDE FLANGE SHAPES

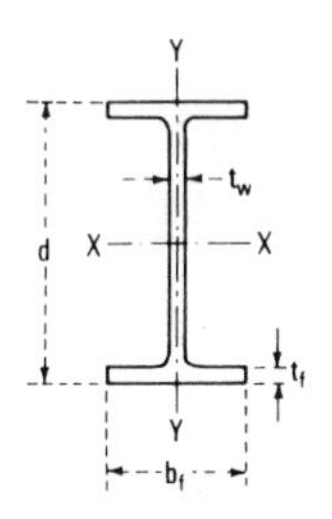


Theoretical Dimensions and Properties for **Designing**

| Section Number | Weight per Foot | Area of Section | Depth of Section | Flange Width | Flange Thickness | Web Thickness | $I_x$ | $Z_x$ | $r_x$ | $I_y$ | $Z_y$ | $r_y$ |
|---|---|---|---|---|---|---|---|---|---|---|---|---|
| | | A | d | $b_f$ | $t_f$ | $t_w$ | | | | | | |
| | lb | in.² | in. | in. | in. | in. | in.⁴ | in.³ | in. | in.⁴ | in.³ | in. |
| **W14 x** | **730***  | 215 | 22.42 | 17.890 | 4.910 | 3.070 | 14300 | 1280 | 8.17 | 4720 | 527 | 4.69 |
| | **665*** | 196 | 21.64 | 17.650 | 4.520 | 2.830 | 12400 | 1150 | 7.98 | 4170 | 472 | 4.62 |
| | **605*** | 178 | 20.92 | 17.415 | 4.160 | 2.595 | 10800 | 1040 | 7.80 | 3680 | 423 | 4.55 |
| | **550*** | 162 | 20.24 | 17.200 | 3.820 | 2.380 | 9430 | 931 | 7.63 | 3250 | 378 | 4.49 |
| | **500*** | 147 | 19.60 | 17.010 | 3.500 | 2.190 | 8210 | 838 | 7.48 | 2880 | 339 | 4.43 |
| | **455*** | 134 | 19.02 | 16.835 | 3.210 | 2.015 | 7190 | 756 | 7.33 | 2560 | 304 | 4.38 |
| **W14 x** | **426** | 125 | 18.67 | 16.695 | 3.035 | 1.875 | 6600 | 707 | 7.26 | 2360 | 283 | 4.34 |
| | **398** | 117 | 18.29 | 16.590 | 2.845 | 1.770 | 6000 | 656 | 7.16 | 2170 | 262 | 4.31 |
| | **370** | 109 | 17.92 | 16.475 | 2.660 | 1.655 | 5440 | 607 | 7.07 | 1990 | 241 | 4.27 |
| | **342** | 101 | 17.54 | 16.360 | 2.470 | 1.540 | 4900 | 559 | 6.98 | 1810 | 221 | 4.24 |
| | **311** | 91.4 | 17.12 | 16.230 | 2.260 | 1.410 | 4330 | 506 | 6.88 | 1610 | 199 | 4.20 |
| | **283** | 83.3 | 16.74 | 16.110 | 2.070 | 1.290 | 3840 | 459 | 6.79 | 1440 | 179 | 4.17 |
| | **257** | 75.6 | 16.38 | 15.995 | 1.890 | 1.175 | 3400 | 415 | 6.71 | 1290 | 161 | 4.13 |
| | **233** | 68.5 | 16.04 | 15.890 | 1.720 | 1.070 | 3010 | 375 | 6.63 | 1150 | 145 | 4.10 |
| | **211** | 62.0 | 15.72 | 15.800 | 1.560 | 0.980 | 2660 | 338 | 6.55 | 1030 | 130 | 4.07 |
| | **193** | 56.8 | 15.48 | 15.710 | 1.440 | 0.890 | 2400 | 310 | 6.50 | 931 | 119 | 4.05 |
| | **176** | 51.8 | 15.22 | 15.650 | 1.310 | 0.830 | 2140 | 281 | 6.43 | 838 | 107 | 4.02 |
| | **159** | 46.7 | 14.98 | 15.565 | 1.190 | 0.745 | 1900 | 254 | 6.38 | 748 | 96.2 | 4.00 |
| | **145** | 42.7 | 14.78 | 15.500 | 1.090 | 0.680 | 1710 | 232 | 6.33 | 677 | 87.3 | 3.98 |
| **W14 x** | **132** | 38.8 | 14.66 | 14.725 | 1.030 | 0.645 | 1530 | 209 | 6.28 | 548 | 74.5 | 3.76 |
| | **120** | 35.3 | 14.48 | 14.670 | 0.940 | 0.590 | 1380 | 190 | 6.24 | 495 | 67.5 | 3.74 |
| | **109** | 32.0 | 14.32 | 14.605 | 0.860 | 0.525 | 1240 | 173 | 6.22 | 447 | 61.2 | 3.73 |
| | **99** | 29.1 | 14.16 | 14.565 | 0.780 | 0.485 | 1110 | 157 | 6.17 | 402 | 55.2 | 3.71 |
| | **90** | 26.5 | 14.02 | 14.520 | 0.710 | 0.440 | 999 | 143 | 6.14 | 362 | 49.9 | 3.70 |
| **W14 x** | **82** | 24.1 | 14.31 | 10.130 | 0.855 | 0.510 | 882 | 123 | 6.05 | 148 | 29.3 | 2.48 |
| | **74** | 21.8 | 14.17 | 10.070 | 0.785 | 0.450 | 796 | 112 | 6.04 | 134 | 26.6 | 2.48 |
| | **68** | 20.0 | 14.04 | 10.035 | 0.720 | 0.415 | 723 | 103 | 6.01 | 121 | 24.2 | 2.46 |
| | **61** | 17.9 | 13.89 | 9.995 | 0.645 | 0.375 | 640 | 92.2 | 5.98 | 107 | 21.5 | 2.45 |
| **W14 x** | **53** | 15.6 | 13.92 | 8.060 | 0.660 | 0.370 | 541 | 77.8 | 5.89 | 57.7 | 14.3 | 1.92 |
| | **48** | 14.1 | 13.79 | 8.030 | 0.595 | 0.340 | 485 | 70.3 | 5.85 | 51.4 | 12.8 | 1.91 |
| | **43** | 12.6 | 13.66 | 7.995 | 0.530 | 0.305 | 428 | 62.7 | 5.82 | 45.2 | 11.3 | 1.89 |

*These shapes have a 1°-00′ (1.75%) flange slope. Flange thicknesses shown are average thicknesses. Properties shown are for a parallel flange section.

All other shapes on these pages have parallel-faced flanges.

# WIDE FLANGE SHAPES

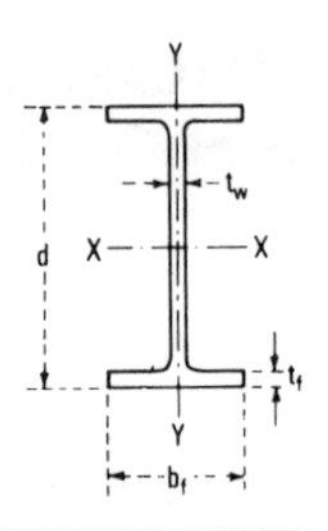


Theoretical Dimensions and Properties for **Designing**

| Section Number | Weight per Foot | Area of Section A | Depth of Section d | Flange Width $b_f$ | Flange Thickness $t_f$ | Web Thickness $t_w$ | $I_x$ | $Z_x$ | $r_x$ | $I_y$ | $Z_y$ | $r_y$ |
|---|---|---|---|---|---|---|---|---|---|---|---|---|
| | lb | in.² | in. | in. | in. | in. | in.⁴ | in.³ | in. | in.⁴ | in.³ | in. |
| W14 x | 38 | 11.2 | 14.10 | 6.770 | 0.515 | 0.310 | 385 | 54.6 | 5.88 | 26.7 | 7.88 | 1.55 |
| | 34 | 10.0 | 13.98 | 6.745 | 0.455 | 0.285 | 340 | 48.6 | 5.83 | 23.3 | 6.91 | 1.53 |
| | 30 | 8.85 | 13.84 | 6.730 | 0.385 | 0.270 | 291 | 42.0 | 5.73 | 19.6 | 5.82 | 1.49 |
| W14 x | 26 | 7.69 | 13.91 | 5.025 | 0.420 | 0.255 | 245 | 35.3 | 5.65 | 8.91 | 3.54 | 1.08 |
| | 22 | 6.49 | 13.74 | 5.000 | 0.335 | 0.230 | 199 | 29.0 | 5.54 | 7.00 | 2.80 | 1.04 |
| W12 x | 190 | 55.8 | 14.38 | 12.670 | 1.735 | 1.060 | 1890 | 263 | 5.82 | 589 | 93.0 | 3.25 |
| | 170 | 50.0 | 14.03 | 12.570 | 1.560 | 0.960 | 1650 | 235 | 5.74 | 517 | 82.3 | 3.22 |
| | 152 | 44.7 | 13.71 | 12.480 | 1.400 | 0.870 | 1430 | 209 | 5.66 | 454 | 72.8 | 3.19 |
| | 136 | 39.9 | 13.41 | 12.400 | 1.250 | 0.790 | 1240 | 186 | 5.58 | 398 | 64.2 | 3.16 |
| | 120 | 35.3 | 13.12 | 12.320 | 1.105 | 0.710 | 1070 | 163 | 5.51 | 345 | 56.0 | 3.13 |
| | 106 | 31.2 | 12.89 | 12.220 | 0.990 | 0.610 | 933 | 145 | 5.47 | 301 | 49.3 | 3.11 |
| | 96 | 28.2 | 12.71 | 12.160 | 0.900 | 0.550 | 833 | 131 | 5.44 | 270 | 44.4 | 3.09 |
| | 87 | 25.6 | 12.53 | 12.125 | 0.810 | 0.515 | 740 | 118 | 5.38 | 241 | 39.7 | 3.07 |
| | 79 | 23.2 | 12.38 | 12.080 | 0.735 | 0.470 | 662 | 107 | 5.34 | 216 | 35.8 | 3.05 |
| | 72 | 21.1 | 12.25 | 12.040 | 0.670 | 0.430 | 597 | 97.4 | 5.31 | 195 | 32.4 | 3.04 |
| | 65 | 19.1 | 12.12 | 12.000 | 0.605 | 0.390 | 533 | 87.9 | 5.28 | 174 | 29.1 | 3.02 |
| W12 x | 58 | 17.0 | 12.19 | 10.010 | 0.640 | 0.360 | 475 | 78.0 | 5.28 | 107 | 21.4 | 2.51 |
| | 53 | 15.6 | 12.06 | 9.995 | 0.575 | 0.345 | 425 | 70.6 | 5.23 | 95.8 | 19.2 | 2.48 |
| W12 x | 50 | 14.7 | 12.19 | 8.080 | 0.640 | 0.370 | 394 | 64.7 | 5.18 | 56.3 | 13.9 | 1.96 |
| | 45 | 13.2 | 12.06 | 8.045 | 0.575 | 0.335 | 350 | 58.1 | 5.15 | 50.0 | 12.4 | 1.94 |
| | 40 | 11.8 | 11.94 | 8.005 | 0.515 | 0.295 | 310 | 51.9 | 5.13 | 44.1 | 11.0 | 1.93 |
| W12 x | 35 | 10.3 | 12.50 | 6.560 | 0.520 | 0.300 | 285 | 45.6 | 5.25 | 24.5 | 7.47 | 1.54 |
| | 30 | 8.79 | 12.34 | 6.520 | 0.440 | 0.260 | 238 | 38.6 | 5.21 | 20.3 | 6.24 | 1.52 |
| | 26 | 7.65 | 12.22 | 6.490 | 0.380 | 0.230 | 204 | 33.4 | 5.17 | 17.3 | 5.34 | 1.51 |
| W12 x | 22 | 6.48 | 12.31 | 4.030 | 0.425 | 0.260 | 156 | 25.4 | 4.91 | 4.66 | 2.31 | 0.848 |
| | 19 | 5.57 | 12.16 | 4.005 | 0.350 | 0.235 | 130 | 21.3 | 4.82 | 3.76 | 1.88 | 0.822 |
| | 16 | 4.71 | 11.99 | 3.990 | 0.265 | 0.220 | 103 | 17.1 | 4.67 | 2.82 | 1.41 | 0.773 |
| | 14 | 4.16 | 11.91 | 3.970 | 0.225 | 0.200 | 88.6 | 14.9 | 4.62 | 2.36 | 1.19 | 0.753 |

All shapes on these pages have parallel-faced flanges.

# WIDE FLANGE SHAPES

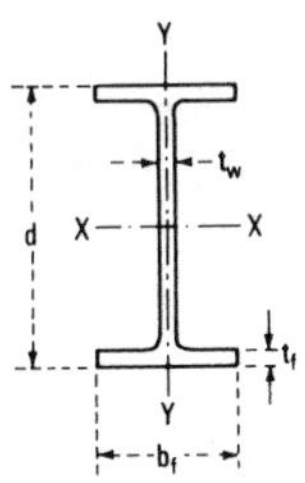


Theoretical Dimensions and Properties for **Designing**

| Section Number | Weight per Foot | Area of Section A | Depth of Section d | Flange Width $b_f$ | Flange Thickness $t_f$ | Web Thickness $t_w$ | $I_x$ | $Z_x$ | $r_x$ | $I_y$ | $Z_y$ | $r_y$ |
|---|---|---|---|---|---|---|---|---|---|---|---|---|
| | lb | in.$^2$ | in. | in. | in. | in. | in.$^4$ | in.$^3$ | in. | in.$^4$ | in.$^3$ | in. |
| W10 x | 112 | 32.9 | 11.36 | 10.415 | 1.250 | 0.755 | 716 | 126 | 4.66 | 236 | 45.3 | 2.68 |
| | 100 | 29.4 | 11.10 | 10.340 | 1.120 | 0.680 | 623 | 112 | 4.60 | 207 | 40.0 | 2.65 |
| | 88 | 25.9 | 10.84 | 10.265 | 0.990 | 0.605 | 534 | 98.5 | 4.54 | 179 | 34.8 | 2.63 |
| | 77 | 22.6 | 10.60 | 10.190 | 0.870 | 0.530 | 455 | 85.9 | 4.49 | 154 | 30.1 | 2.60 |
| | 68 | 20.0 | 10.40 | 10.130 | 0.770 | 0.470 | 394 | 75.7 | 4.44 | 134 | 26.4 | 2.59 |
| | 60 | 17.6 | 10.22 | 10.080 | 0.680 | 0.420 | 341 | 66.7 | 4.39 | 116 | 23.0 | 2.57 |
| | 54 | 15.8 | 10.09 | 10.030 | 0.615 | 0.370 | 303 | 60.0 | 4.37 | 103 | 20.6 | 2.56 |
| | 49 | 14.4 | 9.98 | 10.000 | 0.560 | 0.340 | 272 | 54.6 | 4.35 | 93.4 | 18.7 | 2.54 |
| W10 x | 45 | 13.3 | 10.10 | 8.020 | 0.620 | 0.350 | 248 | 49.1 | 4.33 | 53.4 | 13.3 | 2.01 |
| | 39 | 11.5 | 9.92 | 7.985 | 0.530 | 0.315 | 209 | 42.1 | 4.27 | 45.0 | 11.3 | 1.98 |
| | 33 | 9.71 | 9.73 | 7.960 | 0.435 | 0.290 | 170 | 35.0 | 4.19 | 36.6 | 9.20 | 1.94 |
| W10 x | 30 | 8.84 | 10.47 | 5.810 | 0.510 | 0.300 | 170 | 32.4 | 4.38 | 16.7 | 5.75 | 1.37 |
| | 26 | 7.61 | 10.33 | 5.770 | 0.440 | 0.260 | 144 | 27.9 | 4.35 | 14.1 | 4.89 | 1.36 |
| | 22 | 6.49 | 10.17 | 5.750 | 0.360 | 0.240 | 118 | 23.2 | 4.27 | 11.4 | 3.97 | 1.33 |
| W10 x | 19 | 5.62 | 10.24 | 4.020 | 0.395 | 0.250 | 96.3 | 18.8 | 4.14 | 4.29 | 2.14 | 0.874 |
| | 17 | 4.99 | 10.11 | 4.010 | 0.330 | 0.240 | 81.9 | 16.2 | 4.05 | 3.56 | 1.78 | 0.845 |
| | 15 | 4.41 | 9.99 | 4.000 | 0.270 | 0.230 | 68.9 | 13.8 | 3.95 | 2.89 | 1.45 | 0.810 |
| | 12 | 3.54 | 9.87 | 3.960 | 0.210 | 0.190 | 53.8 | 10.9 | 3.90 | 2.18 | 1.10 | 0.785 |
| W8 x | 67 | 19.7 | 9.00 | 8.280 | 0.935 | 0.570 | 272 | 60.4 | 3.72 | 88.6 | 21.4 | 2.12 |
| | 58 | 17.1 | 8.75 | 8.220 | 0.810 | 0.510 | 228 | 52.0 | 3.65 | 75.1 | 18.3 | 2.10 |
| | 48 | 14.1 | 8.50 | 8.110 | 0.685 | 0.400 | 184 | 43.3 | 3.61 | 60.9 | 15.0 | 2.08 |
| | 40 | 11.7 | 8.25 | 8.070 | 0.560 | 0.360 | 146 | 35.5 | 3.53 | 49.1 | 12.2 | 2.04 |
| | 35 | 10.3 | 8.12 | 8.020 | 0.495 | 0.310 | 127 | 31.2 | 3.51 | 42.6 | 10.6 | 2.03 |
| | 31 | 9.13 | 8.00 | 7.995 | 0.435 | 0.285 | 110 | 27.5 | 3.47 | 37.1 | 9.27 | 2.02 |
| W8 x | 28 | 8.25 | 8.06 | 6.535 | 0.465 | 0.285 | 98.0 | 24.3 | 3.45 | 21.7 | 6.63 | 1.62 |
| | 24 | 7.08 | 7.93 | 6.495 | 0.400 | 0.245 | 82.8 | 20.9 | 3.42 | 18.3 | 5.63 | 1.61 |
| W8 x | 21 | 6.16 | 8.28 | 5.270 | 0.400 | 0.250 | 75.3 | 18.2 | 3.49 | 9.77 | 3.71 | 1.26 |
| | 18 | 5.26 | 8.14 | 5.250 | 0.330 | 0.230 | 61.9 | 15.2 | 3.43 | 7.97 | 3.04 | 1.23 |
| W8 x | 15 | 4.44 | 8.11 | 4.015 | 0.315 | 0.245 | 48.0 | 11.8 | 3.29 | 3.41 | 1.70 | 0.876 |
| | 13 | 3.84 | 7.99 | 4.000 | 0.255 | 0.230 | 39.6 | 9.91 | 3.21 | 2.73 | 1.37 | 0.843 |
| | 10 | 2.96 | 7.89 | 3.940 | 0.205 | 0.170 | 30.8 | 7.81 | 3.22 | 2.09 | 1.06 | 0.841 |

# WIDE FLANGE SHAPES

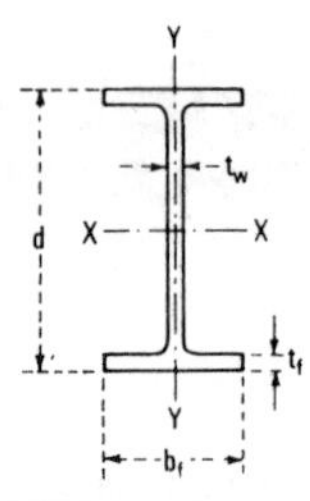


Theoretical Dimensions and Properties for **Designing**

| Section Number | Weight per Foot | Area of Section | Depth of Section | Flange | | Web Thickness | | | | | | |
|---|---|---|---|---|---|---|---|---|---|---|---|---|
| | | | | Width | Thickness | | $I_x$ | $Z_x$ | $r_x$ | $I_y$ | $Z_y$ | $r_y$ |
| | | A | d | $b_f$ | $t_f$ | $t_w$ | | | | | | |
| | lb | in.² | in. | in. | in. | in. | in.⁴ | in.³ | in. | in.⁴ | in.³ | in. |
| **W6 x** | **25** | 7.34 | 6.38 | 6.080 | 0.455 | 0.320 | 53.4 | 16.7 | 2.70 | 17.1 | 5.61 | 1.52 |
| | **20** | 5.87 | 6.20 | 6.020 | 0.365 | 0.260 | 41.4 | 13.4 | 2.66 | 13.3 | 4.41 | 1.50 |
| | **15** | 4.43 | 5.99 | 5.990 | 0.260 | 0.230 | 29.1 | 9.72 | 2.56 | 9.32 | 3.11 | 1.45 |
| **W6 x** | **16** | 4.74 | 6.28 | 4.030 | 0.405 | 0.260 | 32.1 | 10.2 | 2.60 | 4.43 | 2.20 | 0.967 |
| | **12** | 3.55 | 6.03 | 4.000 | 0.280 | 0.230 | 22.1 | 7.31 | 2.49 | 2.99 | 1.50 | 0.918 |
| | **9** | 2.68 | 5.90 | 3.940 | 0.215 | 0.170 | 16.4 | 5.56 | 2.47 | 2.20 | 1.11 | 0.905 |
| **W5 x** | **19** | 5.54 | 5.15 | 5.030 | 0.430 | 0.270 | 26.2 | 10.2 | 2.17 | 9.13 | 3.63 | 1.28 |
| | **16** | 4.68 | 5.01 | 5.000 | 0.360 | 0.240 | 21.3 | 8.51 | 2.13 | 7.51 | 3.00 | 1.27 |
| **W4 x** | **13** | 3.83 | 4.16 | 4.060 | 0.345 | 0.280 | 11.3 | 5.46 | 1.72 | 3.86 | 1.90 | 1.00 |

# AMERICAN STANDARD SHAPES

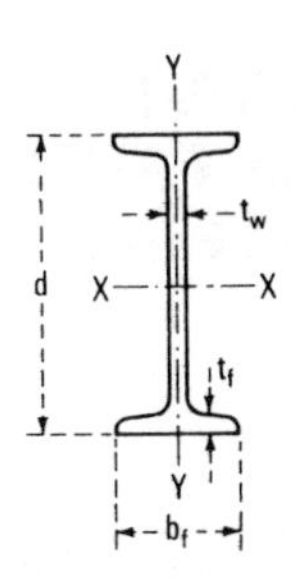


Theoretical Dimensions and Properties for **Designing**

| Section Number | Weight per Foot | Area of Section A | Depth of Section d | Flange Width $b_f$ | Flange Average Thickness $t_f$ | Web Thickness $t_w$ | $I_x$ | $Z_x$ | $r_x$ | $I_y$ | $Z_y$ | $r_y$ |
|---|---|---|---|---|---|---|---|---|---|---|---|---|
| | lb | in.$^2$ | in. | in. | in. | in. | in.$^4$ | in.$^3$ | in. | in.$^4$ | in.$^3$ | in. |
| **S24 x** | **121.0** | 35.6 | 24.50 | 8.050 | 1.090 | 0.800 | 3160 | 258 | 9.43 | 83.3 | 20.7 | 1.53 |
| | **106.0** | 31.2 | 24.50 | 7.870 | 1.090 | 0.620 | 2940 | 240 | 9.71 | 77.1 | 19.6 | 1.57 |
| **S24 x** | **100.0** | 29.3 | 24.00 | 7.245 | 0.870 | 0.745 | 2390 | 199 | 9.02 | 47.7 | 13.2 | 1.27 |
| | **90.0** | 26.5 | 24.00 | 7.125 | 0.870 | 0.625 | 2250 | 187 | 9.21 | 44.9 | 12.6 | 1.30 |
| | **80.0** | 23.5 | 24.00 | 7.000 | 0.870 | 0.500 | 2100 | 175 | 9.47 | 42.2 | 12.1 | 1.34 |
| **S20 x** | **96.0** | 28.2 | 20.30 | 7.200 | 0.920 | 0.800 | 1670 | 165 | 7.71 | 50.2 | 13.9 | 1.33 |
| | **86.0** | 25.3 | 20.30 | 7.060 | 0.920 | 0.660 | 1580 | 155 | 7.89 | 46.8 | 13.3 | 1.36 |
| **S20 x** | **75.0** | 22.0 | 20.00 | 6.385 | 0.795 | 0.635 | 1280 | 128 | 7.62 | 29.8 | 9.32 | 1.16 |
| | **66.0** | 19.4 | 20.00 | 6.255 | 0.795 | 0.505 | 1190 | 119 | 7.83 | 27.7 | 8.85 | 1.19 |
| **S18 x** | **70.0** | 20.6 | 18.00 | 6.251 | 0.691 | 0.711 | 926 | 103 | 6.71 | 24.1 | 7.72 | 1.08 |
| | **54.7** | 16.1 | 18.00 | 6.001 | 0.691 | 0.461 | 804 | 89.4 | 7.07 | 20.8 | 6.94 | 1.14 |
| **S15 x** | **50.0** | 14.7 | 15.00 | 5.640 | 0.622 | 0.550 | 486 | 64.8 | 5.75 | 15.7 | 5.57 | 1.03 |
| | **42.9** | 12.6 | 15.00 | 5.501 | 0.622 | 0.411 | 447 | 59.6 | 5.95 | 14.4 | 5.23 | 1.07 |
| **S12 x** | **50.0** | 14.7 | 12.00 | 5.477 | 0.659 | 0.687 | 305 | 50.8 | 4.55 | 15.7 | 5.74 | 1.03 |
| | **40.8** | 12.0 | 12.00 | 5.252 | 0.659 | 0.462 | 272 | 45.4 | 4.77 | 13.6 | 5.16 | 1.06 |
| **S12 x** | **35.0** | 10.3 | 12.00 | 5.078 | 0.544 | 0.428 | 229 | 38.2 | 4.72 | 9.87 | 3.89 | 0.980 |
| | **31.8** | 9.35 | 12.00 | 5.000 | 0.544 | 0.350 | 218 | 36.4 | 4.83 | 9.36 | 3.74 | 1.00 |

All shapes on these pages have a flange slope of 16⅔ pct.

# AMERICAN STANDARD SHAPES

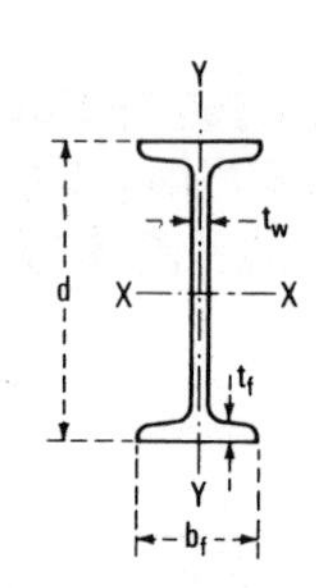


Theoretical Dimensions and Properties for **Designing**

| Section Number | Weight per Foot | Area of Section A | Depth of Section d | Flange Width $b_f$ | Flange Average Thickness $t_f$ | Web Thickness $t_w$ | $I_x$ | $Z_x$ | $r_x$ | $I_y$ | $Z_y$ | $r_y$ |
|---|---|---|---|---|---|---|---|---|---|---|---|---|
| | lb | in.$^2$ | in. | in. | in. | in. | in.$^4$ | in.$^3$ | in. | in.$^4$ | in.$^3$ | in. |
| **S10 x** | **35.0** | 10.3 | 10.00 | 4.944 | 0.491 | 0.594 | 147 | 29.4 | 3.78 | 8.36 | 3.38 | 0.901 |
| | **25.4** | 7.46 | 10.00 | 4.661 | 0.491 | 0.311 | 124 | 24.7 | 4.07 | 6.79 | 2.91 | 0.954 |
| **S8 x** | **23.0** | 6.77 | 8.00 | 4.171 | 0.425 | 0.441 | 64.9 | 16.2 | 3.10 | 4.31 | 2.07 | 0.798 |
| | **18.4** | 5.41 | 8.00 | 4.001 | 0.425 | 0.271 | 57.6 | 14.4 | 3.26 | 3.73 | 1.86 | 0.831 |
| **S7 x** | **15.3** | 4.50 | 7.00 | 3.662 | 0.392 | 0.252 | 36.7 | 10.5 | 2.86 | 2.64 | 1.44 | 0.766 |
| **S6 x** | **17.25** | 5.07 | 6.00 | 3.565 | 0.359 | 0.465 | 26.3 | 8.77 | 2.28 | 2.31 | 1.30 | 0.675 |
| | **12.5** | 3.67 | 6.00 | 3.332 | 0.359 | 0.232 | 22.1 | 7.37 | 2.45 | 1.82 | 1.09 | 0.705 |
| **S5 x** | **10.0** | 2.94 | 5.00 | 3.004 | 0.326 | 0.214 | 12.3 | 4.92 | 2.05 | 1.22 | 0.809 | 0.643 |
| **S4 x** | [y]**9.5** | 2.79 | 4.00 | 2.796 | 0.293 | 0.326 | 6.79 | 3.39 | 1.56 | 0.903 | 0.646 | 0.569 |
| | **7.7** | 2.26 | 4.00 | 2.663 | 0.293 | 0.193 | 6.08 | 3.04 | 1.64 | 0.764 | 0.574 | 0.581 |
| **S3 x** | **7.5** | 2.21 | 3.00 | 2.509 | 0.260 | 0.349 | 2.93 | 1.95 | 1.15 | 0.586 | 0.468 | 0.516 |
| | **5.7** | 1.67 | 3.00 | 2.330 | 0.260 | 0.170 | 2.52 | 1.68 | 1.23 | 0.455 | 0.390 | 0.522 |

All shapes on these pages have a flange slope of 16⅔ pct.
[y]Available subject to inquiry.

# AMERICAN STANDARD CHANNELS

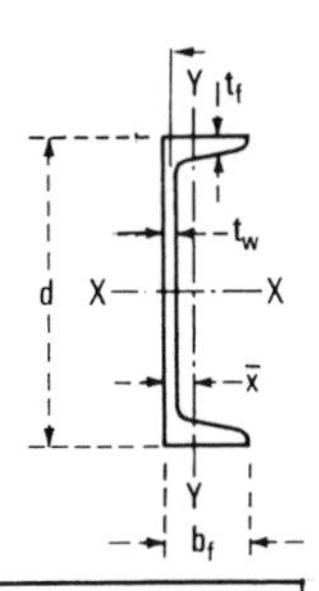


Theoretical Dimensions and Properties for **Designing**

| Section Number | Weight per Foot | Area of Section | Depth of Section | Flange Width | Flange Average Thickness | Web Thickness | $I_x$ | $Z_x$ | $r_x$ | $I_y$ | $Z_y$ | $r_y$ | $\bar{x}$ |
|---|---|---|---|---|---|---|---|---|---|---|---|---|---|
| | | A | d | $b_f$ | $t_f$ | $t_w$ | | | | | | | |
| | lb | in.² | in. | in. | in. | in. | in.⁴ | in.³ | in. | in.⁴ | in.³ | in. | in. |
| C15 x | **50.0** | 14.7 | 15.00 | 3.716 | 0.650 | 0.716 | 404 | 53.8 | 5.24 | 11.0 | 3.78 | 0.867 | 0.799 |
| | **40.0** | 11.8 | 15.00 | 3.520 | 0.650 | 0.520 | 349 | 46.5 | 5.44 | 9.23 | 3.36 | 0.886 | 0.778 |
| | **33.9** | 9.96 | 15.00 | 3.400 | 0.650 | 0.400 | 315 | 42.0 | 5.62 | 8.13 | 3.11 | 0.904 | 0.787 |
| C12 x | **30.0** | 8.82 | 12.00 | 3.170 | 0.501 | 0.510 | 162 | 27.0 | 4.29 | 5.14 | 2.06 | 0.763 | 0.674 |
| | **25.0** | 7.35 | 12.00 | 3.047 | 0.501 | 0.387 | 144 | 24.1 | 4.43 | 4.47 | 1.88 | 0.780 | 0.674 |
| | **20.7** | 6.09 | 12.00 | 2.942 | 0.501 | 0.282 | 129 | 21.5 | 4.61 | 3.88 | 1.73 | 0.799 | 0.698 |
| C10 x | **30.0** | 8.82 | 10.00 | 3.033 | 0.436 | 0.673 | 103 | 20.7 | 3.42 | 3.94 | 1.65 | 0.669 | 0.649 |
| | **25.0** | 7.35 | 10.00 | 2.886 | 0.436 | 0.526 | 91.2 | 18.2 | 3.52 | 3.36 | 1.48 | 0.676 | 0.617 |
| | **20.0** | 5.88 | 10.00 | 2.739 | 0.436 | 0.379 | 78.9 | 15.8 | 3.66 | 2.81 | 1.32 | 0.691 | 0.606 |
| | **15.3** | 4.49 | 10.00 | 2.600 | 0.436 | 0.240 | 67.4 | 13.5 | 3.87 | 2.28 | 1.16 | 0.713 | 0.634 |
| C9 x | **15.0** | 4.41 | 9.00 | 2.485 | 0.413 | 0.285 | 51.0 | 11.3 | 3.40 | 1.93 | 1.01 | 0.661 | 0.586 |
| | **13.4** | 3.94 | 9.00 | 2.433 | 0.413 | 0.233 | 47.9 | 10.6 | 3.48 | 1.76 | 0.962 | 0.668 | 0.601 |
| C8 x | **18.75** | 5.51 | 8.00 | 2.527 | 0.390 | 0.487 | 44.0 | 11.0 | 2.82 | 1.98 | 1.01 | 0.599 | 0.565 |
| | **13.75** | 4.04 | 8.00 | 2.343 | 0.390 | 0.303 | 36.1 | 9.03 | 2.99 | 1.53 | 0.853 | 0.615 | 0.553 |
| | **11.5** | 3.38 | 8.00 | 2.260 | 0.390 | 0.220 | 32.6 | 8.14 | 3.11 | 1.32 | 0.781 | 0.625 | 0.571 |
| C7 x | **12.25** | 3.60 | 7.00 | 2.194 | 0.366 | 0.314 | 24.2 | 6.93 | 2.60 | 1.17 | 0.702 | 0.571 | 0.525 |
| | **9.8** | 2.87 | 7.00 | 2.090 | 0.366 | 0.210 | 21.3 | 6.08 | 2.72 | 0.968 | 0.625 | 0.581 | 0.541 |
| C6 x | **13.0** | 3.83 | 6.00 | 2.157 | 0.343 | 0.437 | 17.4 | 5.80 | 2.13 | 1.05 | 0.642 | 0.525 | 0.514 |
| | **10.5** | 3.09 | 6.00 | 2.034 | 0.343 | 0.314 | 15.2 | 5.06 | 2.22 | 0.865 | 0.564 | 0.529 | 0.500 |
| | **8.2** | 2.40 | 6.00 | 1.920 | 0.343 | 0.200 | 13.1 | 4.38 | 2.34 | 0.692 | 0.492 | 0.537 | 0.512 |
| C5 x | **9.0** | 2.64 | 5.00 | 1.885 | 0.320 | 0.325 | 8.90 | 3.56 | 1.83 | 0.632 | 0.449 | 0.489 | 0.478 |
| | **6.7** | 1.97 | 5.00 | 1.750 | 0.320 | 0.190 | 7.49 | 3.00 | 1.95 | 0.478 | 0.378 | 0.493 | 0.484 |
| C4 x | **7.25** | 2.13 | 4.00 | 1.721 | 0.296 | 0.321 | 4.59 | 2.29 | 1.47 | 0.432 | 0.343 | 0.450 | 0.459 |
| | **5.4** | 1.59 | 4.00 | 1.584 | 0.296 | 0.184 | 3.85 | 1.93 | 1.56 | 0.319 | 0.283 | 0.449 | 0.458 |
| C3 x | **5.0** | 1.47 | 3.00 | 1.498 | 0.273 | 0.258 | 1.85 | 1.24 | 1.12 | 0.247 | 0.233 | 0.410 | 0.438 |
| | **4.1** | 1.21 | 3.00 | 1.410 | 0.273 | 0.170 | 1.66 | 1.10 | 1.17 | 0.197 | 0.202 | 0.404 | 0.437 |

All shapes on these pages have a flange slope of 16⅔ pct.

# STRUCTURAL TEES (Cut from W Shapes)

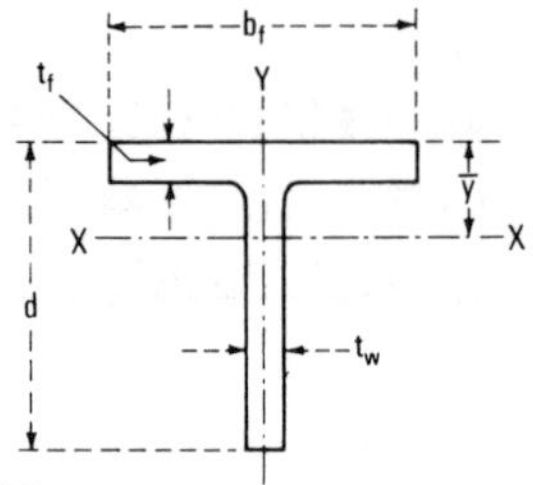


Theoretical Dimensions and Properties for **Designing**

| Section Number | Weight per Foot | Area of Section A | Depth of Section d | Flange Width $b_f$ | Flange Thickness $t_f$ | Stem Thickness $t_w$ | $I_x$ | $Z_x$ | $r_x$ | $\bar{y}$ | $I_y$ | $Z_y$ | $r_y$ |
|---|---|---|---|---|---|---|---|---|---|---|---|---|---|
| | lb | in.$^2$ | in. | in. | in. | in. | in.$^4$ | in.$^3$ | in. | in.$^4$ | in.$^3$ | in.$^3$ | in. |
| **WT18 x** | **150** | 44.1 | 18.370 | 16.655 | 1.680 | 0.945 | 1230 | 86.1 | 5.27 | 4.13 | 648 | 77.8 | 3.83 |
| | **140** | 41.2 | 18.260 | 16.595 | 1.570 | 0.885 | 1140 | 80.0 | 5.25 | 4.07 | 599 | 72.2 | 3.81 |
| | **130** | 38.2 | 18.130 | 16.550 | 1.440 | 0.840 | 1060 | 75.1 | 5.26 | 4.05 | 545 | 65.9 | 3.78 |
| | **122.5** | 36.0 | 18.040 | 16.510 | 1.350 | 0.800 | 995 | 71.0 | 5.26 | 4.03 | 507 | 61.4 | 3.75 |
| | **115** | 33.8 | 17.950 | 16.470 | 1.260 | 0.760 | 934 | 67.0 | 5.25 | 4.01 | 470 | 57.1 | 3.73 |
| **WT18 x** | **105** | 30.9 | 18.345 | 12.180 | 1.360 | 0.830 | 985 | 73.1 | 5.65 | 4.87 | 206 | 33.8 | 2.58 |
| | **97** | 28.5 | 18.245 | 12.115 | 1.260 | 0.765 | 901 | 67.0 | 5.62 | 4.80 | 187 | 30.9 | 2.56 |
| | **91** | 26.8 | 18.165 | 12.075 | 1.180 | 0.725 | 845 | 63.1 | 5.62 | 4.77 | 174 | 28.8 | 2.55 |
| | **85** | 25.0 | 18.085 | 12.030 | 1.100 | 0.680 | 786 | 58.9 | 5.61 | 4.73 | 160 | 26.6 | 2.53 |
| | **80** | 23.5 | 18.005 | 12.000 | 1.020 | 0.650 | 740 | 55.8 | 5.61 | 4.74 | 147 | 24.6 | 2.50 |
| | **75** | 22.1 | 17.925 | 11.975 | 0.940 | 0.625 | 698 | 53.1 | 5.62 | 4.78 | 135 | 22.5 | 2.47 |
| | **67.5** | 19.9 | 17.775 | 11.950 | 0.790 | 0.600 | 636 | 49.7 | 5.66 | 4.96 | 113 | 18.9 | 2.38 |
| **WT16.5 x** | **120.5** | 35.4 | 17.090 | 15.860 | 1.400 | 0.830 | 871 | 65.8 | 4.96 | 3.85 | 466 | 58.8 | 3.63 |
| | **110.5** | 32.5 | 16.965 | 15.805 | 1.275 | 0.775 | 799 | 60.8 | 4.96 | 3.81 | 420 | 53.2 | 3.59 |
| | **100.5** | 29.5 | 16.840 | 15.745 | 1.150 | 0.715 | 725 | 55.5 | 4.95 | 3.78 | 375 | 47.6 | 3.56 |
| **WT16.5 x** | **76** | 22.4 | 16.745 | 11.565 | 1.055 | 0.635 | 592 | 47.4 | 5.14 | 4.26 | 136 | 23.6 | 2.47 |
| | **70.5** | 20.8 | 16.650 | 11.535 | 0.960 | 0.605 | 552 | 44.7 | 5.15 | 4.29 | 123 | 21.3 | 2.43 |
| | **65** | 19.2 | 16.545 | 11.510 | 0.855 | 0.580 | 513 | 42.1 | 5.18 | 4.36 | 109 | 18.9 | 2.39 |
| | **59** | 17.3 | 16.430 | 11.480 | 0.740 | 0.550 | 469 | 39.2 | 5.20 | 4.47 | 93.6 | 16.3 | 2.32 |
| **WT15 x** | **105.5** | 31.0 | 15.470 | 15.105 | 1.315 | 0.775 | 610 | 50.5 | 4.43 | 3.40 | 378 | 50.1 | 3.49 |
| | **95.5** | 28.1 | 15.340 | 15.040 | 1.185 | 0.710 | 549 | 45.7 | 4.42 | 3.35 | 336 | 44.7 | 3.46 |
| | **86.5** | 25.4 | 15.220 | 14.985 | 1.065 | 0.655 | 497 | 41.7 | 4.42 | 3.31 | 299 | 39.9 | 3.43 |
| **WT15 x** | **66** | 19.4 | 15.155 | 10.545 | 1.000 | 0.615 | 421 | 37.4 | 4.66 | 3.90 | 98.0 | 18.6 | 2.25 |
| | **62** | 18.2 | 15.085 | 10.515 | 0.930 | 0.585 | 396 | 35.3 | 4.66 | 3.90 | 90.4 | 17.2 | 2.23 |
| | **58** | 17.1 | 15.005 | 10.495 | 0.850 | 0.565 | 373 | 33.7 | 4.67 | 3.94 | 82.1 | 15.7 | 2.19 |
| | **54** | 15.9 | 14.915 | 10.475 | 0.760 | 0.545 | 349 | 32.0 | 4.69 | 4.01 | 73.0 | 13.9 | 2.15 |
| | **49.5** | 14.5 | 14.825 | 10.450 | 0.670 | 0.520 | 322 | 30.0 | 4.71 | 4.09 | 63.9 | 12.2 | 2.10 |
| **WT13.5 x** | **89** | 26.1 | 13.905 | 14.085 | 1.190 | 0.725 | 414 | 38.2 | 3.98 | 3.05 | 278 | 39.4 | 3.26 |
| | **80.5** | 23.7 | 13.795 | 14.020 | 1.080 | 0.660 | 372 | 34.4 | 3.96 | 2.99 | 248 | 35.4 | 3.24 |
| | **73** | 21.5 | 13.690 | 13.965 | 0.975 | 0.605 | 336 | 31.2 | 3.95 | 2.95 | 222 | 31.7 | 3.21 |
| **WT13.5 x** | **57** | 16.8 | 13.645 | 10.070 | 0.930 | 0.570 | 289 | 28.3 | 4.15 | 3.42 | 79.4 | 15.8 | 2.18 |
| | **51** | 15.0 | 13.545 | 10.015 | 0.830 | 0.515 | 258 | 25.3 | 4.14 | 3.37 | 69.6 | 13.9 | 2.15 |
| | **47** | 13.8 | 13.460 | 9.990 | 0.745 | 0.490 | 239 | 23.8 | 4.16 | 3.41 | 62.0 | 12.4 | 2.12 |
| | **42** | 12.4 | 13.355 | 9.960 | 0.640 | 0.460 | 216 | 21.9 | 4.18 | 3.48 | 52.8 | 10.6 | 2.07 |

Properties shown in this table are for the full center split section.

# STRUCTURAL TEES (Cut from W Shapes)

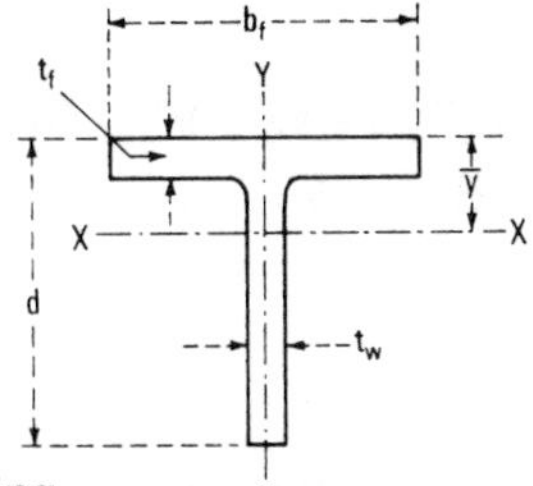


Theoretical Dimensions and Properties for **Designing**

| Section Number | Weight per Foot | Area of Section | Depth of Section | Flange | | Stem Thickness | | | | | | | |
|---|---|---|---|---|---|---|---|---|---|---|---|---|---|
| | | A | d | Width $b_f$ | Thickness $t_f$ | $t_w$ | $I_x$ | $Z_x$ | $r_x$ | $\bar{y}$ | $I_y$ | $Z_y$ | $r_y$ |
| | lb | in.² | in. | in. | in. | in. | in.⁴ | in.³ | in. | in.⁴ | in.³ | in.³ | in. |
| WT12 x | 81 | 23.9 | 12.500 | 12.955 | 1.220 | 0.705 | 293 | 29.9 | 3.50 | 2.70 | 221 | 34.2 | 3.05 |
| | 73 | 21.5 | 12.370 | 12.900 | 1.090 | 0.650 | 264 | 27.2 | 3.50 | 2.66 | 195 | 30.3 | 3.01 |
| | 65.5 | 19.3 | 12.240 | 12.855 | 0.960 | 0.605 | 238 | 24.8 | 3.52 | 2.65 | 170 | 26.5 | 2.97 |
| | 58.5 | 17.2 | 12.130 | 12.800 | 0.850 | 0.550 | 212 | 22.3 | 3.51 | 2.62 | 149 | 23.2 | 2.94 |
| | 52 | 15.3 | 12.030 | 12.750 | 0.750 | 0.500 | 189 | 20.0 | 3.51 | 2.59 | 130 | 20.3 | 2.91 |
| WT12 x | 47 | 13.8 | 12.155 | 9.065 | 0.875 | 0.515 | 186 | 20.3 | 3.67 | 2.99 | 54.5 | 12.0 | 1.98 |
| | 42 | 12.4 | 12.050 | 9.020 | 0.770 | 0.470 | 166 | 18.3 | 3.67 | 2.97 | 47.2 | 10.5 | 1.95 |
| | 38 | 11.2 | 11.960 | 8.990 | 0.680 | 0.440 | 151 | 16.9 | 3.68 | 3.00 | 41.3 | 9.18 | 1.92 |
| | 34 | 10.0 | 11.865 | 8.965 | 0.585 | 0.415 | 137 | 15.6 | 3.70 | 3.06 | 35.2 | 7.85 | 1.87 |
| WT12 x | 31 | 9.11 | 11.870 | 7.040 | 0.590 | 0.430 | 131 | 15.6 | 3.79 | 3.46 | 17.2 | 4.90 | 1.38 |
| | 27.5 | 8.10 | 11.785 | 7.005 | 0.505 | 0.395 | 117 | 14.1 | 3.80 | 3.50 | 14.5 | 4.15 | 1.34 |
| WT10.5 x | 73.5 | 21.6 | 11.030 | 12.510 | 1.150 | 0.720 | 204 | 23.7 | 3.08 | 2.39 | 188 | 30.0 | 2.95 |
| | 66 | 19.4 | 10.915 | 12.440 | 1.035 | 0.650 | 181 | 21.1 | 3.06 | 2.33 | 166 | 26.7 | 2.93 |
| | 61 | 17.9 | 10.840 | 12.390 | 0.960 | 0.600 | 166 | 19.3 | 3.04 | 2.28 | 152 | 24.6 | 2.92 |
| | 55.5 | 16.3 | 10.755 | 12.340 | 0.875 | 0.550 | 150 | 17.5 | 3.03 | 2.23 | 137 | 22.2 | 2.90 |
| | 50.5 | 14.9 | 10.680 | 12.290 | 0.800 | 0.500 | 135 | 15.8 | 3.01 | 2.18 | 124 | 20.2 | 2.89 |
| WT10.5 x | 46.5 | 13.7 | 10.810 | 8.420 | 0.930 | 0.580 | 144 | 17.9 | 3.25 | 2.74 | 46.4 | 11.0 | 1.84 |
| | 41.5 | 12.2 | 10.715 | 8.355 | 0.835 | 0.515 | 127 | 15.7 | 3.22 | 2.66 | 40.7 | 9.75 | 1.83 |
| | 36.5 | 10.7 | 10.620 | 8.295 | 0.740 | 0.455 | 110 | 13.8 | 3.21 | 2.60 | 35.3 | 8.51 | 1.81 |
| | 34 | 10.0 | 10.565 | 8.270 | 0.685 | 0.430 | 103 | 12.9 | 3.20 | 2.59 | 32.4 | 7.83 | 1.80 |
| | 31 | 9.13 | 10.495 | 8.240 | 0.615 | 0.400 | 93.8 | 11.9 | 3.21 | 2.58 | 28.7 | 6.97 | 1.77 |
| WT10.5 x | 28.5 | 8.37 | 10.530 | 6.555 | 0.650 | 0.405 | 90.4 | 11.8 | 3.29 | 2.85 | 15.3 | 4.67 | 1.35 |
| | 25 | 7.36 | 10.415 | 6.530 | 0.535 | 0.380 | 80.3 | 10.7 | 3.30 | 2.93 | 12.5 | 3.82 | 1.30 |
| | 22 | 6.49 | 10.330 | 6.500 | 0.450 | 0.350 | 71.1 | 9.68 | 3.31 | 2.98 | 10.3 | 3.18 | 1.26 |
| WT9 x | 59.5 | 17.5 | 9.485 | 11.265 | 1.060 | 0.655 | 119 | 15.9 | 2.60 | 2.03 | 126 | 22.5 | 2.69 |
| | 53 | 15.6 | 9.365 | 11.200 | 0.940 | 0.590 | 104 | 14.1 | 2.59 | 1.97 | 110 | 19.7 | 2.66 |
| | 48.5 | 14.3 | 9.295 | 11.145 | 0.870 | 0.535 | 93.8 | 12.7 | 2.56 | 1.91 | 100 | 18.0 | 2.65 |
| | 43 | 12.7 | 9.195 | 11.090 | 0.770 | 0.480 | 82.4 | 11.2 | 2.55 | 1.86 | 87.6 | 15.8 | 2.63 |
| | 38 | 11.2 | 9.105 | 11.035 | 0.680 | 0.425 | 71.8 | 9.83 | 2.54 | 1.80 | 76.2 | 13.8 | 2.61 |
| WT9 x | 35.5 | 10.4 | 9.235 | 7.635 | 0.810 | 0.495 | 78.2 | 11.2 | 2.74 | 2.26 | 30.1 | 7.89 | 1.70 |
| | 32.5 | 9.55 | 9.175 | 7.590 | 0.750 | 0.450 | 70.7 | 10.1 | 2.72 | 2.20 | 27.4 | 7.22 | 1.69 |
| | 30 | 8.82 | 9.120 | 7.555 | 0.695 | 0.415 | 64.7 | 9.29 | 2.71 | 2.16 | 25.0 | 6.63 | 1.69 |
| | 27.5 | 8.10 | 9.055 | 7.530 | 0.630 | 0.390 | 59.5 | 8.63 | 2.71 | 2.16 | 22.5 | 5.97 | 1.67 |
| | 25 | 7.33 | 8.995 | 7.495 | 0.570 | 0.355 | 53.5 | 7.79 | 2.70 | 2.12 | 20.0 | 5.35 | 1.65 |
| WT9 x | 23 | 6.77 | 9.030 | 6.060 | 0.605 | 0.360 | 52.1 | 7.77 | 2.77 | 2.33 | 11.3 | 3.72 | 1.29 |
| | 20 | 5.88 | 8.950 | 6.015 | 0.525 | 0.315 | 44.8 | 6.73 | 2.76 | 2.29 | 9.55 | 3.17 | 1.27 |
| | 17.5 | 5.15 | 8.850 | 6.000 | 0.425 | 0.300 | 40.1 | 6.21 | 2.79 | 2.39 | 7.67 | 2.56 | 1.22 |

Properties shown in this table are for the full center split section.

# STRUCTURAL TEES (Cut from W Shapes)

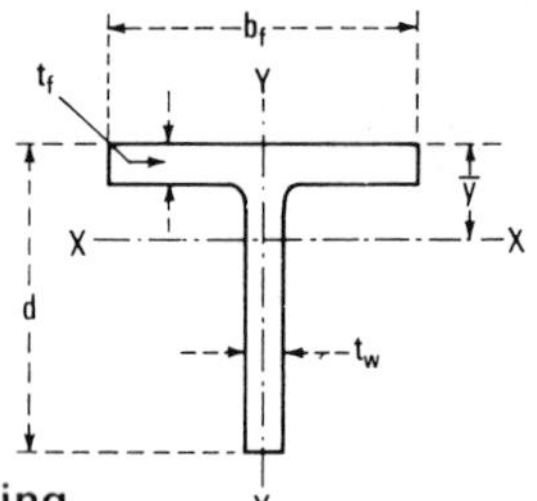


Theoretical Dimensions and Properties for **Designing**

| Section Number | Weight per Foot | Area of Section A | Depth of Section d | Flange Width $b_f$ | Flange Thickness $t_f$ | Stem Thickness $t_w$ | | | | | | | |
|---|---|---|---|---|---|---|---|---|---|---|---|---|---|
| | | | | | | | $I_x$ | $Z_x$ | $r_x$ | $\bar{y}$ | $I_y$ | $Z_y$ | $r_y$ |
| | lb | in.$^2$ | in. | in. | in. | in. | in.$^4$ | in.$^3$ | in. | in. | in.$^4$ | in.$^3$ | in. |
| **WT8 x** | **50** | 14.7 | 8.485 | 10.425 | 0.985 | 0.585 | 76.8 | 11.4 | 2.28 | 1.76 | 93.1 | 17.9 | 2.52 |
| | **44.5** | 13.1 | 8.375 | 10.365 | 0.875 | 0.525 | 67.2 | 10.1 | 2.27 | 1.70 | 81.3 | 15.7 | 2.49 |
| | **38.5** | 11.3 | 8.260 | 10.295 | 0.760 | 0.455 | 56.9 | 8.59 | 2.24 | 1.63 | 69.2 | 13.4 | 2.47 |
| | **33.5** | 9.84 | 8.165 | 10.235 | 0.665 | 0.395 | 48.6 | 7.36 | 2.22 | 1.56 | 59.5 | 11.6 | 2.46 |
| **WT8 x** | **28.5** | 8.38 | 8.215 | 7.120 | 0.715 | 0.430 | 48.7 | 7.77 | 2.41 | 1.94 | 21.6 | 6.06 | 1.60 |
| | **25** | 7.37 | 8.130 | 7.070 | 0.630 | 0.380 | 42.3 | 6.78 | 2.40 | 1.89 | 18.6 | 5.26 | 1.59 |
| | **22.5** | 6.63 | 8.065 | 7.035 | 0.565 | 0.345 | 37.8 | 6.10 | 2.39 | 1.86 | 16.4 | 4.67 | 1.57 |
| | **20** | 5.89 | 8.005 | 6.995 | 0.505 | 0.305 | 33.1 | 5.35 | 2.37 | 1.81 | 14.4 | 4.12 | 1.57 |
| | **18** | 5.28 | 7.930 | 6.985 | 0.430 | 0.295 | 30.6 | 5.05 | 2.41 | 1.88 | 12.2 | 3.50 | 1.52 |
| **WT8 x** | **15.5** | 4.56 | 7.940 | 5.525 | 0.440 | 0.275 | 27.4 | 4.64 | 2.45 | 2.02 | 6.20 | 2.24 | 1.17 |
| | **13** | 3.84 | 7.845 | 5.500 | 0.345 | 0.250 | 23.5 | 4.09 | 2.47 | 2.09 | 4.80 | 1.74 | 1.12 |
| **WT7 x** | **365** | 107 | 11.210 | 17.890 | 4.910 | 3.070 | 739 | 95.4 | 2.62 | 3.47 | 2360 | 264 | 4.69 |
| | **332.5** | 97.8 | 10.820 | 17.650 | 4.520 | 2.830 | 622 | 82.1 | 2.52 | 3.25 | 2080 | 236 | 4.62 |
| | **302.5** | 88.9 | 10.460 | 17.415 | 4.160 | 2.595 | 524 | 70.6 | 2.43 | 3.05 | 1840 | 211 | 4.55 |
| | **275** | 80.9 | 10.120 | 17.200 | 3.820 | 2.380 | 442 | 60.9 | 2.34 | 2.85 | 1630 | 189 | 4.49 |
| | **250** | 73.5 | 9.800 | 17.010 | 3.500 | 2.190 | 375 | 52.7 | 2.26 | 2.67 | 1440 | 169 | 4.43 |
| | **227.5** | 66.9 | 9.510 | 16.835 | 3.210 | 2.015 | 321 | 45.9 | 2.19 | 2.51 | 1280 | 152 | 4.38 |
| **WT7 x** | **213** | 62.6 | 9.335 | 16.695 | 3.035 | 1.875 | 287 | 41.4 | 2.14 | 2.40 | 1180 | 141 | 4.34 |
| | **199** | 58.5 | 9.145 | 16.590 | 2.845 | 1.770 | 257 | 37.6 | 2.10 | 2.30 | 1090 | 131 | 4.31 |
| | **185** | 54.4 | 8.960 | 16.475 | 2.660 | 1.655 | 229 | 33.9 | 2.05 | 2.19 | 994 | 121 | 4.27 |
| | **171** | 50.3 | 8.770 | 16.360 | 2.470 | 1.540 | 203 | 30.4 | 2.01 | 2.09 | 903 | 110 | 4.24 |
| | **155.5** | 45.7 | 8.560 | 16.230 | 2.260 | 1.410 | 176 | 26.7 | 1.96 | 1.97 | 807 | 99.4 | 4.20 |
| | **141.5** | 41.6 | 8.370 | 16.110 | 2.070 | 1.290 | 153 | 23.5 | 1.92 | 1.86 | 722 | 89.7 | 4.17 |
| | **128.5** | 37.8 | 8.190 | 15.995 | 1.890 | 1.175 | 133 | 20.7 | 1.88 | 1.75 | 645 | 80.7 | 4.13 |
| | **116.5** | 34.2 | 8.020 | 15.890 | 1.720 | 1.070 | 116 | 18.2 | 1.84 | 1.65 | 576 | 72.5 | 4.10 |
| | **105.5** | 31.0 | 7.860 | 15.800 | 1.560 | 0.098 | 102 | 16.2 | 1.81 | 1.57 | 513 | 65.0 | 4.07 |
| | **96.5** | 28.4 | 7.740 | 15.710 | 1.440 | 0.890 | 89.8 | 14.4 | 1.78 | 1.49 | 466 | 59.3 | 4.05 |
| | **88** | 25.9 | 7.610 | 15.650 | 1.310 | 0.830 | 80.5 | 13.0 | 1.76 | 1.43 | 419 | 53.5 | 4.02 |
| | **79.5** | 23.4 | 7.490 | 15.565 | 1.190 | 0.745 | 70.2 | 11.4 | 1.73 | 1.35 | 374 | 48.1 | 4.00 |
| | **72.5** | 21.3 | 7.390 | 15.500 | 1.090 | 0.680 | 62.5 | 10.2 | 1.71 | 1.29 | 338 | 43.7 | 3.98 |

Properties shown in this table are for the full center split section.

# STRUCTURAL TEES (Cut from W Shapes)

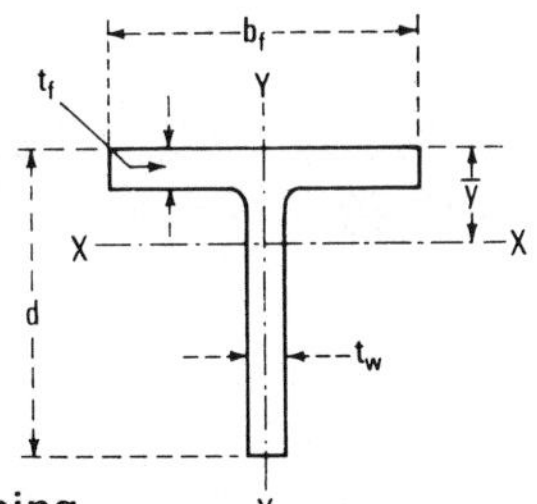


Theoretical Dimensions and Properties for **Designing**

| Section Number | Weight per Foot | Area of Section $A$ | Depth of Section $d$ | Flange Width $b_f$ | Flange Thickness $t_f$ | Stem Thickness $t_w$ | $I_x$ | $Z_x$ | $r_x$ | $\bar{y}$ | $I_y$ | $Z_y$ | $r_y$ |
|---|---|---|---|---|---|---|---|---|---|---|---|---|---|
| | lb | in.² | in. | in. | in. | in. | in.⁴ | in.³ | in. | in. | in.⁴ | in.³ | in. |
| **WT7 x** | **66** | 19.4 | 7.330 | 14.725 | 1.030 | 0.645 | 57.8 | 9.57 | 1.73 | 1.29 | 274 | 37.2 | 3.76 |
| | **60** | 17.7 | 7.240 | 14.670 | 0.940 | 0.590 | 51.7 | 8.61 | 1.71 | 1.24 | 247 | 33.7 | 3.74 |
| | **54.5** | 16.0 | 7.160 | 14.605 | 0.860 | 0.525 | 45.3 | 7.56 | 1.68 | 1.17 | 223 | 30.6 | 3.73 |
| | **49.5** | 14.6 | 7.080 | 14.565 | 0.780 | 0.485 | 40.9 | 6.88 | 1.67 | 1.14 | 201 | 27.6 | 3.71 |
| | **45** | 13.2 | 7.010 | 14.520 | 0.710 | 0.440 | 36.4 | 6.16 | 1.66 | 1.09 | 181 | 25.0 | 3.70 |
| **WT7 x** | **41** | 12.0 | 7.155 | 10.130 | 0.855 | 0.510 | 41.2 | 7.14 | 1.85 | 1.39 | 74.2 | 14.6 | 2.48 |
| | **37** | 10.9 | 7.085 | 10.070 | 0.785 | 0.450 | 36.0 | 6.25 | 1.82 | 1.32 | 66.9 | 13.3 | 2.48 |
| | **34** | 9.99 | 7.020 | 10.035 | 0.720 | 0.415 | 32.6 | 5.69 | 1.81 | 1.29 | 60.7 | 12.1 | 2.46 |
| | **30.5** | 8.96 | 6.945 | 9.995 | 0.645 | 0.375 | 28.9 | 5.07 | 1.80 | 1.25 | 53.7 | 10.7 | 2.45 |
| **WT7 x** | **26.5** | 7.81 | 6.960 | 8.060 | 0.660 | 0.370 | 27.6 | 4.94 | 1.88 | 1.38 | 28.8 | 7.16 | 1.92 |
| | **24** | 7.07 | 6.895 | 8.030 | 0.595 | 0.340 | 24.9 | 4.48 | 1.87 | 1.35 | 25.7 | 6.40 | 1.91 |
| | **21.5** | 6.31 | 6.830 | 7.995 | 0.530 | 0.305 | 21.9 | 3.98 | 1.86 | 1.31 | 22.6 | 5.65 | 1.89 |
| **WT7 x** | **19** | 5.58 | 7.050 | 6.770 | 0.515 | 0.310 | 23.3 | 4.22 | 2.04 | 1.54 | 13.3 | 3.94 | 1.55 |
| | **17** | 5.00 | 6.990 | 6.745 | 0.455 | 0.285 | 20.9 | 3.83 | 2.04 | 1.53 | 11.7 | 3.45 | 1.53 |
| | **15** | 4.42 | 6.920 | 6.730 | 0.385 | 0.270 | 19.0 | 3.55 | 2.07 | 1.58 | 9.79 | 2.91 | 1.49 |
| **WT7 x** | **13** | 3.85 | 6.955 | 5.025 | 0.420 | 0.255 | 17.3 | 3.31 | 2.12 | 1.72 | 4.45 | 1.77 | 1.08 |
| | **11** | 3.25 | 6.870 | 5.000 | 0.335 | 0.230 | 14.8 | 2.91 | 2.14 | 1.76 | 3.50 | 1.40 | 1.04 |
| **WT6 x** | **95** | 27.9 | 7.190 | 12.670 | 1.735 | 1.060 | 79.0 | 14.2 | 1.68 | 1.62 | 295 | 46.5 | 3.25 |
| | **85** | 25.0 | 7.015 | 12.570 | 1.560 | 0.960 | 67.8 | 12.3 | 1.65 | 1.52 | 259 | 41.2 | 3.22 |
| | **76** | 22.4 | 6.855 | 12.480 | 1.400 | 0.870 | 58.5 | 10.8 | 1.62 | 1.43 | 227 | 36.4 | 3.19 |
| | **68** | 20.0 | 6.705 | 12.400 | 1.250 | 0.790 | 50.6 | 9.46 | 1.59 | 1.35 | 199 | 32.1 | 3.16 |
| | **60** | 17.6 | 6.560 | 12.320 | 1.105 | 0.710 | 43.4 | 8.22 | 1.57 | 1.28 | 172 | 28.0 | 3.13 |
| | **53** | 15.6 | 6.445 | 12.220 | 0.990 | 0.610 | 36.3 | 6.91 | 1.53 | 1.19 | 151 | 24.7 | 3.11 |
| | **48** | 14.1 | 6.355 | 12.160 | 0.900 | 0.550 | 32.0 | 6.12 | 1.51 | 1.13 | 135 | 22.2 | 3.09 |
| | **43.5** | 12.8 | 6.265 | 12.125 | 0.810 | 0.515 | 28.9 | 5.60 | 1.50 | 1.10 | 120 | 19.9 | 3.07 |
| | **39.5** | 11.6 | 6.190 | 12.080 | 0.735 | 0.470 | 25.8 | 5.03 | 1.49 | 1.06 | 108 | 17.9 | 3.05 |
| | **36** | 10.6 | 6.125 | 12.040 | 0.670 | 0.430 | 23.2 | 4.54 | 1.48 | 1.02 | 97.5 | 16.2 | 3.04 |
| | **32.5** | 9.54 | 0.060 | 12.000 | 0.605 | 0.390 | 20.6 | 4.06 | 1.47 | 0.985 | 87.2 | 14.5 | 3.02 |

Properties shown in this table are for the full center split section.

# STRUCTURAL TEES (Cut from W Shapes)

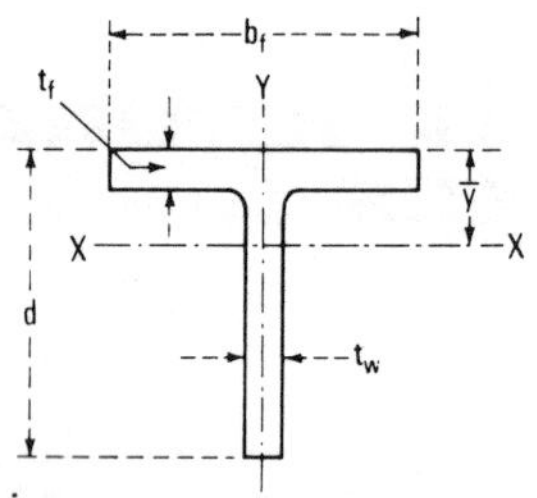


Theoretical Dimensions and Properties for **Designing**

| Section Number | Weight per Foot | Area of Section A | Depth of Section d | Flange Width $b_f$ | Flange Thick-ness $t_f$ | Stem Thick-ness $t_w$ | $I_x$ | $Z_x$ | $r_x$ | $\bar{y}$ | $I_y$ | $Z_y$ | $r_y$ |
|---|---|---|---|---|---|---|---|---|---|---|---|---|---|
| | lb | in.² | in. | in. | in. | in. | in.⁴ | in.³ | in. | in. | in.⁴ | in.³ | in. |
| WT6 x | 29 | 8.52 | 6.095 | 10.010 | 0.640 | 0.360 | 19.1 | 3.76 | 1.50 | 1.03 | 53.5 | 10.7 | 2.51 |
| | 26.5 | 7.78 | 6.030 | 9.995 | 0.575 | 0.345 | 17.7 | 3.54 | 1.51 | 1.02 | 47.9 | 9.58 | 2.48 |
| WT6 x | 25 | 7.34 | 6.095 | 8.080 | 0.640 | 0.370 | 18.7 | 3.79 | 1.60 | 1.17 | 28.2 | 6.97 | 1.96 |
| | 22.5 | 6.61 | 6.030 | 8.045 | 0.575 | 0.335 | 16.6 | 3.39 | 1.58 | 1.13 | 25.0 | 6.21 | 1.94 |
| | 20 | 5.89 | 5.970 | 8.005 | 0.515 | 0.295 | 14.4 | 2.95 | 1.57 | 1.08 | 22.0 | 5.51 | 1.93 |
| WT6 x | 17.5 | 5.17 | 6.250 | 6.560 | 0.520 | 0.300 | 16.0 | 3.23 | 1.76 | 1.30 | 12.2 | 3.73 | 1.54 |
| | 15 | 4.40 | 6.170 | 6.520 | 0.440 | 0.260 | 13.5 | 2.75 | 1.75 | 1.27 | 10.2 | 3.12 | 1.52 |
| | 13 | 3.82 | 6.110 | 6.490 | 0.380 | 0.230 | 11.7 | 2.40 | 1.75 | 1.25 | 8.66 | 2.67 | 1.51 |
| WT6 x | 11 | 3.24 | 6.155 | 4.030 | 0.425 | 0.260 | 11.7 | 2.59 | 1.90 | 1.63 | 2.33 | 1.16 | 0.848 |
| | 9.5 | 2.79 | 6.080 | 4.005 | 0.350 | 0.235 | 10.1 | 2.28 | 1.90 | 1.65 | 1.88 | 0.939 | 0.822 |
| | 8 | 2.36 | 5.995 | 3.990 | 0.265 | 0.220 | 8.70 | 2.04 | 1.92 | 1.74 | 1.41 | 0.706 | 0.773 |
| | 7 | 2.08 | 5.955 | 3.970 | 0.225 | 0.200 | 7.67 | 1.83 | 1.92 | 1.76 | 1.18 | 0.594 | 0.753 |
| WT5 x | 56 | 16.5 | 5.680 | 10.415 | 1.250 | 0.755 | 28.6 | 6.40 | 1.32 | 1.21 | 118 | 22.6 | 2.68 |
| | 50 | 14.7 | 5.550 | 10.340 | 1.120 | 0.680 | 24.5 | 5.56 | 1.29 | 1.13 | 103 | 20.0 | 2.65 |
| | 44 | 12.9 | 5.420 | 10.265 | 0.990 | 0.605 | 20.8 | 4.77 | 1.27 | 1.06 | 89.3 | 17.4 | 2.63 |
| | 38.5 | 11.3 | 5.300 | 10.190 | 0.870 | 0.530 | 17.4 | 4.05 | 1.24 | 0.990 | 76.8 | 15.1 | 2.60 |
| | 34 | 9.99 | 5.200 | 10.130 | 0.770 | 0.470 | 14.9 | 3.49 | 1.22 | 0.932 | 66.8 | 13.2 | 2.59 |
| | 30 | 8.82 | 5.110 | 10.080 | 0.680 | 0.420 | 12.9 | 3.04 | 1.21 | 0.884 | 58.1 | 11.5 | 2.57 |
| | 27 | 7.91 | 5.045 | 10.030 | 0.615 | 0.370 | 11.1 | 2.64 | 1.19 | 0.836 | 51.7 | 10.3 | 2.56 |
| | 24.5 | 7.21 | 4.990 | 10.000 | 0.560 | 0.340 | 10.0 | 2.39 | 1.18 | 0.807 | 46.7 | 9.34 | 2.54 |
| WT5 x | 22.5 | 6.63 | 5.050 | 8.020 | 0.620 | 0.350 | 10.2 | 2.47 | 1.24 | 0.907 | 26.7 | 6.65 | 2.01 |
| | 19.5 | 5.73 | 4.960 | 7.985 | 0.530 | 0.315 | 8.84 | 2.16 | 1.24 | 0.876 | 22.5 | 5.64 | 1.98 |
| | 16.5 | 4.85 | 4.865 | 7.960 | 0.435 | 0.290 | 7.71 | 1.93 | 1.26 | 0.869 | 18.3 | 4.60 | 1.94 |
| WT5 x | 15 | 4.42 | 5.235 | 5.810 | 0.510 | 0.300 | 9.28 | 2.24 | 1.45 | 1.10 | 8.35 | 2.87 | 1.37 |
| | 13 | 3.81 | 5.165 | 5.770 | 0.440 | 0.260 | 7.86 | 1.91 | 1.44 | 1.06 | 7.05 | 2.44 | 1.36 |
| | 11 | 3.24 | 5.085 | 5.750 | 0.360 | 0.240 | 6.88 | 1.72 | 1.46 | 1.07 | 5.71 | 1.99 | 1.33 |
| WT5 x | 9.5 | 2.81 | 5.120 | 4.020 | 0.395 | 0.250 | 6.68 | 1.74 | 1.54 | 1.28 | 2.15 | 1.07 | 0.874 |
| | 8.5 | 2.50 | 5.055 | 4.010 | 0.330 | 0.240 | 6.06 | 1.62 | 1.56 | 1.32 | 1.78 | 0.888 | 0.845 |
| | 7.5 | 2.21 | 4.995 | 4.000 | 0.270 | 0.230 | 5.45 | 1.50 | 1.57 | 1.37 | 1.45 | 0.723 | 0.810 |
| | 6 | 1.77 | 4.935 | 3.960 | 0.210 | 0.190 | 4.35 | 1.22 | 1.57 | 1.36 | 1.09 | 0.551 | 0.785 |

Properties shown in this table are for the full center split section.

# STRUCTURAL TEES (Cut from W Shapes)

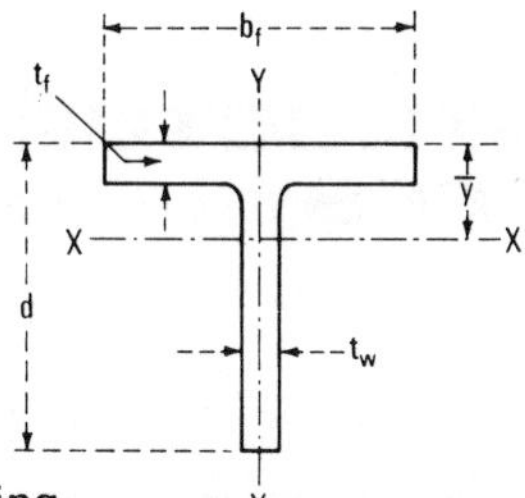


Theoretical Dimensions and Properties for **Designing**

| Section Number | Weight per Foot | Area of Section | Depth of Section | Flange | | Stem Thickness | | | | | | | |
|---|---|---|---|---|---|---|---|---|---|---|---|---|---|
| | | A | d | Width $b_f$ | Thickness $t_f$ | $t_w$ | $I_x$ | $Z_x$ | $r_x$ | $\bar{y}$ | $I_y$ | $Z_y$ | $r_y$ |
| | lb | in.$^2$ | in. | in. | in. | in. | in.$^4$ | in.$^3$ | in. | in. | in.$^4$ | in.$^3$ | in. |
| **WT4 x** | **33.5** | 9.84 | 4.500 | 8.280 | 0.935 | 0.570 | 10.9 | 3.05 | 1.05 | 0.936 | 44.3 | 10.7 | 2.12 |
| | **29** | 8.55 | 4.375 | 8.220 | 0.810 | 0.510 | 9.12 | 2.61 | 1.03 | 0.874 | 37.5 | 9.13 | 2.10 |
| | **24** | 7.05 | 4.250 | 8.110 | 0.685 | 0.400 | 6.85 | 1.97 | 0.986 | 0.777 | 30.5 | 7.52 | 2.08 |
| | **20** | 5.87 | 4.125 | 8.070 | 0.560 | 0.360 | 5.73 | 1.69 | 0.988 | 0.735 | 24.5 | 6.08 | 2.04 |
| | **17.5** | 5.14 | 4.060 | 8.020 | 0.495 | 0.310 | 4.81 | 1.43 | 0.968 | 0.688 | 21.3 | 5.31 | 2.03 |
| | **15.5** | 4.56 | 4.000 | 7.995 | 0.435 | 0.285 | 4.28 | 1.28 | 0.968 | 0.668 | 18.5 | 4.64 | 2.02 |
| **WT4 x** | **14** | 4.12 | 4.030 | 6.535 | 0.465 | 0.285 | 4.22 | 1.28 | 1.01 | 0.734 | 10.8 | 3.31 | 1.62 |
| | **12** | 3.54 | 3.965 | 6.495 | 0.400 | 0.245 | 3.53 | 1.08 | 0.999 | 0.695 | 9.14 | 2.81 | 1.61 |
| **WT4 x** | **10.5** | 3.08 | 4.140 | 5.270 | 0.400 | 0.250 | 3.90 | 1.18 | 1.12 | 0.831 | 4.89 | 1.85 | 1.26 |
| | **9** | 2.63 | 4.070 | 5.250 | 0.330 | 0.230 | 3.41 | 1.05 | 1.14 | 0.834 | 3.98 | 1.52 | 1.23 |
| **WT4 x** | **7.5** | 2.22 | 4.055 | 4.015 | 0.315 | 0.245 | 3.28 | 1.07 | 1.22 | 0.998 | 1.70 | 0.849 | 0.876 |
| | **6.5** | 1.92 | 3.995 | 4.000 | 0.255 | 0.230 | 2.89 | 0.974 | 1.23 | 1.03 | 1.37 | 0.683 | 0.844 |
| | **5** | 1.48 | 3.945 | 3.940 | 0.205 | 0.170 | 2.15 | 0.717 | 1.20 | 0.953 | 1.05 | 0.532 | 0.841 |
| **WT3 x** | **12.5** | 3.67 | 3.190 | 6.080 | 0.455 | 0.320 | 2.29 | 0.886 | 0.789 | 0.610 | 8.53 | 2.81 | 1.52 |
| | **10** | 2.94 | 3.100 | 6.020 | 0.365 | 0.260 | 1.76 | 0.693 | 0.774 | 0.560 | 6.64 | 2.21 | 1.50 |
| | **7.5** | 2.21 | 2.995 | 5.990 | 0.260 | 0.230 | 1.41 | 0.577 | 0.797 | 0.558 | 4.66 | 1.56 | 1.45 |
| **WT3 x** | **8** | 2.37 | 3.140 | 4.030 | 0.405 | 0.260 | 1.69 | 0.685 | 0.844 | 0.677 | 2.21 | 1.10 | 0.967 |
| | **6** | 1.78 | 3.015 | 4.000 | 0.280 | 0.230 | 1.32 | 0.564 | 0.862 | 0.677 | 1.50 | 0.748 | 0.918 |
| | **4.5** | 1.34 | 2.950 | 3.940 | 0.215 | 0.170 | 0.950 | 0.408 | 0.842 | 0.623 | 1.10 | 0.557 | 0.905 |
| **WT2.5 x** | **9.5** | 2.77 | 2.575 | 5.030 | 0.430 | 0.270 | 1.01 | 0.485 | 0.605 | 0.487 | 4.56 | 1.82 | 1.28 |
| | **8** | 2.34 | 2.505 | 5.000 | 0.360 | 0.240 | 0.845 | 0.413 | 0.601 | 0.458 | 3.75 | 1.50 | 1.27 |
| **WT2 x** | **6.5** | 1.91 | 2.080 | 4.060 | 0.345 | 0.280 | 0.526 | 0.321 | 0.524 | 0.440 | 1.93 | 0.950 | 1.00 |

Properties shown in this table are for the full center split section.

# STRUCTURAL TEES (Cut from S Shapes)

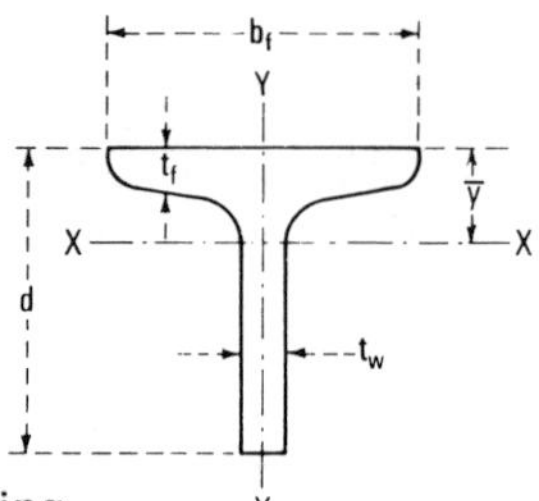


Theoretical Dimensions and Properties for **Designing**

| Section Number | Weight per Foot | Area of Section A | Depth of Tee $d$ | Flange Width $b_f$ | Flange Average Thickness $t_f$ | Stem Thickness $t_w$ | $I_x$ | $Z_x$ | $r_x$ | $\bar{y}$ | $I_y$ | $Z_y$ | $r_y$ |
|---|---|---|---|---|---|---|---|---|---|---|---|---|---|
| | lb | in.$^2$ | in. | in. | in. | in. | in.$^4$ | in.$^3$ | in. | in. | in.$^4$ | in.$^3$ | in. |
| **ST12 x** | **60.5** | 17.8 | 12.250 | 8.050 | 1.090 | 0.800 | 259 | 30.1 | 3.82 | 3.63 | 41.7 | 10.4 | 1.53 |
| | **53** | 15.6 | 12.250 | 7.870 | 1.090 | 0.620 | 216 | 24.1 | 3.72 | 3.28 | 38.5 | 9.80 | 1.57 |
| **ST12 x** | **50** | 14.7 | 12.000 | 7.245 | 0.870 | 0.745 | 215 | 26.3 | 3.83 | 3.84 | 23.8 | 6.58 | 1.27 |
| | **45** | 13.2 | 12.000 | 7.125 | 0.870 | 0.625 | 190 | 22.6 | 3.79 | 3.60 | 22.5 | 6.31 | 1.30 |
| | **40** | 11.7 | 12.000 | 7.000 | 0.870 | 0.500 | 162 | 18.7 | 3.72 | 3.29 | 21.1 | 6.04 | 1.34 |
| **ST10 x** | **48** | 14.1 | 10.150 | 7.200 | 0.920 | 0.800 | 143 | 20.3 | 3.18 | 3.13 | 25.1 | 6.97 | 1.33 |
| | **43** | 12.7 | 10.150 | 7.060 | 0.920 | 0.660 | 125 | 17.2 | 3.14 | 2.91 | 23.4 | 6.63 | 1.36 |
| **ST10 x** | **37.5** | 11.0 | 10.000 | 6.385 | 0.795 | 0.635 | 109 | 15.8 | 3.15 | 3.07 | 14.9 | 4.66 | 1.16 |
| | **33** | 9.70 | 10.000 | 6.255 | 0.795 | 0.505 | 93.1 | 12.9 | 3.10 | 2.81 | 13.8 | 4.43 | 1.19 |
| **ST9 x** | **35** | 10.3 | 9.000 | 6.251 | 0.691 | 0.711 | 84.7 | 14.0 | 2.87 | 2.94 | 12.1 | 3.86 | 1.08 |
| | **27.35** | 8.04 | 9.000 | 6.001 | 0.691 | 0.461 | 62.4 | 9.61 | 2.79 | 2.50 | 10.4 | 3.47 | 1.14 |
| **ST7.5 x** | **25** | 7.35 | 7.500 | 5.640 | 0.622 | 0.550 | 40.6 | 7.73 | 2.35 | 2.25 | 7.85 | 2.78 | 1.03 |
| | **21.45** | 6.31 | 7.500 | 5.501 | 0.622 | 0.411 | 33.0 | 6.00 | 2.29 | 2.01 | 7.19 | 2.61 | 1.07 |
| **ST6 x** | **25** | 7.35 | 6.000 | 5.477 | 0.659 | 0.687 | 25.2 | 6.05 | 1.85 | 1.84 | 7.85 | 2.87 | 1.03 |
| | **20.4** | 6.00 | 6.000 | 5.252 | 0.659 | 0.462 | 18.9 | 4.28 | 1.78 | 1.58 | 6.78 | 2.58 | 1.06 |
| **ST6 x** | **17.5** | 5.14 | 6.000 | 5.078 | 0.544 | 0.428 | 17.2 | 3.95 | 1.83 | 1.65 | 4.93 | 1.94 | 0.980 |
| | **15.9** | 4.68 | 6.000 | 5.000 | 0.544 | 0.350 | 14.9 | 3.31 | 1.78 | 1.51 | 4.68 | 1.87 | 1.00 |
| **ST5 x** | **17.5** | 5.15 | 5.000 | 4.944 | 0.491 | 0.594 | 12.5 | 3.63 | 1.56 | 1.56 | 4.18 | 1.69 | 0.901 |
| | **12.7** | 3.73 | 5.000 | 4.661 | 0.491 | 0.311 | 7.83 | 2.06 | 1.45 | 1.20 | 3.39 | 1.46 | 0.954 |
| **ST4 x** | **11.5** | 3.38 | 4.000 | 4.171 | 0.425 | 0.441 | 5.03 | 1.77 | 1.22 | 1.15 | 2.15 | 1.03 | 0.798 |
| | **9.2** | 2.70 | 4.000 | 4.001 | 0.425 | 0.271 | 3.51 | 1.15 | 1.14 | 0.941 | 1.86 | 0.932 | 0.831 |
| **ST3.5 x** | **7.65** | 2.25 | 3.500 | 3.662 | 0.392 | 0.252 | 2.19 | 0.816 | 0.987 | 0.817 | 1.32 | 0.720 | 0.766 |
| **ST3 x** | **8.625** | 2.53 | 3.000 | 3.565 | 0.359 | 0.465 | 2.13 | 1.02 | 0.917 | 0.914 | 1.15 | 0.648 | 0.675 |
| | **6.25** | 1.83 | 3.000 | 3.332 | 0.359 | 0.232 | 1.27 | 0.552 | 0.833 | 0.691 | 0.911 | 0.547 | 0.705 |
| **ST2.5 x** | **5** | 1.47 | 2.500 | 3.004 | 0.326 | 0.214 | 0.681 | 0.353 | 0.681 | 0.569 | 0.608 | 0.405 | 0.643 |
| **ST2 x** | **4.75** | 1.40 | 2.000 | 2.796 | 0.293 | 0.326 | 0.470 | 0.325 | 0.580 | 0.553 | 0.451 | 0.323 | 0.569 |
| | **3.85** | 1.13 | 2.000 | 2.663 | 0.293 | 0.193 | 0.316 | 0.203 | 0.528 | 0.448 | 0.382 | 0.287 | 0.581 |

Properties shown in this table are for the full center split section.

# ANGLES
equal legs

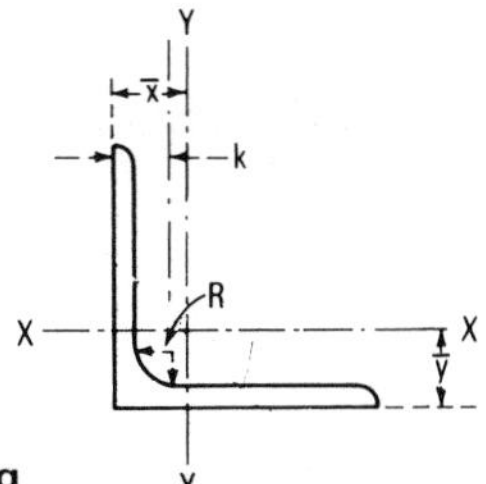


Theoretical Dimensions and Properties for **Designing**

| Section Number and Size | Thickness | Weight per Foot | Area of Section | k | $I_{x,y}$ | $Z_{x,y}$ | $r_{x,y}$ | $\bar{y}$ or $\bar{x}$ |
|---|---|---|---|---|---|---|---|---|
| | | | | | Axis X-X and Axis Y-Y | | | |
| in. | in. | lb | in.² | in. | in.⁴ | in.³ | in. | in. |
| **L8 x 8 x** | **1⅛** | 56.9 | 16.7 | 1¾ | 98.0 | 17.5 | 2.42 | 2.41 |
| R=⅝ | **1** | 51.0 | 15.0 | 1⅝ | 89.0 | 15.8 | 2.44 | 2.37 |
| | **⅞** | 45.0 | 13.2 | 1½ | 79.6 | 14.0 | 2.45 | 2.32 |
| | **¾** | 38.9 | 11.4 | 1⅜ | 69.7 | 12.2 | 2.47 | 2.28 |
| | **⅝** | 32.7 | 9.61 | 1¼ | 59.4 | 10.3 | 2.49 | 2.23 |
| | [y]9/16 | 29.6 | 8.68 | 1 3/16 | 54.1 | 9.34 | 2.50 | 2.21 |
| | ½ | 26.4 | 7.75 | 1⅛ | 48.6 | 8.36 | 2.50 | 2.19 |
| **L6 x 6 x** | **1** | 37.4 | 11.0 | 1½ | 35.5 | 8.57 | 1.80 | 1.86 |
| R=½ | ⅞ | 33.1 | 9.73 | 1⅜ | 31.9 | 7.63 | 1.81 | 1.82 |
| | ¾ | 28.7 | 8.44 | 1¼ | 28.2 | 6.66 | 1.83 | 1.78 |
| | ⅝ | 24.2 | 7.11 | 1⅛ | 24.2 | 5.66 | 1.84 | 1.73 |
| | [y]9/16 | 21.9 | 6.43 | 1 1/16 | 22.1 | 5.14 | 1.85 | 1.71 |
| | ½ | 19.6 | 5.75 | 1 | 19.9 | 4.61 | 1.86 | 1.68 |
| | [y]7/16 | 17.2 | 5.06 | 15/16 | 17.7 | 4.08 | 1.87 | 1.66 |
| | ⅜ | 14.9 | 4.36 | ⅞ | 15.4 | 3.53 | 1.88 | 1.64 |
| **L5 x 5 x** | ⅞ | 27.2 | 7.98 | 1⅜ | 17.8 | 5.17 | 1.49 | 1.57 |
| R=½ | ¾ | 23.6 | 6.94 | 1¼ | 15.7 | 4.53 | 1.51 | 1.52 |
| | [x][y]⅝ | 20.0 | 5.86 | 1⅛ | 13.6 | 3.86 | 1.52 | 1.48 |
| | ½ | 16.2 | 4.75 | 1 | 11.3 | 3.16 | 1.54 | 1.43 |
| | [y]7/16 | 14.3 | 4.18 | 15/16 | 10.0 | 2.79 | 1.55 | 1.41 |
| | ⅜ | 12.3 | 3.61 | ⅞ | 8.74 | 2.42 | 1.56 | 1.39 |
| | 5/16 | 10.3 | 3.03 | 13/16 | 7.42 | 2.04 | 1.57 | 1.37 |
| **L4 x 4 x** | ¾ | 18.5 | 5.44 | 1⅛ | 7.67 | 2.81 | 1.19 | 1.27 |
| R=⅜ | ⅝ | 15.7 | 4.61 | 1 | 6.66 | 2.40 | 1.20 | 1.23 |
| | ½ | 12.8 | 3.75 | ⅞ | 5.56 | 1.97 | 1.22 | 1.18 |
| | [y]7/16 | 11.3 | 3.31 | 13/16 | 4.97 | 1.75 | 1.23 | 1.16 |
| | ⅜ | 9.8 | 2.86 | ¾ | 4.36 | 1.52 | 1.23 | 1.14 |
| | 5/16 | 8.2 | 2.40 | 11/16 | 3.71 | 1.29 | 1.24 | 1.12 |
| | ¼ | 6.6 | 1.94 | ⅝ | 3.04 | 1.05 | 1.25 | 1.09 |
| **L3½ x 3½ x** | ⅜ | 8.5 | 2.48 | ¾ | 2.87 | 1.15 | 1.07 | 1.01 |
| R=⅜ | 5/16 | 7.2 | 2.09 | 11/16 | 2.45 | 0.976 | 1.08 | 0.990 |
| | ¼ | 5.8 | 1.69 | ⅝ | 2.01 | 0.794 | 1.09 | 0.968 |

[x]Price—Subject to Inquiry.
[y]Availability—Subject to Inquiry.

# ANGLES
## equal legs

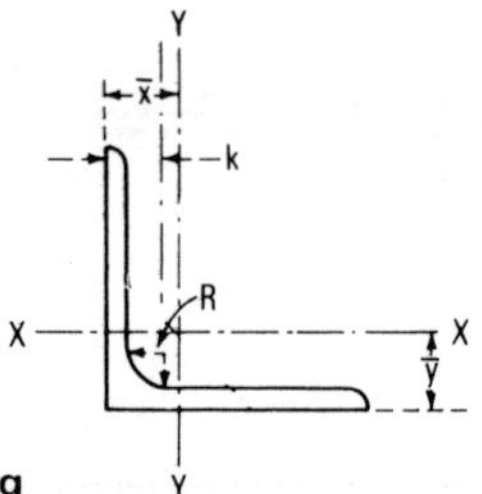


Theoretical Dimensions and Properties for **Designing**

| Section Number and Size | Thick-ness | Weight per Foot | Area of Section | k | Axis X-X and Axis Y-Y | | | |
|---|---|---|---|---|---|---|---|---|
| | | | | | $I_{x,y}$ | $Z_{x,y}$ | $r_{x\,y}$ | $\bar{y}$ or $\bar{x}$ |
| in. | in. | lb | in.$^2$ | in. | in.$^4$ | in.$^3$ | in. | in. |
| **L3 x 3 x** | 1/2 | 9.4 | 2.75 | 13/16 | 2.22 | 1.07 | 0.898 | 0.932 |
| R=5/16 | [x] [y]7/16 | 8.3 | 2.43 | 3/4 | 1.99 | 0.954 | 0.905 | 0.910 |
| | 3/8 | 7.2 | 2.11 | 11/16 | 1.76 | 0.833 | 0.913 | 0.888 |
| | 5/16 | 6.1 | 1.78 | 5/8 | 1.51 | 0.707 | 0.922 | 0.865 |
| | 1/4 | 4.9 | 1.44 | 9/16 | 1.24 | 0.577 | 0.930 | 0.842 |
| | 3/16 | 3.71 | 1.09 | 1/2 | 0.962 | 0.441 | 0.939 | 0.820 |
| **L2½ x 2½ x** | [z]1/2 | 7.7 | 2.25 | 13/16 | 1.23 | 0.724 | 0.739 | 0.806 |
| R=1/4 | [z]3/8 | 5.9 | 1.73 | 11/16 | 0.984 | 0.566 | 0.753 | 0.762 |
| | [z]5/16 | 5.0 | 1.46 | 5/8 | 0.849 | 0.482 | 0.761 | 0.740 |
| | [z]1/4 | 4.1 | 1.19 | 9/16 | 0.703 | 0.394 | 0.769 | 0.717 |
| | [z]3/16 | 3.07 | 0.902 | 1/2 | 0.547 | 0.303 | 0.778 | 0.694 |
| **L2 x 2 x** | [z]3/8 | 4.7 | 1.36 | 11/16 | 0.479 | 0.351 | 0.594 | 0.636 |
| R=9/32 | [z]5/16 | 3.92 | 1.15 | 5/8 | 0.416 | 0.300 | 0.601 | 0.614 |
| | [z]1/4 | 3.19 | 0.938 | 9/16 | 0.348 | 0.247 | 0.609 | 0.592 |
| | [z]3/16 | 2.44 | 0.715 | 1/2 | 0.272 | 0.190 | 0.617 | 0.569 |
| | [z]1/8 | 1.65 | 0.484 | 7/16 | 0.190 | 0.131 | 0.626 | 0.546 |
| **L1¾ x 1¾ x** | [z]1/4 | 2.77 | 0.813 | 1/2 | 0.227 | 0.186 | 0.529 | 0.529 |
| R=1/4 | [z]3/16 | 2.12 | 0.621 | 7/16 | 0.179 | 0.144 | 0.537 | 0.506 |
| | [z]1/8 | 1.44 | 0.422 | 3/8 | 0.126 | 0.099 | 0.546 | 0.484 |
| **L1½ x 1½ x** | [z]1/4 | 2.34 | 0.688 | 7/16 | 0.139 | 0.134 | 0.449 | 0.466 |
| R=3/16 | [z]3/16 | 1.80 | 0.527 | 3/8 | 0.110 | 0.104 | 0.457 | 0.444 |
| | [z]5/32 | 1.52 | 0.444 | 3/8 | 0.094 | 0.088 | 0.461 | 0.433 |
| | [z]1/8 | 1.23 | 0.359 | 5/16 | 0.078 | 0.072 | 0.465 | 0.421 |
| **L1¼ x 1¼ x** | [z]1/4 | 1.92 | 0.563 | 7/16 | 0.077 | 0.091 | 0.369 | 0.403 |
| R=3/16 | [z]3/16 | 1.48 | 0.434 | 3/8 | 0.061 | 0.071 | 0.377 | 0.381 |
| | [z]1/8 | 1.01 | 0.297 | 5/16 | 0.044 | 0.049 | 0.385 | 0.359 |
| **L1 x 1 x** | [z]1/4 | 1.49 | 0.438 | 3/8 | 0.037 | 0.056 | 0.290 | 0.339 |
| R=1/8 | [z]3/16 | 1.16 | 0.340 | 5/16 | 0.030 | 0.044 | 0.297 | 0.318 |
| | [z]1/8 | 0.80 | 0.234 | 1/4 | 0.022 | 0.031 | 0.304 | 0.296 |

[x]Price—Subject to Inquiry.
[y]Availability—Subject to Inquiry.
[z]West Coast Mills Only.

# ANGLES

unequal legs

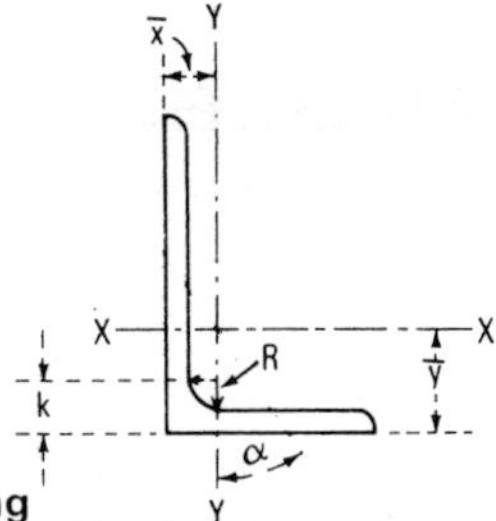


Theoretical Dimensions and Properties for **Designing**

| Section Number and Size | Thickness | Weight per Foot | Area of Section | k | | | | | | | | |
|---|---|---|---|---|---|---|---|---|---|---|---|---|
| | | | | | $I_x$ | $Z_x$ | $r_x$ | $\bar{y}$ | $I_y$ | $Z_y$ | $r_y$ | $\bar{x}$ |
| in. | in. | lb | in.$^2$ | in. | in.$^4$ | in.$^3$ | in. | in. | in.$^4$ | in.$^3$ | in. | in. |
| **L9 x 4 x** | [z]5/8 | 26.3 | 7.73 | 1 1/8 | 64.9 | 11.5 | 2.90 | 3.36 | 8.32 | 2.65 | 0.858 | 1.04 |
| R=1/2 | [z]9/16 | 23.8 | 7.00 | 1 1/16 | 59.1 | 10.4 | 2.91 | 3.33 | 7.63 | 2.41 | 0.834 | 1.04 |
| | [z]1/2 | 21.3 | 6.25 | 1 | 53.2 | 9.34 | 2.92 | 3.31 | 6.92 | 2.17 | 0.810 | 1.05 |
| **L8 x 6 x** | 1 | 44.2 | 13.0 | 1 1/2 | 80.8 | 15.1 | 2.49 | 2.65 | 38.8 | 8.92 | 1.73 | 1.65 |
| R=1/2 | [x y]7/8 | 39.1 | 11.5 | 1 3/8 | 72.3 | 13.4 | 2.51 | 2.61 | 34.9 | 7.94 | 1.74 | 1.61 |
| | 3/4 | 33.8 | 9.94 | 1 1/4 | 63.4 | 11.7 | 2.53 | 2.56 | 30.7 | 6.92 | 1.76 | 1.56 |
| | 9/16 | 25.7 | 7.56 | 1 1/16 | 49.3 | 8.95 | 2.55 | 2.50 | 24.0 | 5.34 | 1.78 | 1.50 |
| | 1/2 | 23.0 | 6.75 | 1 | 44.3 | 8.02 | 2.56 | 2.47 | 21.7 | 4.79 | 1.79 | 1.47 |
| | [y]7/16 | 20.2 | 5.93 | 15/16 | 39.2 | 7.07 | 2.57 | 2.45 | 19.3 | 4.23 | 1.80 | 1.45 |
| **L8 x 4 x** | 1 | 37.4 | 11.0 | 1 1/2 | 69.6 | 14.1 | 2.52 | 3.05 | 11.6 | 3.94 | 1.03 | 1.05 |
| R=1/2 | [x y]7/8 | 33.1 | 9.73 | 1 3/8 | 62.5 | 12.5 | 2.53 | 3.00 | 10.5 | 3.51 | 1.04 | 0.999 |
| | 3/4 | 28.7 | 8.44 | 1 1/4 | 54.9 | 10.9 | 2.55 | 2.95 | 9.36 | 3.07 | 1.05 | 0.953 |
| | [x y]5/8 | 24.2 | 7.11 | 1 1/8 | 46.9 | 9.21 | 2.57 | 2.91 | 8.10 | 2.62 | 1.07 | 0.906 |
| | 9/16 | 21.9 | 6.43 | 1 1/16 | 42.8 | 8.35 | 2.58 | 2.88 | 7.43 | 2.38 | 1.07 | 0.882 |
| | 1/2 | 19.6 | 5.75 | 1 | 38.5 | 7.49 | 2.59 | 2.86 | 6.74 | 2.15 | 1.08 | 0.859 |
| | [y]7/16 | 17.2 | 5.06 | 15/16 | 34.1 | 6.60 | 2.60 | 2.83 | 6.02 | 1.90 | 1.09 | 0.835 |
| **L7 x 4 x** | 3/4 | 26.2 | 7.69 | 1 1/4 | 37.8 | 8.42 | 2.22 | 2.51 | 9.05 | 3.03 | 1.09 | 1.01 |
| R=1/2 | [x y]5/8 | 22.1 | 6.48 | 1 1/8 | 32.4 | 7.14 | 2.24 | 2.46 | 7.84 | 2.58 | 1.10 | 0.963 |
| | 1/2 | 17.9 | 5.25 | 1 | 26.7 | 5.81 | 2.25 | 2.42 | 6.53 | 2.12 | 1.11 | 0.917 |
| | [y]7/16 | 15.8 | 4.62 | 15/16 | 23.7 | 5.13 | 2.26 | 2.39 | 5.83 | 1.88 | 1.12 | 0.893 |
| | 3/8 | 13.6 | 3.98 | 7/8 | 20.6 | 4.44 | 2.27 | 2.37 | 5.10 | 1.63 | 1.13 | 0.870 |

[x]Price—Subject to Inquiry
[y]Availability—Subject to Inquiry.
[z]West Coast Mills Only.

# ANGLES

## unequal legs

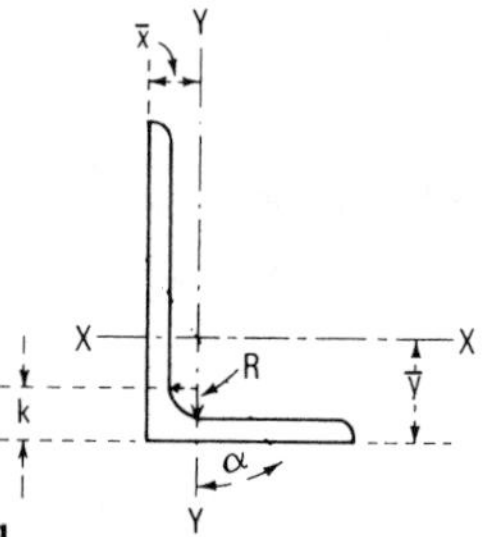


Theoretical Dimensions and Properties for **Designing**

| Section Number and Size | Thickness | Weight per Foot | Area of Section | k | $I_x$ | $Z_x$ | $r_x$ | $\bar{y}$ | $I_y$ | $Z_y$ | $r_y$ | $\bar{x}$ |
|---|---|---|---|---|---|---|---|---|---|---|---|---|
| in. | in. | lb | in.² | in. | in.⁴ | in.³ | in. | in. | in.⁴ | in.³ | in. | in. |
| **L6 x 4 x** | 3/4 | 23.6 | 6.94 | 1 1/4 | 24.5 | 6.25 | 1.88 | 2.08 | 8.68 | 2.97 | 1.12 | 1.08 |
| R=1/2 | 5/8 | 20.0 | 5.86 | 1 1/8 | 21.1 | 5.31 | 1.90 | 2.03 | 7.52 | 2.54 | 1.13 | 1.03 |
| | [x,y] 9/16 | 18.1 | 5.31 | 1 1/16 | 19.3 | 4.83 | 1.90 | 2.01 | 6.91 | 2.31 | 1.14 | 1.01 |
| | 1/2 | 16.2 | 4.75 | 1 | 17.4 | 4.33 | 1.91 | 1.99 | 6.27 | 2.08 | 1.15 | 0.987 |
| | [y] 7/16 | 14.3 | 4.18 | 15/16 | 15.5 | 3.83 | 1.92 | 1.96 | 5.60 | 1.85 | 1.16 | 0.964 |
| | 3/8 | 12.3 | 3.61 | 7/8 | 13.5 | 3.32 | 1.93 | 1.94 | 4.90 | 1.60 | 1.17 | 0.941 |
| | [y] 5/16 | 10.3 | 3.03 | 13/16 | 11.4 | 2.79 | 1.94 | 1.92 | 4.18 | 1.35 | 1.17 | 0.918 |
| **L6 x 3 1/2 x** | 3/8 | 11.7 | 3.42 | 7/8 | 12.9 | 3.24 | 1.94 | 2.04 | 3.34 | 1.23 | 0.988 | 0.787 |
| R=1/2 | 5/16 | 9.8 | 2.87 | 13/16 | 10.9 | 2.73 | 1.95 | 2.01 | 2.85 | 1.04 | 0.996 | 0.763 |
| | [y] 1/4 | 7.9 | 2.31 | 3/4 | 8.86 | 2.21 | 1.96 | 1.99 | 2.34 | 0.847 | 1.01 | 0.740 |
| **L5 x 3 1/2 x** | 3/4 | 19.8 | 5.81 | 1 1/4 | 13.9 | 4.28 | 1.55 | 1.75 | 5.55 | 2.22 | 0.977 | 0.996 |
| R=7/16 | [x,y] 5/8 | 16.8 | 4.92 | 1 1/8 | 12.0 | 3.65 | 1.56 | 1.70 | 4.83 | 1.90 | 0.991 | 0.951 |
| | 1/2 | 13.6 | 4.00 | 1 | 9.99 | 2.99 | 1.58 | 1.66 | 4.05 | 1.56 | 1.01 | 0.906 |
| | [y] 7/16 | 12.0 | 3.53 | 15/16 | 8.90 | 2.64 | 1.59 | 1.63 | 3.63 | 1.39 | 1.01 | 0.883 |
| | 3/8 | 10.4 | 3.05 | 7/8 | 7.78 | 2.29 | 1.60 | 1.61 | 3.18 | 1.21 | 1.02 | 0.861 |
| | 5/16 | 8.7 | 2.56 | 13/16 | 6.60 | 1.94 | 1.61 | 1.59 | 2.72 | 1.02 | 1.03 | 0.838 |
| | [y] 1/4 | 7.0 | 2.06 | 3/4 | 5.39 | 1.57 | 1.62 | 1.56 | 2.23 | 0.830 | 1.04 | 0.814 |
| **L5 x 3 x** | 1/2 | 12.8 | 3.75 | 1 | 9.45 | 2.91 | 1.59 | 1.75 | 2.58 | 1.15 | 0.829 | 0.750 |
| R=3/8 | [x,y] 7/16 | 11.3 | 3.31 | 15/16 | 8.43 | 2.58 | 1.60 | 1.73 | 2.32 | 1.02 | 0.837 | 0.727 |
| | 3/8 | 9.8 | 2.86 | 7/8 | 7.37 | 2.24 | 1.61 | 1.70 | 2.04 | 0.888 | 0.845 | 0.704 |
| | 5/16 | 8.2 | 2.40 | 13/16 | 6.26 | 1.89 | 1.61 | 1.68 | 1.75 | 0.753 | 0.853 | 0.681 |
| | 1/4 | 6.6 | 1.94 | 3/4 | 5.11 | 1.53 | 1.62 | 1.66 | 1.44 | 0.614 | 0.861 | 0.657 |
| **L4 x 3 1/2 x** | 1/2 | 11.9 | 3.50 | 15/16 | 5.32 | 1.94 | 1.23 | 1.25 | 3.79 | 1.52 | 1.04 | 1.00 |
| R=3/8 | [x,y] 7/16 | 10.6 | 3.09 | 7/8 | 4.76 | 1.72 | 1.24 | 1.23 | 3.40 | 1.35 | 1.05 | 0.978 |
| | 3/8 | 9.1 | 2.67 | 13/16 | 4.18 | 1.49 | 1.25 | 1.21 | 2.95 | 1.17 | 1.06 | 0.955 |
| | 5/16 | 7.7 | 2.25 | 3/4 | 3.56 | 1.26 | 1.26 | 1.18 | 2.55 | 0.994 | 1.07 | 0.932 |
| | 1/4 | 6.2 | 1.81 | 11/16 | 2.91 | 1.03 | 1.27 | 1.16 | 2.09 | 0.808 | 1.07 | 0.909 |

[x] Price—Subject to Inquiry.
[y] Availability—Subject to Inquiry.

# ANGLES

unequal legs

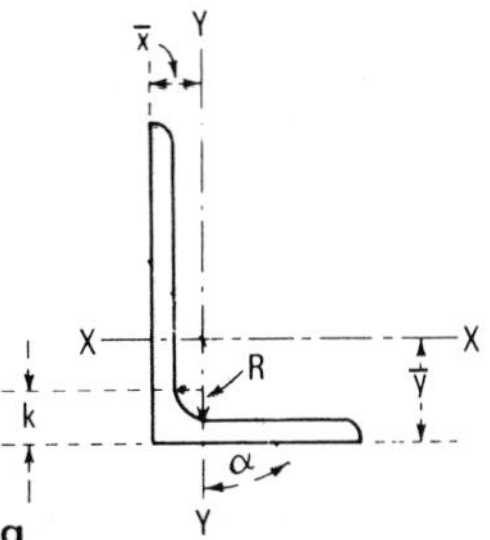


Theoretical Dimensions and Properties for **Designing**

| Section Number and Size | Thickness | Weight per Foot | Area of Section | k | $I_x$ | $Z_x$ | $r_x$ | $\bar{y}$ | $I_y$ | $Z_y$ | $r_y$ | $\bar{x}$ |
|---|---|---|---|---|---|---|---|---|---|---|---|---|
| in. | in. | lb | in.² | in. | in.⁴ | in.³ | in. | in. | in.⁴ | in.³ | in. | in. |
| **L4 x 3 x** | ½ | 11.1 | 3.25 | 15/16 | 5.05 | 1.89 | 1.25 | 1.33 | 2.42 | 1.12 | 0.864 | 0.827 |
| R=⅜ | [x,y]7/16 | 9.8 | 2.87 | ⅞ | 4.52 | 1.68 | 1.25 | 1.30 | 2.18 | 0.992 | 0.871 | 0.804 |
| | ⅜ | 8.5 | 2.48 | 13/16 | 3.96 | 1.46 | 1.26 | 1.28 | 1.92 | 0.866 | 0.879 | 0.782 |
| | 5/16 | 7.2 | 2.09 | ¾ | 3.38 | 1.23 | 1.27 | 1.26 | 1.65 | 0.734 | 0.887 | 0.759 |
| | ¼ | 5.8 | 1.69 | 11/16 | 2.77 | 1.00 | 1.28 | 1.24 | 1.36 | 0.599 | 0.896 | 0.736 |
| **L3½ x 3 x** | ⅜ | 7.9 | 2.30 | 13/16 | 2.72 | 1.13 | 1.09 | 1.08 | 1.85 | 0.851 | 0.897 | 0.830 |
| R=⅜ | 5/16 | 6.6 | 1.93 | ¾ | 2.33 | 0.954 | 1.10 | 1.06 | 1.58 | 0.722 | 0.905 | 0.808 |
| | ¼ | 5.4 | 1.56 | 11/16 | 1.91 | 0.776 | 1.11 | 1.04 | 1.30 | 0.589 | 0.914 | 0.785 |
| **L3½ x 2½ x** | ⅜ | 7.2 | 2.11 | 13/16 | 2.56 | 1.09 | 1.10 | 1.16 | 1.09 | 0.592 | 0.719 | 0.660 |
| R=5/16 | 5/16 | 6.1 | 1.78 | ¾ | 2.19 | 0.927 | 1.11 | 1.14 | 0.939 | 0.504 | 0.727 | 0.637 |
| | ¼ | 4.9 | 1.44 | 11/16 | 1.80 | 0.755 | 1.12 | 1.11 | 0.777 | 0.412 | 0.735 | 0.614 |
| **L3 x 2½ x** | [z]⅜ | 6.6 | 1.92 | ¾ | 1.66 | 0.810 | 0.928 | 0.956 | 1.04 | 0.581 | 0.736 | 0.706 |
| R=5/16 | [z]5/16 | 5.6 | 1.62 | 11/16 | 1.42 | 0.688 | 0.937 | 0.933 | 0.898 | 0.494 | 0.744 | 0.683 |
| | [z]¼ | 4.5 | 1.31 | ⅝ | 1.17 | 0.561 | 0.945 | 0.911 | 0.743 | 0.404 | 0.753 | 0.661 |
| **L3 x 2 x** | [z]⅜ | 5.9 | 1.73 | 11/16 | 1.53 | 0.781 | 0.940 | 1.04 | 0.543 | 0.371 | 0.559 | 0.539 |
| R=5/16 | [z]5/16 | 5.0 | 1.46 | ⅝ | 1.32 | 0.664 | 0.948 | 1.02 | 0.470 | 0.317 | 0.567 | 0.516 |
| | [z]¼ | 4.1 | 1.19 | 9/16 | 1.09 | 0.542 | 0.957 | 0.993 | 0.392 | 0.260 | 0.574 | 0.493 |
| | [y,z]3/16 | 3.07 | 0.902 | ½ | 0.842 | 0.415 | 0.966 | 0.970 | 0.307 | 0.200 | 0.583 | 0.470 |
| **L2½ x 2 x** | [z]⅜ | 5.3 | 1.55 | 11/16 | 0.912 | 0.547 | 0.768 | 0.831 | 0.514 | 0.363 | 0.577 | 0.581 |
| R=¼ | [z]5/16 | 4.5 | 1.31 | ⅝ | 0.788 | 0.466 | 0.776 | 0.809 | 0.446 | 0.310 | 0.584 | 0.559 |
| | [z]¼ | 3.62 | 1.06 | 9/16 | 0.654 | 0.381 | 0.784 | 0.787 | 0.372 | 0.254 | 0.592 | 0.537 |
| | [z]3/16 | 2.75 | 0.809 | ½ | 0.509 | 0.293 | 0.793 | 0.764 | 0.291 | 0.196 | 0.600 | 0.514 |

[x]Price—Subject to Inquiry.
[y]Availability—Subject to Inquiry.
[z]West Coast Mills Only.

# ANGLES

## unequal legs

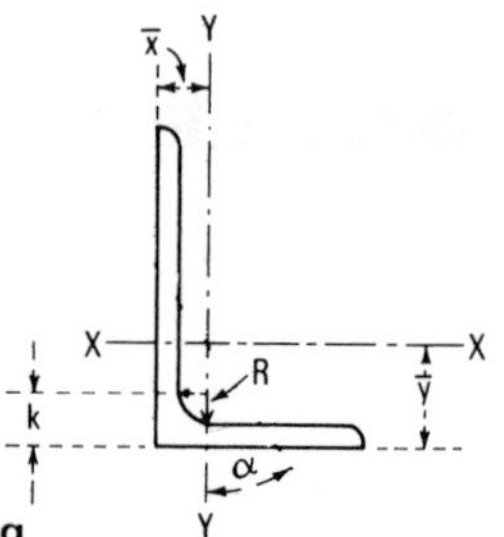


Theoretical Dimensions and Properties for **Designing**

| Section Number and Size | Thickness | Weight per Foot | Area of Section | k | | | | | | | | |
|---|---|---|---|---|---|---|---|---|---|---|---|---|
| | | | | | $I_x$ | $Z_x$ | $r_x$ | $\bar{y}$ | $I_y$ | $Z_y$ | $r_y$ | $\bar{x}$ |
| in. | in. | lb | in.$^2$ | in. | in.$^4$ | in.$^3$ | in. | in. | in.$^4$ | in.$^3$ | in. | in. |
| **L2½ x 1½ x** | [z] 5/16 | 3.92 | 1.15 | 5/8 | 0.711 | 0.444 | 0.785 | 0.898 | 0.191 | 0.174 | 0.408 | 0.398 |
| R=3/16 | [z] ¼ | 3.19 | 0.938 | 9/16 | 0.591 | 0.364 | 0.794 | 0.875 | 0.161 | 0.143 | 0.415 | 0.375 |
| | [z] 3/16 | 2.44 | 0.715 | ½ | 0.461 | 0.279 | 0.803 | 0.852 | 0.127 | 0.111 | 0.422 | 0.352 |
| **L2 x 1½ x** | [z] ¼ | 2.77 | 0.813 | ½ | 0.316 | 0.236 | 0.623 | 0.663 | 0.151 | 0.139 | 0.432 | 0.413 |
| R=3/16 | [z] 3/16 | 2.12 | 0.621 | 7/16 | 0.248 | 0.182 | 0.632 | 0.641 | 0.120 | 0.108 | 0.440 | 0.391 |
| | [z] ⅛ | 1.44 | 0.422 | 3/8 | 0.173 | 0.125 | 0.641 | 0.618 | 0.085 | 0.075 | 0.448 | 0.368 |
| **L1¾ x 1¼ x** | [z] ¼ | 2.34 | 0.688 | 7/16 | 0.202 | 0.176 | 0.543 | 0.602 | 0.085 | 0.095 | 0.352 | 0.352 |
| R=3/16 | [z] 3/16 | 1.80 | 0.527 | 3/8 | 0.160 | 0.137 | 0.551 | 0.580 | 0.068 | 0.074 | 0.359 | 0.330 |
| | [z] ⅛ | 1.23 | 0.359 | 5/16 | 0.113 | 0.094 | 0.560 | 0.557 | 0.049 | 0.051 | 0.368 | 0.307 |

[z]West Coast Mills Only.

# APPENDIX B

## UNIT CONVERSIONS BETWEEN U.S. CUSTOMARY AND SI UNITS

*Length*

inches (in.) × 25.4 = millimeters (mm)
inches (in.) × 2.54 = centimeters (cm)
inches (in.) × 0.0254 = meters (m)
feet (ft) × 0.3048 = meters (m)
meters (m) × 39.4 = inches (in.)
meters (m) × 3.28 = feet (ft)
inches (in.) × 12 = feet (ft)
feet (ft) × 3 = yards (yd)
yards (yd) × 0.3333 = feet (ft)
yards (yd) × 36 = inches (in.)
yards (yd) × 0.9144 = meters (m)
meters (m) × 1.0936 = yards (yd)

*Area*

square inches ($in.^2$) × 645.2 = square millimeters ($mm^2$)
square feet ($ft^2$) × 0.0929 = square meters ($m^2$)
square meters ($m^2$) × 10.76 = square feet ($ft^2$)
square millimeters ($mm^2$) × $1.55 \times 10^{-3}$ = square inches ($in.^2$)

*Force*

pounds (lb) × 4.448 = newtons (N)
pounds (lb) × 1000 = kilopounds (kips)
newtons (N) × 0.225 = pounds (lb)
kilopounds (kips) × 0.001 = pounds (lb)
pounds (lb) × 2000 = tons (T)

*Moment*

pound feet (lb ft) × 1.356 = newton meters (Nm)
newton meters (Nm) × 0.7375 = pound feet (lb ft)

*Area Moment of Inertia*

quartic inches ($in.^4$) × $4.162 \times 10^5$ = quartic millimeters ($mm^4$)
quartic feet ($ft^4$) × $8.631 \times 10^{-3}$ = quartic meters ($m^4$)
quartic millimeters ($mm^4$) × $2.403 \times 10^{-6}$ = quartic inches ($in.^4$)
quartic meters ($m^4$) × 115.86 = quartic feet ($ft^4$)

# Answers to Even-Numbered Problems

## Chapter 1

1.2 (a) 90°; (b) 20°; (c) 60°; (d) 130°
1.4 $\alpha_1 = 150°$; $\alpha_2 = 70°$; $\alpha_3 = 140°$
1.6 $\alpha = 100°$
1.8 $\sin A = 0.640$; $\cos A = 0.768$; $\tan A = 0.833$
$\sin B = 0.768$; $\cos B = 0.640$; $\tan B = 1.2$
1.10 (a) $A = B = 0°$; $C = 180°$
(b) $A = 96.2°$; $B = 53.8°$; $c = 12.4''$
(c) $A = 80°$; $b = 10.1$; $C = 2.60°$
(d) $B = 90°$; $a = 3.1$; $c = 11.6$
(e) $B = 80°$; $a = 19$; $b = 22.8$
1.12 111.5′
1.14 (a) 1, 0.5; (b) 2, −2.667; (c) 42.66, 2.34; (d) 1, 0.5
1.16 (a) 2; (b) $33x^2 - 42x$; (c) $x^6 + x^5 + 1$; (d) $12x^{11} - 6x^5 + 3x^2 + 0.5x$

## Chapter 2

2.4 (a) $R = 95$ lb, $\theta = 49°$; (b) $R = 1747$ lb, $\theta = 97°$;
(c) $R = 610$ N, $\theta = 18°$; (d) $R = 5246$ lb, $\theta = 62°$
2.6 (a) $F_x = 196$ lb; $F_y = 784$ lb; $F_z = 588$ lb
(b) $F_x = 534$ lb; $F_y = -133$ lb; $F_z = 0$
(c) $F_x = 2.67$ kips; $F_y = -5.35$ kips; $F_z = 8.02$ kips
(d) $F_x = 6.06$ kN; $F_y = 0$; $F_z = 24.25$ kN
2.8 $R = 1400$ lb

## Chapter 3

3.2 −25 kNm
3.4 −1414 lb ft
3.6 2250 lb in.
3.8 500 kNm
3.10 $M_A = 500$ kNm; $M_B = 100$ kNm

### Chapter 4

4.10 115 lb
4.12 $A = 13.3$ kips; $B = 16.7$ kips
4.14 $A_x = 10$ kN; $A_y = 21.2$ kN; $B = 11.1$ kN
4.16 $A_x = 0$; $A_y = 0$; $B = 1500$ N
4.18 $F = 250$ lb

### Chapter 5

5.2 $\bar{x} = 9.93''$; $\bar{y} = 7.93''$
5.4 $x = 80$ mm; $y = 70$ mm
5.6 $\bar{x} = 13.19''$; $\bar{y} = 8.93''$
5.8 $\bar{x} = 2.63''$; $\bar{y} = 2.38''$
5.10 $\bar{x} = 5.02$; $\bar{y} = 4.69''$
5.12 $\bar{y} = 23.66''$
5.14 $\bar{x} = 5.23''$; $\bar{y} = 6.81''$

### Chapter 6

(C) = *compression*; (T) = *tension*.

6.2 $AB = BC = 3500$ lb (C); $AE = CD = 3030$ lb (T)
$BE = BD = 2740$ lb (T); $ED = 577$ lb (T)
6.4 $AB = 3.89$ k (C); $BC = 3.5$ k (C); $CD = 4.6$ k (C); $DE = 3.25$ k (T)
$EA = 5.75$ k (T); $BE = 1.06$ k (T); $CE = 4.6$ k (T)
6.6 $AB = 8.3$ kN (C); $BD = 13$ kN (T); $DE = 13.3$ kN (T); $EF = 5$ kN (T)
$EH = 10$ kN (T); $AE = 13$ kN (C); $AD = 10$ kN (C); $CD = 5$ kN (T)
$BC = AH = HG = EG = FG = 0$
6.8 $BC = 8$ kN (C); $CE = 10$ kN (C); $DE = 10.4$ kN (T)
6.10 $AB = 1000$ lb (C); $BC = 1414$ lb (T); $CD = 1414$ lb (C)
6.12 $BD$
6.14 $BL$, $CK$, $DJ$, $FH$, $IF$
6.16 $EG$, $EH$, $DH$, $CH$, $CI$, $CJ$, $BJ$, $AJ$
6.18 $B_x = 390$ lb; $B_y = 390$ lb; $C_x = 240$ lb; $C_y = 390$ lb; $D_x = D_y = 390$ lb
6.20 $A_x = 0$; $A_y = 1635$ N; $B_x = 0$; $B_y = 1635$ N; $C_x = 0$; $C_y = 3270$ N;
$D_y = 1635$ N

### Chapter 7

7.2
$$\left.\begin{aligned} V &= -500 \\ M &= -500x \end{aligned}\right\} 0 < x < 10'$$

$$\left.\begin{aligned} V &= -1000 \\ M &= -500x - 500(x - 10) \end{aligned}\right\} 10' < x < 20'$$

$V_{max} = -1000$ lb
$M_{max} = -15{,}000$ lb ft

7.4 $\left.\begin{aligned} V &= -1000x \\ M &= -500x^2 \end{aligned}\right\} 0 < x < 5 \text{ m}$

$\left.\begin{aligned} V &= -5000 \text{ N} \\ M &= -5000(x - 2.5) \end{aligned}\right\} 5 \text{ m} < x < 10 \text{ m}$

$V_{max} = -5000$ N
$M_{max} = -37{,}500$ Nm

7.6 $\left.\begin{aligned} V &= 0 \\ M &= -300 \text{ lb ft} \end{aligned}\right\} 0 < x < 10'$

$\left.\begin{aligned} V &= -500 \text{ lb} \\ M &= -300 - 500(x - 10) \end{aligned}\right\} 10' < x < 20'$

$V_{max} = -500$ lb
$M_{max} = -5300$ lb ft

7.8 $\left.\begin{aligned} V &= -10000 \text{ N} \\ M &= -10000\, x \end{aligned}\right\} 0 < x < 4 \text{ m}$

$\left.\begin{aligned} V &= 4375 - 100(x - 4) \\ M &= -10{,}000\, x + 14{,}375(x - 4) - 50(x - 4)^2 \end{aligned}\right\} 4 \text{ m} < x < 9 \text{ m}$

$\left.\begin{aligned} V &= 3875 \text{ N} \\ M &= -10{,}000\, x + 14{,}375(x - 4) - 500(x - 6.5) \end{aligned}\right\} 9 \text{ m} < x < 13 \text{ m}$

$V_{max} = -10{,}000$ N
$M_{max} = -40{,}000$ Nm

7.10 $\left.\begin{aligned} V &= 5050x - 50x^2 \\ M &= 5050x - 50/3x^3 \end{aligned}\right\} 0 < x < 12'$

$\left.\begin{aligned} V &= -2150 \text{ lb} \\ M &= 5050x - 7200(x - 8) \end{aligned}\right\} 12' < x < 18'$

$$\left.\begin{aligned} V &= -3150 \\ M &= 5050x - 7200(x - 8) - 1000(x - 18) \end{aligned}\right\} 18' < x < 24'$$

$V_{max} = -3150$ lb
$M_{max} = -33{,}835$ lb ft

7.12
$$\left.\begin{aligned} V &= 1500 \text{ lb to } 0 \text{ linear} \\ M &= 0 \text{ to } 3750 \text{ lb ft parabolic} \end{aligned}\right\} 0 < x < 5'$$

$$\left.\begin{aligned} V &= 0 \\ M &= 3750 \text{ lb ft} \end{aligned}\right\} 0 < x < 15'$$

$$\left.\begin{aligned} V &= 0 \text{ to } -1500 \text{ lb linear} \\ M &= 3750 \text{ lb ft to } 0 \text{ parabolic} \end{aligned}\right\} 0 < x < 20'$$

7.14 $V = 1622$ lb to 1222 lb linear
$M = 0$ to 5688 lb ft parabolic
$V = 1222$ lb to 22 lb linear
$M = 5688$ lb ft to 9420 lb ft parabolic
$V = 22$ lb to $-2378$ lb linear
$M = 9420$ lb ft to 0 parabolic
$V_{max} = -2378$ lb
$M_{max} = 9420$ lb ft at $x = 10.07'$

7.16 $M = 0$ to $-2000$ lb ft linear $\quad 0 < x < 20'$
$M = 0$ to $-1500$ lb ft linear $\quad 5' < x < 20'$
$M = 0$ to $-1000$ lb ft linear $\quad 10' < x < 20'$
$M = 0$ to $-500$ lb ft linear $\quad 15' < x < 20'$
$M_{max} = -5000$ lb ft at $x = 20'$

7.18 $M = 0$ to $-5000$ lb ft $\quad 0 < x < 50'$
$M = 0$ to $-80{,}000$ lb ft $\quad 10' < x < 50'$
$M = 0$ to $-20{,}000$ lb ft $\quad 30' < x < 50'$
$M_{max} = -105{,}000$ lb ft at $x = 50'$

7.20 $M = 200$ lb ft $\quad 0 < x < 16'$
$M = 0$ to $-3600$ lb ft $\quad 4' < x < 16'$
$M = 0$ to $-8000$ lb ft $\quad 8' < x < 16'$
$M = 0$ to $+ 800$ lb ft $\quad 4' < x < 16'$
$M_{max} = -10{,}600$ lb ft at $x = 16'$

7.22 $N = 3$ kN $(0 < x < 5')$
$N = 2$ kN $(5' < x < 15')$
$N = 0$ $(15' < x < 20')$

## Chapter 8

8.2 $I_x = 1624\ \text{cm}^4$; $I_y = 887\ \text{cm}^4$
8.4 $I_x = 342\ \text{in.}^4$; $I_y = 336.5\ \text{in.}^4$
8.6 $I_x = 974\ \text{in.}^4$; $I_y = 97\ \text{in.}^4$
8.8 $\bar{I}_x = 155\ \text{in.}^4$; $\bar{I}_y = 48.5\ \text{in.}^4$
8.10 $\bar{I}_x = 109\ \text{in.}^4$; $\bar{I}_y = 49\ \text{in.}^4$
8.12 $\bar{I}_x = 130\ \text{in.}^4$; $\bar{I}_y = 14.2\ \text{in.}^4$
8.14 $\bar{I}_x = 166\ \text{in.}^4$; $\bar{I}_y = 456\ \text{in.}^4$
8.16 $\bar{I}_x = 155\ \text{in.}^4$; $\bar{I}_y = 48.5\ \text{in.}^4$; $\bar{I}_{xy} = 0$
8.18 $\bar{I}_x = 109\ \text{in.}^4$; $\bar{I}_y = 49\ \text{in.}^4$; $\bar{I}_{xy} = -38\ \text{in.}^4$
8.20 $I_{\max} = 155\ \text{in.}^4$; $I_{\min} = 48.5\ \text{in.}^4$; $\alpha_m = 0°$
8.22 $I_{\max} = 127.4\ \text{in.}^4$; $I_{\min} = 30.6\ \text{in.}^4$; $\alpha_m = 26°$

# INDEX

# INDEX